Mammalian Endocrinology and Male Reproductive Biology

Mammalian Endocrinology and Male Reproductive Biology

Edited by

Shio Kumar Singh

CRC Press
Taylor & Francis Group
Boca Raton London New York

CRC Press is an imprint of the
Taylor & Francis Group, an **informa** business

CRC Press
Taylor & Francis Group
6000 Broken Sound Parkway NW, Suite 300
Boca Raton, FL 33487-2742

First issued in paperback 2019

ISBN-13: 978-1-4987-2735-8 (hbk)
ISBN-13: 978-0-367-37745-8 (pbk)

Visit the Taylor & Francis Web site at
http://www.taylorandfrancis.com

and the CRC Press Web site at
http://www.crcpress.com

This book is dedicated to the memory of my revered teacher, late Professor C J Dominic, who introduced me to the field of mammalian endocrinology and male reproductive biology.

Contents

Preface ... ix
Editor ... xi
Contributors ... xiii

1 Sexual Dimorphism in the Central Nervous System ... 1
DES GILMORE

2 Mammalian Gonadotropin-Releasing Hormone ... 31
EDATHIL VIJAYAN

3 Gonadotropins: Past, Present and Future ... 43
KAMBADUR MURALIDHAR, RAJESH CHAUDHURI, SHAH SADDAD HUSSAIN AND NIKKI KUMARI

4 Endocrine Regulation of Spermatogenesis in Mammals ... 77
HANUMANTHAPPA KRISHNAMURTHY AND M. RAM SAIRAM

5 Role of Apoptosis in Spermatogenesis ... 97
AMIYA P. SINHA HIKIM

6 The Epididymis: Structure and Function ... 115
MOHAMMAD A. AKBARSHA, KUNNATHODI FAISAL AND ARUMUGAM RADHA

7 Regulation of Growth and Function of Epididymides ... 167
NIRMALA SINGH YADUVANSHI, SHAYU N. DESHPANDE, VIJAYAKUMAR GOVINDARAJ AND A. JAGANNADHA RAO

8 Role of Sperm Surface Molecules in Motility Regulation ... 197
GOPAL C. MAJUMDER, SUDIPTA SAHA, KAUSHIK DAS, DEBJANI NATH, ARUNIMA MAITI, SOUVIK DEY, DEBARUN ROY, CHINMOY SANKAR DEY, SUTAPA MITRA, AJAY RANA, JITAMANYU CHAKRABARTY, SUJOY DAS, ARPITA BHOUMIK, SASWATI BANERJEE, MAHITOSH MANDAL, BIJAY SHANKAR JAISWAL, PRASANTA GHOSH, ABHI DAS, DEBDAS BHATTACHARYYA AND SANDHYA REKHA DUNGDUNG

9 Male Accessory Sex Glands: Structure and Function 245
MARGOT FLINT, DEBRA A. McALISTER, ASHOK AGARWAL AND STEFAN S. DU PLESSIS

10 Reproductive Endocrinology and Contraception in the Male 259
MANNENCHERIL RAJALAKSHMI, PRAMOD CHANDRA PAL AND RADHEY SHYAM SHARMA

11 Molecular Mechanism of Androgen Action 285
MAHENDRA KUMAR THAKUR

12 Genomics of Male Infertility 299
KIRAN SINGH AND RAJIVA RAMAN

Index 319

Preface

The compiling of this book was taken up to create a reference book for readers interested in mammalian endocrinology and male reproductive biology. This discipline has witnessed considerable progress over the past several years. Keeping this in view, an attempt has been made to provide up-to-date information on the recent developments in this field. Unlike a book written by a single author, this book is the result of a multiauthor effort and provides comprehensive systematic coverage of topics in mammalian endocrinology and male reproductive biology.

There are 12 chapters in this book. Chapter 1 reviews the evidence for sexual dimorphism in the central nervous system and considers the mechanisms by which this is brought about. Chapter 2 on mammalian gonadotropin-releasing hormone (GnRH) presents an account of its structure, control of release and potential functions. Chapter 3 on gonadotropins – past, present and future – provides a coherent view of gonadotropins of pituitary origin and their role in the regulation of gonadal functions. In this chapter, selected topics concerning pituitary protein hormones and their influence on gonadal processes are discussed, and these include their purification, structural highlights, microheterogeneity, *in vivo* and *in vitro* bioassay, regulation of biosynthesis and secretion, their receptors in gonadal tissue, and clinical usage. Chapter 4 presents an account of endocrine regulation of spermatogenesis. This chapter illustrates how hormones of the hypothalamo–pituitary–gonadal axis regulate the process of spermatogenesis. Chapter 5 discusses the role of apoptosis in spermatogenesis and highlights the signal transduction pathways involved in apoptosis of germ cells in the testis. Chapter 6 provides a detailed account of structure and function of the epididymis, which plays a major role in physiological maturation of spermatozoa, and this maturation process is a prerequisite for the sperm to acquire the ability to be motile and to fertilise the oocyte. Chapter 7 deals with regulation of growth and function of epididymides. Hormones such as follicle-stimulating hormone (FSH), androgen and oestrogen are known to play role in regulation of epididymal growth and function, so their role in this regulation is discussed. Chapter 8 describes sperm forward motility initiating and regulating molecules located on surface of sperm cell in the epididymis. The authors have focussed on the sperm cell surface molecules with special reference to external domain. In Chapter 9, structure and function of the male accessory sex glands are described. The major male accessory sex glands are the seminal vesicles, prostate glands and Cowper's glands, and this chapter provides details of their structure and function, with associated pathologies. Chapter 10 focuses on recent advances in male contraception along with a resume of the male reproductive processes and male reproductive endocrinology, which form the basis for the development of male-based contraceptive regimens. Chapter 11 describes molecular mechanism of androgen action. In this article, the author has reviewed current progress in understanding the molecular mechanism of androgen action with emphasis on coregulators and their role in physiology and pathology, particularly with a focus on the brain. Finally, Chapter 12 deals with genomics of male infertility. It is well to remember that infertility due to male factor contributes approximately 50% of the infertility cases in humans. A number of autosomal as well as sex chromosomal genes have been identified that are believed to play a pivotal role in the production of viable, normal and functional sperm. In this chapter, the authors discuss diverse aspects of male infertility and the role of genome.

There are inevitably some overlaps, but these are essential for flow and clarity of the subject matter, and efforts have been made to restrict these to the minimum. The chapters have been written by authors who are acknowledged leaders in their respective fields. I wish to express my personal gratitude to all of them for their valuable contribution and overwhelming support and encouragement. This book could not have been completed without their fullest cooperation. My thanks are also due to my research students for their help, and I wish to record my special appreciation to Ms. Deepanshu Joshi for her invaluable help in many ways. I will be failing in my duty if I do not acknowledge my wife for the sustained support in her own way.

Thanks are also due to colleagues at CRC Press/Taylor & Francis, especially to Dr. Chuck Crumly, Hayley Ruggieri and Amanda Parida for their efforts in ensuring the successful publication of this book.

Finally, I accomplished this task with the Holy Blessings of my Gurumaa and Gurudev. I hope this book will be useful to all those interested in the area of mammalian endocrinology and male reproductive biology be they students, teachers or researchers.

Shio Kumar Singh

Editor

Shio Kumar Singh, PhD, is a professor in the Department of Zoology at Banaras Hindu University, Varanasi, India. Born in the Kaimur district of the state of Bihar (India), he obtained his BSc (1973), MSc (1975), and PhD (1980) from Banaras Hindu University. Dr. Singh is on faculty of Banaras Hindu University since 1980. He is the recipient of the direct fellowship of the Japan Society for the Promotion of Science (JSPS) and the fellowship in Reproductive Biology of the Rockefeller Foundation (USA), for postdoctoral research at Hokkaido University School of Medicine, Sapporo (Japan) and the General Hospital and the University, Salzburg (Austria), respectively. Dr. Singh is also the recipient of the Science Academy Medal for Young Scientists of the Indian National Science Academy (INSA) and the ISCA Young Scientists Award of the Indian Science Congress Association for the year 1984.

Dr. Singh's research interests centre on male reproductive biology, regulation of male fertility and male reproductive toxicology. He has published numerous research papers in journals of repute, contributed chapters in books, and guest edited (by invitation) in 2005 a special issue on Male Reproductive Biology of the *Indian Journal of Experimental Biology* (*IJEB*), a NISCAIR publication, Government of India. He is a referee for many pioneer journals in his area of research and served/serves as member of the editorial board of the *Indian Journal of Experimental Biology (IJEB)/World Journal of Biological Chemistry* (*WJBC*). Dr. Singh has successfully supervised the PhD research work of several students, and a foreign faculty also received postdoctoral training in his laboratory. He has delivered invited/plenary talks and chaired sessions in several national and international conferences. He is a life member of the Indian Society for the Study of Reproduction and Fertility and the Society for Reproductive Biology and Comparative Endocrinology, India.

Contributors

Ashok Agarwal
Center for Reproductive Medicine
Cleveland Clinic
Cleveland, Ohio

Mohammad A. Akbarsha
Department of Animal Science
Bharathidasan University
Tiruchirappalli, Tamil Nadu, India

Saswati Banerjee
Department of Cell Biology and Physiology
Indian Institute of Chemical Biology
Kolkata, West Bengal, India

Debdas Bhattacharyya
Division of Cryobiology
Centre for Rural and Cryogenic Technologies
Jadavpur University
Kolkata, West Bengal, India

Arpita Bhoumik
Division of Cryobiology
Centre for Rural and Cryogenic Technologies
Jadavpur University
and
Department of Cell Biology and Physiology
Indian Institute of Chemical Biology
Kolkata, West Bengal, India

Jitamanyu Chakrabarty
National Institute of Technology
Durgapur, West Bengal, India

Rajesh Chaudhuri
Department of Biomedical Science
University of Delhi
New Delhi, India

Abhi Das
Department of Cell Biology and Physiology
Indian Institute of Chemical Biology
Kolkata, West Bengal, India

Kaushik Das
Burdwan Model School
Burdwan, West Bengal, India

Sujoy Das
Department of Cell Biology and Physiology
Indian Institute of Chemical Biology
Kolkata, West Bengal, India

Shayu N. Deshpande
Department of Developmental Biology
Memorial Sloan Kettering Cancer Research Centre
New York, New York

Chinmoy Sankar Dey
School of Biological Sciences
Indian Institute of Technology Delhi
New Delhi, India

Souvik Dey
Division of Cryobiology
Centre for Rural and Cryogenic Technologies
Jadavpur University
Kolkata, West Bengal, India

Sandhya Rekha Dungdung
Department of Cell Biology and Physiology
Indian Institute of Chemical Biology
Kolkata, West Bengal, India

Stefan S. du Plessis
Division of Medical Physiology
Stellenbosch University
Tygerberg, South Africa

Kunnathodi Faisal
Department of Animal Science
Bharathidasan University
Tiruchirappalli, Tamil Nadu, India

Margot Flint
Division of Medical Physiology
Stellenbosch University
Tygerberg, South Africa

Prasanta Ghosh
Department of Cell Biology and Physiology
Indian Institute of Chemical Biology
Kolkata, West Bengal, India

Des Gilmore
Laboratory of Human Anatomy
University of Glasgow
Glasgow, Scotland

Vijayakumar Govindaraj
Department of Biochemistry
Indian Institute of Science
Bangalore, Karnataka, India

Amiya P. Sinha Hikim
Division of Endocrinology, Metabolism and Molecular Medicine
Department of Internal Medicine
Charles R. Drew University of Medicine and Science
Los Angeles, California

Shah Saddad Hussain
Hormone Research Laboratory
Department of Zoology
University of Delhi
New Delhi, India

Bijay Shankar Jaiswal
Department of Molecular Biology
Genentech Inc.
South San Francisco, California

Hanumanthappa Krishnamurthy
National Centre for Biological Sciences
Tata Institute for Fundamental Research
Bangalore, Karnataka, India

Nikki Kumari
Hormone Research Laboratory
Department of Zoology
University of Delhi
New Delhi, India

Arunima Maiti
Indian Institute of Chemical Biology
Kolkata, West Bengal, India

Gopal C. Majumder
Division of Cryobiology
Centre for Rural and Cryogenic Technologies
Jadavpur University
Kolkata, West Bengal, India

Mahitosh Mandal
Indian Institute of Technology
Kharagpur, West Bengal, India

Debra A. McAlister
Division of Medical Physiology
Stellenbosch University
Tygerberg, South Africa

Sutapa Mitra
Department of Molecular Diagnostics
DiagnoRite Innovative Healthcare Pvt. Ltd.
Pune, Maharashtra, India

Kambadur Muralidhar
South Asian University
New Delhi, India

Debjani Nath
Department of Zoology
University of Kalyani
Kalyani, West Bengal, India

Pramod Chandra Pal
Department of Biochemistry
All India Institute of Medical Sciences
New Delhi, India

Arumugam Radha
Department of Animal Science
Bharathidasan University
Tiruchirappalli, Tamil Nadu, India

Mannencheril Rajalakshmi
Academic Cell
Medical Council of India
Dwarka, New Delhi, India

Rajiva Raman
Cytogenetics Laboratory
Department of Zoology
Banaras Hindu University
Varanasi, Uttar Pradesh, India

Ajay Rana
Department of Pharmacology
Loyola University Medical Center
Maywood, Illinois

A. Jagannadha Rao
Department of Biochemistry
Indian Institute of Science
Bangalore, Karnataka, India

Debarun Roy
Division of Cryobiology
Centre for Rural and Cryogenic Technologies
Jadavpur University
Kolkata, West Bengal, India

Sudipta Saha
Department of Anatomy
Chang Gung University
Taoyuan City, Taiwan

M. Ram Sairam
Molecular Endocrinology Laboratory
Clinical Research Institute of Montreal
Montreal, Quebec, Canada

Radhey Shyam Sharma
Indian Council of Medical Research
New Delhi, India

Kiran Singh
Department of Molecular and Human Genetics
Banaras Hindu University
Varanasi, Uttar Pradesh, India

Mahendra Kumar Thakur
Biochemistry and Molecular Biology Laboratory
Department of Zoology
Banaras Hindu University
Varanasi, Uttar Pradesh, India

Edathil Vijayan
Center for Neuroscience, Department of Biotechnology
Cochin University of Science and Technology
Kochi, Kerala, India

Nirmala Singh Yaduvanshi
Department of Biochemistry
Indian Institute of Science
Bangalore, Karnataka, India

Chapter 1

Sexual Dimorphism in the Central Nervous System

Des Gilmore

Contents

1.1 Introduction 2
1.1.1 Major Sexually Dimorphic Regions of the Mammalian CNS 3
1.2 Causes of Sexual Dimorphism in the CNS 3
1.3 Sexually Dimorphic Behaviour 5
1.4 Monoaminergic Systems and Their Participation in Sexual Dimorphism 6
1.5 Serotonin, Dopamine, Noradrenaline and GABA 7
1.6 Timing of Sexual Differentiation and How It Is Achieved 7
1.7 Role of Androgens and Oestrogens in Bringing about Brain Masculinisation 8
1.8 α-Fetoprotein and Sex Hormone Binding Globulin 9
1.9 Testosterone and 5α-Dihydrotestosterone 10
1.10 Endogenous Opioids and Influences of Opiates on the Developing Brain 10
1.11 Sexual Dimorphism in the Spinal Cord 11
1.12 Sexually Dimorphic Structures in the Medial Preoptic Area and the Hypothalamus 12
1.13 Amygdala 15
1.14 Sexual Dimorphism in the Human Brain 16
1.14.1 Non-Sexual Dimorphic Human Behaviour 16
1.14.2 Sexual Dimorphism in the Anterior Hypothalamus 17
1.14.3 Supraoptic and Suprachiasmatic Nuclei 18
1.14.4 Hippocampus 18
1.14.5 Vomeronasal Organ 19

1.14.6 Bed Nucleus of the Stria Terminalis....19
1.14.7 Anterior Commissure....19
1.14.8 Corpus Callosum....20
1.14.9 Sexual Dimorphism in Handedness....20
1.15 Sexual Orientation and the CNS....21
1.16 Conclusions....23
Acknowledgements....23
References....23

1.1 Introduction

The purpose of this chapter is to review the evidence for sexual dimorphism in the mammalian central nervous system (CNS) and to consider the mechanisms by which this is brought about. Although a considerable amount of the research on the causes and consequences of sexual dimorphism has been undertaken on rodents, special attention will be paid to the situation in the human. In mammals sexual dimorphism has been reported in much of the CNS including the weight of the brain, the size of specific regions, the number of neurones present, the pattern of synaptic connections as well as the distribution and content of certain neurotransmitters and neuropeptides (see Gilmore, 2002). As might be suspected, many parts of the CNS showing sexual dimorphism are those associated with reproductive behaviour and affected by gonadal steroid hormones acting both during development and later in adult life (Figure 1.1).

Most of the observations regarding sexual dimorphism in the CNS have been based on post-mortem studies, but more recently in vivo magnetic resonance imaging (MRI) has also been employed as an investigative tool (see Goldstein et al., 2001).

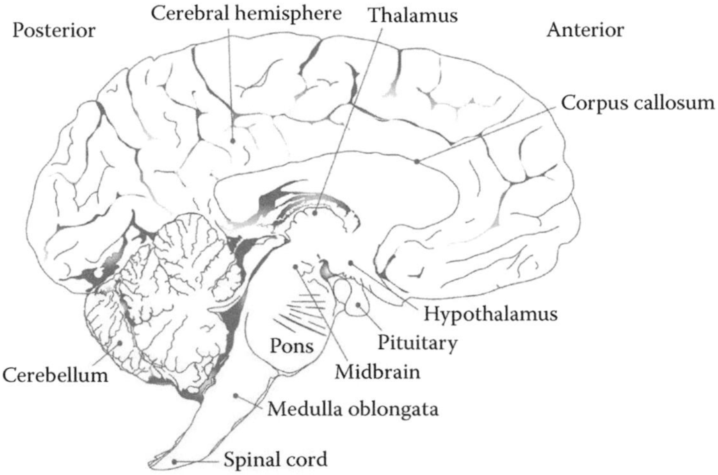

Figure 1.1 Diagram of the human brain.

The developmental pattern of the CNS during the prenatal and neonatal period has been widely investigated with the aid of histofluorescent, immunohistochemical, electrophysiological and spectrofluorescent techniques. Areas shown to be sexually dimorphic have previously been reviewed by a number of authors including Cooke et al. (1998) and Payne (1996). They include certain nuclei in the spinal cord and many parts of the limbic system, which comprises a diverse collection of associated cortical structures including the amygdala, hippocampus, fornix, cingulate gyrus, thalamus and hypothalamus. The hypothalamus can, itself, be subdivided into many regions and has been the focus of an enormous amount of research over the past half century, especially with regard to its role as an endocrine regulatory organ.

1.1.1 Major Sexually Dimorphic Regions of the Mammalian CNS

- Brain
  - Accessory olfactory bulb
  - Amygdala – Posterodorsal nucleus
  - Anterior commissure
  - Arcuate nucleus
  - Bed nucleus of the stria terminalis
  - Corpus callosum
  - Hippocampus
  - Hypothalamus, many areas including the anterior hypothalamus, arcuate, supraoptic, and ventromedial nuclei (see Figure 1.3 for abbreviations)
  - Locus coeruleus
  - Medial preoptic area containing the sexually dimorphic nucleus
  - Vomeronasal organ
- Spinal cord
  - Cremasteric nucleus
  - Dorsolateral nucleus
  - Pelvic autonomic ganglion
  - Spinal nucleus of bulbocavenosus
  - Superior cervical ganglion

1.2 Causes of Sexual Dimorphism in the CNS

Sexual dimorphism in the CNS is brought about by differences in cell division and migration, in programmed cell death, and in growth or synaptogenesis. It is generally assumed that gonadal steroids, acting during the limited so-called critical periods in fetal and/or neonatal life, induce this sexual differentiation, the timing of which is dependent on the maturity of the species at birth; even in the same species certain sexually dimorphic functions may be differentiated by oestrogens and others by androgens. It is only during these critical periods that neuronal tissue is sufficiently plastic to respond irreversibly to the gonadal steroids or to environmental contaminants that mimic their action (see Gilmore, 2002). During early human development testosterone levels are higher in males than females in mid-pregnancy and in the first three months after birth. It is apparently within these two periods that structures and circuits in the brain are permanently programmed, and at puberty become activated by the dramatic increase in circulating levels of the sex hormones. Lombardo et al. (2012) investigated the causes of sexual dimorphism in the human brain using structural MRI. Interestingly, but not surprisingly, they found that the volume of local grey matter in specific brain regions of 8- to 11-year-old boys could be correlated with fetal testosterone levels measured in their amniotic fluid samples removed during amniocentesis at 13–20 weeks gestation.

The organisational action of the gonadal steroids appears to take place at the molecular, ultrastructural, and cellular levels and may either be direct or occur through modulating effects on various neurotransmitter and neuropeptide systems within the CNS. It should be stressed that the critical period is only an empirical concept and does not represent a clearly defined stage of development. Not all sexually differentiated CNS structures are maximally sensitive to gonadal steroids at exactly the same time and thus the critical period may vary temporally for disparate sexually dimorphic traits (see Gilmore, 2002). Androgens act as signal transduction molecules, neurotrophic factors and/or neuromodulators by binding to specific intracellular receptors and so altering gene expression and/or transcriptional factors within the target cell to affect protein synthesis in neuronal tissue (Evans, 1988; McEwen, 1991; Miesfield, 1989). In a similar manner oestrogens have been shown to attach to specific intracellular receptors within the brain to modify them in such a way that they can bind DNA and so alter the expression of oestrogen-sensitive genes (Parker, 1990; Walters, 1985).

Interestingly, evidence also indicates that genes, by directly and differentially affecting patterns of neural development, can bring about sexual dimorphism of the brain (Dewing et al., 2003). This finding would suggest that there are some functional sex differences in the brain that occur independent of hormonal action. Dewing et al. (2003), utilising 10.5 days post coitum mouse embryo brain tissue, identified 51 genes with differential expression between male and female that were implicated in cell differentiation and proliferation as transcriptional regulators or as signalling molecules. Since these genes were present before gonadal formation it was suggested that they might play a role in bringing about sexual differentiation.

If fertilisation is brought about by a Y-chromosome-bearing sperm, then the resultant embryo will normally progress along the male developmental pathway. Although the initial development of embryos is similar, in the male the activation of the sex-determining gene (a transcription factor located on the Y chromosome) regulates the subsequent formation of most sex-specific structures. This includes the development of the testis and the consequent synthesis by it of anti-Müllerian hormone in the Sertoli cells and of testosterone in the Leydig cells which masculinise the genitalia. It was long thought that development of the female reproductive system was the default pathway, but this is now known to be an extreme oversimplification (see Mackay, 2000). Research by Vainio et al. (1999) and others has indicated that a signalling molecule (Wnt-4) not only brings about development of the Müllerian duct but also suppresses Leydig cell differentiation leading to regression of the Wolffian duct (plural), and maintaining post-meiotic oocyte development.

More recently, Chassot et al. (2008), working with mice, described how shortly after the onset of Sry expression the transcription factor Sox9 becomes upregulated inducing the development of the Sertoli cells and leading to the differentiation of the seminiferous tubules. Sox9 goes on to activate anti-Müllerian hormone, which brings about regression of the Müllerian duct (plural). There has been speculation that, as the fetal ovaries contain large amounts of oestrogens, this steroid might have some role in ovarian development including the initiation of meiosis. In a comprehensive review, Bowles and Koopman (2007) described how the active derivative of Vitamin A (retinoic acid) plays a vital role in determining the fate of the germ cells in mammals. In the fetal mouse this signalling molecule has been found to stimulate germ cells to enter meiosis, and it has been suggested retinoic acid may regulate meiosis in the pubertal testis too. Moreover, β-catenin, a protein which can be linked to E-cadherin in cell adhesion but can also act as a signalling molecule affecting gene transcription when free in the cytoplasm, has now been revealed as a pro-ovarian and anti-testis signalling molecule (Chassot et al., 2008; Maatouk et al., 2008).

Amongst the genes that determine male sexual differentiation are ones that bring about production of the androgen receptor protein that is found in most tissues including the brain (see Wilson and Rahman, 2005). Androgen receptors enable testosterone, produced by the fetal Leydig cells, to bind to them forming a testosterone–androgen receptor complex. Interestingly it is not only testosterone but also its two active metabolites 5α-dihydrotestosterone (5α-DHT) and oestradiol-17β that influence the manner in which both the CNS and reproductive system develop (see Figure 1.2). Not unexpectedly, there are differences in the distribution and concentration of gonadal steroid receptors within the CNS between the sexes.

1.3 Sexually Dimorphic Behaviour

Sexual behaviour can be divided into courtship, as well as mounting, intromission and ejaculation by the male and the adoption of the receptive lordosis posture by the female. Visual cues are often important in courtship, for example rodents wriggle their ears and engage in a curious darting run. In many mammals there is grooming and the smelling of the face and genitalia of the partner; this is clearly seen in dogs. The olfactory system provides an important sensory input to the limbic system in sub-primates. Many oestrous females produce a characteristic odour that is attractive to the males and detected by the vomeronasal organ. Almost all sexual behaviour in female non-primates is dependent upon hormone secretion and ovariectomy leads to its abrupt cessation. In contrast, although orchidectomy reduces male sexual

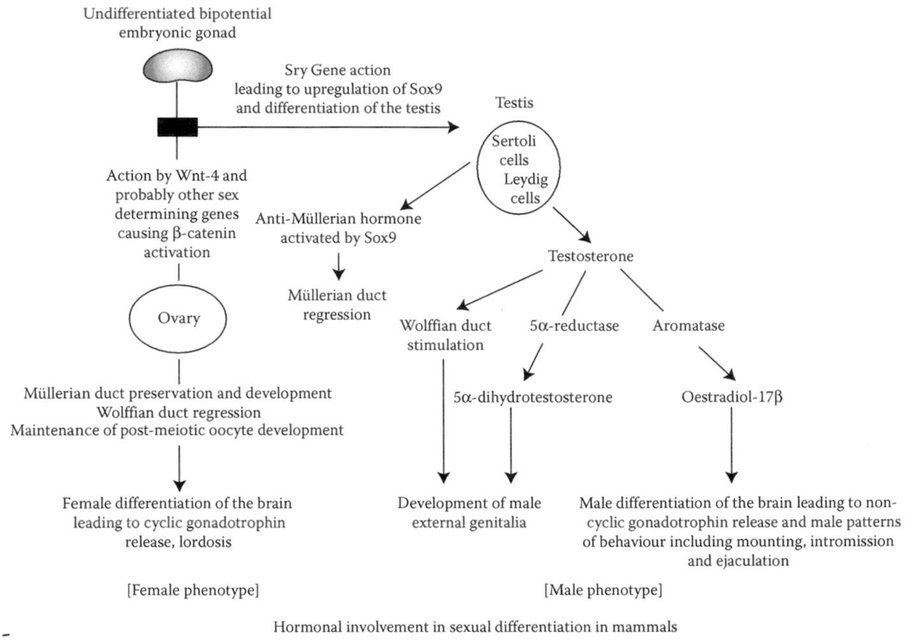

Figure 1.2 Hormonal involvement in sexual differentiation in mammals.

activity the decline is gradual. What is important in this situation is the amount of sexual experience before castration, and replacement therapy with testosterone will eventually in most cases lead to a full restoration of sexual activity.

1.4 Monoaminergic Systems and Their Participation in Sexual Dimorphism

The catecholaminergic and serotoninergic systems play an important role in the control and regulation of many brain functions. There is now strong evidence that the neurotransmitters of these pathways – noradrenaline, adrenaline, dopamine and serotonin (5-HT) – exert a vital role in the regulation of the hypothalamic stimulatory and inhibitory hormones both during fetal and in adult life. Work by Lauder (1990) demonstrated that 5-HT is an important developmental signal in the immature brain. Substantial evidence from numerous sources has now accumulated to indicate that androgens, acting upon these monoaminergic systems in various brain regions during the critical period for sexual differentiation of the CNS, can bring about generalised and long-lasting effects (see Siddiqui and Gilmore, 1988; Siddiqui et al., 1989) leading to sex differences in the content of a number of neurotransmitters in several brain regions. Goldstein et al. (2001) pointed out that unsurprisingly, in those brain areas involved in the control of sexual behaviour, there is co-localisation of gonadal steroid receptors with those of neurotransmitters including the monoamines and the excitatory γ-amino butyric acid (GABA) and with growth factors.

1.5 Serotonin, Dopamine, Noradrenaline and GABA

Research in our laboratory has indicated that when brain 5-HT levels are reduced in male hamsters around the time of sexual differentiation, there is an increase in the amount of female behaviour these animals show when adults (Johnston et al., 1990). This observation implies that 5-HT may be modulating a neural substrate that has already been differentiated by androgens during the perinatal period. The monoamines have also been shown to have direct behavioural effects in the adult. Fabre-Nys (1998) pointed out how this can be stimulatory or inhibitory dependent on the type of receptor implicated, what component of sexual behaviour is involved and the endocrine milieu. Fabre-Nys also reported how increased levels of 5-HT reduce sexual behaviour in both sexes and that dopamine modulates the expression of sexual behaviour, largely through action via the D_2 receptor. Although dopamine stimulates male sexual behaviour, its effects in the female are less clear-cut (Melis and Argiolas, 1995). Fabre-Nys (1998) described how dopamine concentrations in certain brain regions, including the medial preoptic area (MPOA), the paraventricular nucleus and the nucleus accumbens, differ between the sexes in rodents and sheep. This is presumably true for other mammals too. Crowley et al. (1989) reported that the predominant effect of noradrenaline is to stimulate sexual behaviour through its action via α_1 receptors involved in attention mechanisms and sensory processing. In contrast, Pfaus and Everitt (1995) showed that α_2 receptor activation inhibits male sexual behaviour. There is a suggestion that GABA may also be involved in brain sexual development and dimorphism. Tobet et al. (2002), studying mouse embryos, found that GABA has a role to play in differentiation of the ventromedial hypothalamus by providing boundary information. It has been postulated that GABA may affect cell positioning by controlling the guidance of cells by neuronal or glial fibres (see Tobet, 2003).

1.6 Timing of Sexual Differentiation and How It Is Achieved

In the rat and hamster, whereas differentiation of the reproductive system takes place before birth, that of the brain takes place perinatally. One can thus manipulate brain sexual differentiation after that of the genitalia has already taken place. In larger mammals with a long gestation brain sexual differentiation occurs prenatally. In the human one can only make inferences from observations on certain types of pathology as regards the influence of hormones on this process. As a consequence of sexual differentiation of the brain, the organ also becomes programmed to determine patterns of behaviour appropriate to one gender or the other. However, it is only following puberty weeks, months or even years later that the now high levels of circulating gonadal steroids act on the CNS in an excitatory or inhibitory fashion to regulate gonadotrophin secretion and induce the appropriate gender-specific post-pubertal reproductive behaviour.

1.7 Role of Androgens and Oestrogens in Bringing about Brain Masculinisation

There is now overwhelming evidence that the masculinisation of most neural functions involves intraneuronal conversion of testosterone to oestradiol-17β by the P450 aromatase enzyme. This enzyme is only present in certain brain areas, including the medial basal hypothalamus and amygdala. Neurones in these regions contain oestrogen receptors that bind locally produced oestradiol-17β. The cytoplasmic receptor complex is translocated to the cell nucleus where it alters expression of the genome. Aromatisation of androgens to oestrogens appears to be crucial for sexual differentiation of neuroendocrine function and for determining future gender-specific sexual behaviour in many species. These locally produced oestrogens increase neurite growth and establish neural circuitry (see Lephart, 1996). Kudwa et al. (2006) reviewed the roles of the two known oestrogen receptors in this process. Findings, based on work primarily undertaken on knockout mice, indicated there might be separate roles for these. Kudwa et al. suggested that the α receptor is primarily involved in masculinisation of the CNS, whereas the β receptor plays a major role in the defeminisation of sexual behaviour. MacLusky et al. (1985) showed that male rats possess higher levels of neuronal aromatase activity than do females during the time when brain sexual differentiation takes place, and Beyer et al. (1994) later found the same to be true for mice. Treatment of male rats with aromatase inhibitors during the perinatal period increases the capacity for female behaviour after puberty (Fadem and Barfield, 1981; McEwen et al., 1977). Incidentally, Sholl and Kim (1990) working with the rhesus monkey found sex differences in aromatase activity, but not in 5α-reductase activity nor in androgen receptor levels, in the medial basal hypothalamus and amygdala during early development.

Behavioural masculinisation in the ferret, monkey and human, but only to a much lesser extent in the rodent, is generally mediated by both testosterone and 5α-DHT, which bind to many of the same loci as does oestradiol-17β. In rats, where brain sexual differentiation depends on aromatisation of androgens to oestrogens, intracerebral implants of either oestradiol-17β or testosterone were found by Christensen and Gorski (1978) to be equally effective in eliciting masculinisation of reproductive function and sexual behaviour. Anti-androgens are known to block both testosterone- and oestradiol-17β-induced masculinisation (see Cooke et al., 1998). Non-aromatisable androgens such as 5α-DHT appear to be largely ineffective in bringing about brain sexual differentiation (see Cooke et al., 1998). Another indication that androgens act by their conversion to oestrogens, is indicated by the observation that early exposure to the latter affects brain sexual differentiation in much the same way as does early exposure to androgens. As far back as 1963 Gorski demonstrated that treatment of female rats with oestrogens during the perinatal period results in a pattern of anovulatory sterility in adulthood, which resembles that seen after perinatal testosterone administration.

1.8 α-Fetoprotein and Sex Hormone Binding Globulin

In placental mammals the fetus is continually exposed to endogenous oestrogens from the placental and maternal circulations. It thus follows that the brain of the female fetus must somehow be guarded from masculinisation by the effects of the circulating oestrogens upon the neurones. It had long been assumed that this protection, at least in rodents, was brought about by the selective binding with high affinity of oestrogens to α-fetoprotein (AFP) which would consequently prevent them from being able to cross the blood-brain barrier. However, this theory is now under review. AFP is present in large amounts in the plasma of many vertebrates and is produced transiently by the endodermal cells of the visceral yolk sac, by hepatocytes in the fetal liver and in the fetal gastrointestinal tract (see Bakker et al., 2006). Benno and Williams (1978) detected large amounts of AFP within neurones in some regions of the developing rat brain. It has been suggested (see Toran-Allerand, 1985), that AFP may in fact function as a carrier protein actively transporting oestradiol into neurones, thereby controlling the levels available for binding to its receptors and consequently affecting brain sexual differentiation. Recent experiments carried out on Afp $^{-/-}$ mutant mice by Bakker et al. (2006) have indicated that prenatal oestrogen exposure actually defeminises the brain and that AFP protects the female brain from these defeminising effects. Bakker et al. (2006) point out that little if any AFP is found in the limbic area including in the hypothalamus and the amygdala. They suggest that AFP may thus only protect from oestradiol those brain regions involved in reproductive behaviour, while delivering the hormone to other areas and thereby controlling the sexual differentiation of non-reproductive functions. Because AFP does not bind testosterone this androgen is thus free to enter neurones within the brain, where it can be converted to oestradiol-17β and bring about the masculinisation of neural structures. It is not yet possible to state categorically that the same process occurs in species other than the rat and mouse; human AFP for example does not bind oestrogens. In humans and other primates the principal high affinity steroid-binding protein is sex hormone binding globulin (SHBG). Although SHBG strongly binds androgens, it has only a weak affinity for oestrogens. Bakker et al. (2006) thus suggest that SHBG may function to prevent masculinisation of the developing female brain from the effect of the male hormones.

1.9 Testosterone and 5α-Dihydrotestosterone

Testosterone from the fetal testes and more especially 5α-DHT induce the conversion of the Wolffian duct into the male internal genitalia and bring about fusion of the labioscrotal folds. The discovery by Imperato-McGinley et al. (1979) of a genetically determined 5α-reductase enzyme deficiency in a group of related individuals in the Dominican Republic shed considerable light on the process of sexual differentiation in humans. This defect results in the inability to transform testosterone into 5α-DHT. Males are born with marked hypoplasia of the external genitalia, possessing a clitoris-like penis. Internally however, there is complete Müllerian duct inhibition and Wolffian duct stimulation, indicating that it is largely 5α-DHT that is responsible for normal penile and scrotal development in the human fetus. At puberty, however with the tremendous increase in testosterone that then occurs, the penis grows to adult proportions and the enlarged testes descend. Erections and ejaculations can occur. These individuals are reported to be heterosexual, indicating that, due to the effects of the intraneuronal conversion of testosterone to oestradiol-17β, normal male sexual differentiation of the brain has taken place.

The concept that there are specific sex centres in the brain is derived from a great deal of evidence which has now accumulated to show that certain neural pathways are responsive to the gonadal hormones and indispensable for the cyclic release of hormones in the female and for the expression of sexual behaviour in both sexes.

1.10 Endogenous Opioids and Influences of Opiates on the Developing Brain

Work on rats has indicated that the endogenous opioids are important growth-inhibiting, tonically active regulators of prenatal development (McLaughlin et al., 1996) and opiates have been shown to interfere with neural development (Hammer et al., 1989; Ricalde and Hammer, 1990). Exposure of the developing nervous system to opiates can also affect brain sexual differentiation. As mentioned earlier, for complete masculinisation of both the reproductive tract and behaviour, sufficient levels of androgens must be present during limited critical periods. Prenatal exposure of male rats to opiates (Ward et al., 1983) or to stress (Ward, 1983) reduces adult copulatory behaviour and increases feminine behaviour. Stress itself is believed to trigger release of endogenous opioids in the adult (Scallet, 1982), and the same may be true in the fetus. Opiates have been reported to reduce androgen production in the testes of fetal rats (Singh et al., 1980). When the long-acting opiate Duromorph was administered to golden hamsters over the last four days of their pregnancy, and to their male pups during the first four days of life, adult sexual behaviour was altered (Johnston et al., 1992). Although this perinatal opiate exposure had little effect on the female pups' ability to display normal sexual behaviour as adults, feminine sexual behaviour (lordosis) in the males was greatly enhanced.

In a further study in which hamsters were treated with Duromorph throughout pregnancy and into lactation (Johnston et al., 1994) it was discovered, rather unexpectedly, that this chronic opiate exposure increased both feminine and masculine sexual behaviour in the male offspring when adult. Again, no significant changes in either feminine or masculine sexual behaviour were seen in the female offspring, which had received exactly the same treatment. The fact that male sexual behaviour was even higher in the experimental group than in the control males would seem to rule out that the observed changes in feminine behaviour were in fact caused by a reduction in androgen levels during the critical period for brain sexual differentiation, and indicate that another mechanism was involved.

Kent et al. (1982) reported that opioid receptors are first detectable within the rat central nervous system on embryonic day 14. Petrillo et al. (1987) showed that their pattern of development alters during fetal growth and opiate exposure can influence the rate and distribution of opioid receptor ontogeny (Tempel et al., 1988). Kornblum et al. (1987) provided data from membrane-binding studies to indicate that the affinities of the μ-, κ- and δ-opioid receptor sites are similar in both neonatal and adult rats. They found μ- and κ-receptors to be present in significant densities during early neonatal life, but δ-receptors only appeared much later. This finding was supported by Magnan and Tiberi (1989), who presented evidence for the existence of μ and κ, but not δ-opioid sites in the human fetal brain. The chronic exposure to Duromorph in our experiments could have affected opioid receptor distribution and ontogeny in the male hamster offspring and thereby delayed neural development so that the systems involved in the defeminisation process occurred too late in the critical period for this to be completed. This would cause such males to show a higher incidence than expected of lordosis, but would not explain why the same hamsters showed a great increase in male sexual behaviour as well, in comparison to the controls. It is also possible that the opiates affected brain neurotransmitter concentrations and therefore interfered with the signals which these have been shown to provide for neural outgrowth and synaptogenesis (Zagon and McLaughlin, 1987). In the human, exposure of a male fetal brain to opiates, for example, through heroin abuse by the pregnant mother during critical developmental periods, could have serious implications if, as in rodents, it affected sexual orientation along with reproductive and other social behaviour once adolescence was reached.

1.11 Sexual Dimorphism in the Spinal Cord

There are present in the spinal cord certain groups of motoneurones that innervate muscles which are sexually dimorphic. Much of the research on this topic has been undertaken in the rat and hamster. It has been suggested that these muscles have a trophic effect on the neurones that innervate them, increasing their synaptic input (Lowrie and Vrbová, 1992). Sex differences are present in the numbers of neurones comprising the spinal nucleus of the bulbocavernosus (SNB) and the dorsolateral nucleus (DLN) in rodents. The motoneurones present in these nuclei supply perineal muscles associated with penile reflexes and erection, and are believed to be homologous with Onuf's nucleus present in the spinal cord of cats, dogs and primates. There are approximately five times more motoneurones in the SNB of male than in female rats and they are twice as large (Breedlove and Arnold, 1980; Jordan et al., 1982; Tobin and Payne, 1991). Although the ischiocavernosus, bulbocavernosus, levator ani and anal sphincter muscles supplied by the SNB and DLN motoneurones are equal in number in both sexes at birth, they atrophy in the female early in life due to the lack of androgen support (Nordeen et al., 1985). Cowburn and Payne (1992) showed that the manipulation of androgen levels in neonatal rats alters the numbers of motoneurones present in these two nuclei. A little earlier Sengelaub and Arnold (1986) and Nordeen et al. (1985) had demonstrated that no sex difference was apparent in the small number of motoneurones present in the SNB and DLN of fetal rats five days prior to birth. Four days later the migration of neurones into these two nuclei had resulted in their numbers being higher in both sexes than even in the adult male. Cell death early in postnatal life reduced neurone numbers to adult levels, the reduction being much greater in the female. The effect of androgens in preventing muscle atrophy also indirectly preserved the nerves supplying them. Another motoneurone group present in the spinal cord and sexually dimorphic in most mammals is the cremasteric nucleus, which projects via the genitofemoral nerve to innervate the cremaster muscle (Nagy and Senba, 1985; Payne, 1996; Van der Schoot et al., 1999). In females the muscle usually atrophies early in life. In the adult male rat there are approximately 260 cremasteric motoneurones present in the spinal cord; in the female only 70 (Barthold et al., 1994).

1.12 Sexually Dimorphic Structures in the Medial Preoptic Area and the Hypothalamus

Much of our knowledge about the organising action of androgens on the male brain comes from work on the rat by Gorski and his colleagues at the University of California, Los Angeles, who published their initial findings in the 1970s (see Figure 1.3).

The MPOA is a relatively large and complex region involved in the regulation of many functions. However, it is only one of many sites in the CNS for hormone action during development. The ventromedial arcuate region and other areas of the hypothalamus, as well as the cerebral cortex, the amygdala, the hippocampus, and, as already mentioned, certain nuclei in the spinal cord are also targets of the gonadal steroids. These can be permanently altered by androgens acting during critical periods in development. In male rodents the MPOA regulates copulatory behaviour and in females it is involved in the induction of lordosis, maternal

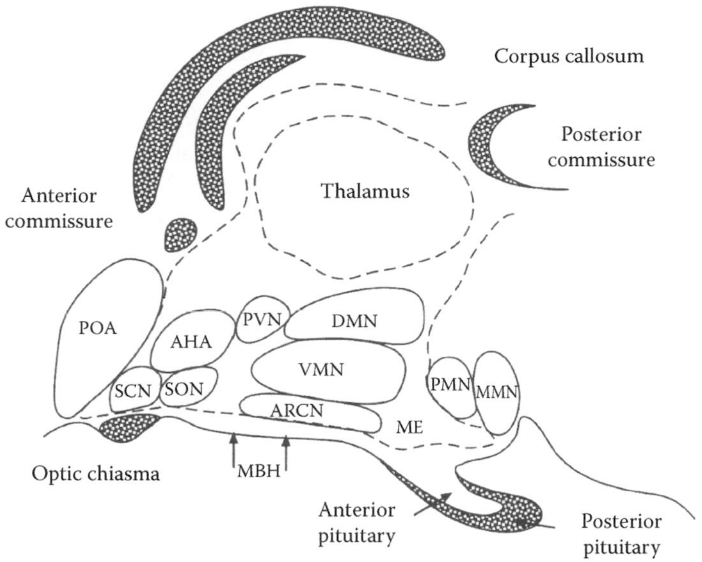

Figure 1.3 Schematic diagram of the rat hypothalamus. AHA = anterior hypothalamic area; ARCN = arcuate nucleus; DMN = dorsomedial nucleus; ME = median eminence; MMN = medial mammillary nucleus; PMN = premammillary nucleus; POA = preoptic area; PVN = paraventricular nucleus; SCN = suprachiasmatic nucleus; SON = supraoptic nucleus; VMN = ventromedial nucleus.

behaviour and cyclic gonadotrophin secretion (see Cooke et al., 1998). Incidentally the ventromedial nucleus has also been implicated in the control of lordosis and maternal behaviour in rats (see Madeira and Lieberman, 1995). Cooke et al. (1998) pointed out that the MPOA receives afferents from the medial amygdala, the bed nucleus of the stria terminalis (BNST), the hippocampus and ascending projections from serotoninergic neurones, as well as being innervated by much of the hypothalamus. Cooke et al. (1998) also described how the MPOA itself projects to all these areas and sends descending inputs to most of the periventricular nucleus of the hypothalamus. Within the MPOA is a group of densely packed neurones that show sexual dimorphism, known as the sexually dimorphic nucleus (SDN). Gorski et al. (1978) discovered that the SDN-MPOA is approximately 1 cubic mm in size in the male rat, being eight times larger than in the female. Since that time sex differences in similar nuclei have been found in other mammals including sheep and humans. There is an ovine SDN that is larger in rams than in ewes, and this nucleus also appears smaller in homosexual than in heterosexual rams (Roselli et al., 2004). Gorski et al. (1980, 1978) were able to show experimentally that the size of the SDN-MPOA in the rat was directly related to androgen levels during brain sexual differentiation and thus an extremely important morphological example of the organising action of androgens on the male brain. Houtsmuller et al. (1994) found that male rats treated with an aromatase inhibitor, thus preventing the conversion of testosterone to oestradiol-17β, had a much smaller SDN-MPOA, which was similar in size to that present in the female.

It is thus now evident that, at least in rodents, gonadal steroids alone can determine the volume of the SDN. The interaction between the gonadal steroids and those processes which lead to the formation of the SDN may be quite complex, even for the small cluster of neurones that comprise the area. Toran-Allerand (1985) postulated five possible mechanisms by which the final number of neurones comprising the SDN could be modified, if not absolutely determined. These are

1. The stimulation of neurogenesis by the steroids.
2. The influence of steroids upon the migration of neurones from their origin in the ependymal lining of the third ventricle to the region of the SDN.
3. Steroids in the vicinity of the SDN promoting the aggregation of these neurones into a distinct nucleus, perhaps by altering a cell surface recognition process.
4. Oestradiol maintaining survival during a phase of neuronal death.
5. Oestradiol influencing the specification of neurones destined to form the SDN. By activating or suppressing certain genes, oestradiol could determine the functional specificity and thus perhaps the migration, aggregation and even survival of more neurones in the male.

It is also possible that oestradiol might prolong the period of mitotic activity that would otherwise be completed.

However, although the SDN has been put forward as a model system for the study of sexual differentiation, there is as yet no clear knowledge of its function. The absolute volume of the nucleus is not correlated with ovulatory function, nor with lordosis responsiveness. Lesions of the SDN do not disrupt male copulatory behaviour, although small lesions just dorsal to it do so (Arendash and Gorski, 1983; De Jonge et al., 1989). When SDN tissue is removed from the brains of newborn male rats and transplanted into their female littermates (stereotaxically inserted into the MPOA) the females, when adult, display enhanced behavioural responses, in terms of both masculine and feminine copulatory behaviour (Arendash and Gorski, 1982). Gorski (1988) also demonstrated that transplant volume is increased by treatment of the recipient with testosterone propionate, indicating a trophic influence of gonadal steroids on the SDN.

Experiments in the newborn female rat have shown that, while androgens act upon the ventromedial hypothalamus to cause anovulatory sterility, their action upon the preoptic area may be the most important factor in the process of sexual differentiation (see Matsumoto and Arai, 1986). Although neonatally androgenised intact female rats display some weak male sexual behaviour as adults, female sexual behaviour prevails. However, following ovariectomy and androgen treatment predominantly male behaviour is observed. Neonatally androgenised females lose the capability of cyclic gonadotrophin secretion and exhibit persistent oestrus, polycystic ovaries without corpora lutea and infertility. In the female guinea pig, a species developmentally much more mature than the rat at birth, it has long been known that androgens must be administered prenatally to produce alterations in sexual behaviour and ovulation (Goy, 1966). As a consequence of prenatal androgen treatment, the external genitalia are masculinised and the newborn animals are pseudohermaphrodites. In the rhesus monkey prenatal androgen treatment also masculinises the genitalia, producing pseudohermaphrodites and altering female-typical behaviour (Pomerantz et al., 1986). Such androgenised females, when adult, display increased mating behaviour and exhibit a pattern of play behaviour more like that of the genetic male. Menarche is significantly delayed, but menstrual cycles, once established, are regular and ovulation occurs.

1.13 Amygdala

It had generally been assumed that once altered during early development sexually dimorphic structures in the CNS remain so thereafter. However, Cooke et al. (1999) reported a sexual dimorphism in the volume of the posterodorsal nucleus of the medial amygdala in adult rats that is affected by levels of circulating androgens. This nucleus is normally larger in adult males, but when these are castrated it shrinks within four weeks to the volume of that seen in females. Conversely, subcutaneous androgen administration of crystalline testosterone to adult females for the same period results in the volume of the nucleus increasing to that seen in males. In a more recent study, Morris et al. (2008) found that the same nucleus in adult mice is also responsive to steroid hormonal manipulation. Johansen et al. (2004) emphasised how such findings suggest that in some instances the androgen-sensitive period for the organisation of certain brain regions, including the posterodorsal nucleus of the medial amygdala, extend into adulthood. Cooke et al. (1999) also recorded a dramatic sex differences in the pattern of immunoreactivity of several neuropeptides in the rat brain. Substance P-immunoreactive perikarya and terminals were identified in the posterodorsal nucleus of the medial amygdala (MePD) and the bed nucleus of the stria terminalis (BNST), respectively. Moreover Substance P staining in the MePD was found to be twice as large in males than in females. This difference was found by Micevych et al. (1987) to be dependent on circulating adult androgen levels. The peptide cholecystokinin (CCK) and the hormone vasopressin also have a sexually dimorphic distribution in the posterodorsal nucleus of the medial amygdala of rodents. Adult male rats possess more CCK-immunoreactive cells there than do females and this also appears related to circulating testosterone levels as the difference disappears after orchidectomy and then reappears after replacement therapy (Simerly and Swanson, 1987). Sex differences in vasopressin immunoreactivity have also been identified in the posterodorsal nucleus of the medial amygdala and the medial posterior portion of the BNST with temporal staining in males being twice as great as in females. Interestingly, Toscano et al. (2009) reported that lesions placed in the amygdala of neonatal female rhesus monkeys affected future maternal behaviour. Damage to this area reduced the extent to which the monkeys as young adults showed interest in the infants of mothers with which they were housed.

1.14 Sexual Dimorphism in the Human Brain

1.14.1 Non-Sexual Dimorphic Human Behaviour

Similar to the situation in rodents, there are many sexually dimorphic areas present in the human brain. In right-handed males language is rigidly segregated in the left hemisphere, while visual spatial skills are contained in the right hemisphere. In right-handed females, the hemispheres are less functionally distinct and more diffusely organised (see Geschwind and Galaburda, 1985). There are numerous examples of human behaviour unrelated to sexual activity that are brought about by gonadal hormones acting on the developing brain (see Wilson and Rahman, 2005). The brains of women are not as lateralised as are those of men. Consequently women are much less likely to be affected by aphasia if they have a stroke restricted to one cerebral hemisphere. There are also many clear sex differences found when certain cognitive tests are carried out. A great deal of evidence has accumulated to confirm that women are more sensitive to touch and have better fine motor coordination than do men. Men are generally better at tasks that require visual spatial skills, reading maps, negotiating mazes, at mathematics and at perceiving and manipulating objects in space (see Huston, 1983; Maccoby and Jacklin, 1974). Moreover, women are significantly better at tasks that require a language ability. It has been suggested (Hiscock et al., 1994, 1995) that males are better at visual spatial tasks because language control may have been displaced from the right to the left hemisphere in favour of spatial processing. An indication of the major influence of testosterone on the CNS during fetal life in the human is seen in individuals with testicular feminisation (androgen insensitivity) syndrome where visual spatial skills are less than those in unaffected XY men. XYY men have been found to have even greater visual spatial skills than normal, whereas XXY (Klinefelter's syndrome) males have normal visual spatial skills, and XO (Turner's syndrome) women tend to be exaggeratedly female with regard to their behaviour. Wilson and Rahman (2005) have suggested that specialisation of the right half of the brain in males for visual spatial tasks might also have occurred at the expense of emotional processing which could account for the greater empathy possessed by women and the higher incidence of Asperger's syndrome that occurs in males.

Even in the infant, sex differences with regard to behaviour are apparent. Baby boys will respond more to what is visually attractive in the environment, for example lights, patterns and three-dimensional objects. Baby girls will respond preferentially to faces of people rather than to inanimate objects. They are more sensitive to sound, vocalise better and are more comforted by speech than are baby boys. Bao and Swaab (2011) suggest that sex differences in play behaviour and artistic expression evident in children arose long ago during evolution and are imprinted by testosterone during fetal life. Aggressive behaviour in men has also been related to fetal testosterone levels (Mazur and Booth, 1998).

Data from post-mortem studies has indicated that the cerebral cortex is larger in adult men than in women by up to 10%. However, some of this difference is accounted for by the greater body and brain size of males (see Raz et al., 2004). Goldstein et al. (2001) used MRI in an attempt to identify sexual dimorphism in the cortical and subcortical regions of the human brain. They found that, relative to total size of the cerebral cortex, women had larger volumes than men especially in the frontal and perimedial cortex, whereas in men the volumes of the frontomedial cortex, amygdala and hypothalamus were larger than those of women. The findings of Goldstein et al. (2001) may at least explain in part some of these differences between the sexes that are seen in cognition.

1.14.2 Sexual Dimorphism in the Anterior Hypothalamus

In 1985, Swaab and Fliers carried out more than 100 autopsies and described a sexually dimorphic cell group in the preoptic area of the young adult human brain; the volume of this nucleus being 2.5 ± 0.6 times as large in men as in women and containing 2.2 ± 0.5 as many cells. A few years later others, also examining human brains, reported sexual dimorphism to be present in two other anterior interstitial hypothalamic nuclei – $INAH_2$ and $INAH_3$ (Allen et al., 1989a; Le Vay, 1991). The $INAH_3$ was almost three times larger in men, the $INAH_2$ less so. The $INAH_3$ lies about 1 mm lateral to the wall of the third ventricle and about 1–2 mm dorsal to the anterior tip of the paraventricular nucleus and, Byne (1998) concluded the $INAH_3$ to be homologous with the SDN-POA in rodents.

Working at the Salk Institute in San Diego, California, Le Vay (1991) decided to investigate whether the size of the $INAH_3$ might be related not only to one's genetic sex, but also to sexual orientation, that is larger in individuals sexually orientated towards women (heterosexual men and lesbians) and smaller in individuals sexually orientated towards men (heterosexual women and gay men). Le Vay obtained the brains of 41 individuals. Nineteen were homosexual men, 16 (presumably) heterosexual men and six (presumably) heterosexual women. The brains were sectioned and stained. One-way ANOVA showed that the volume of the $INAH_3$ was more than twice as large in heterosexual men than in homosexual men (Le Vay, 1991). The whole topic of sexual differentiation in the human hypothalamus in relation to gender and sexual orientation was reviewed by Swaab and Hofman (1995). However, since this time further research, including that by Byne et al. (2001), has been unable to fully replicate Le Vay's results.

Fernández-Guasti et al. (2000) found in men intense androgen receptor immunoreactivity in neurones present in the horizontal limb of the diagonal band of Broca, in the lateromamillary nucleus and in the mediomamillary nucleus. Intermediate immunoreactivity was recorded in the diagonal band of Broca, the SDN-POA, also referred to in the human as the interstitial nucleus of the anterior hypothalamus-1 (ANH_1), and in the paraventricular, supraoptic, ventromedial and infundibular nuclei. Marginally weaker staining was observed in other hypothalamic regions. In most of the same areas immunoreactivity was recorded as being less intense in women and in the lateromamillary and mediomamillary nuclei strong sex differences were present.

1.14.3 Supraoptic and Suprachiasmatic Nuclei

Ishunina et al. (1999) reported that the vasopressinergic neurons of the supraoptic nucleus are more active in young men than in young women. Zhou et al. (1995a) and Swaab et al. (1994) found that the vasopressinergic component of the suprachiasmatic nucleus (SCN) was elongated in women but more spherical in men, and moreover that the vasoactive intestinal peptide (VIP)-containing portion of the SCN was twice as large in men as in women. Incidentally, the SCN primarily governs circadian rhythms but also appears to be involved in the control of sexual behaviour (see Swaab and Hofman, 1995). A little earlier Swaab and Hofman (1990) had published work to indicate that the SCN was almost twice as large in homosexual men as in heterosexual men and contained twice as many cells. No such differences were found in the SDN between the two groups.

1.14.4 Hippocampus

Raz et al. (2004) using MRI examined age-, sex- and hemisphere-related differences in the hippocampus of 200 healthy adults. They found that age-related shrinkage of this area was greater in men than in women from the fifth decade. A little earlier, Maguire et al. (1999) reported how functional brain imaging revealed strong parallels between humans and other animals in the neural involvement in navigation. Sex differences are apparent with males possessing an advantage in tasks requiring navigational skills and they suggest this is related to the function of the right hippocampus. Giedd et al. (1997) using MRI studied brain sexual dimorphism in 121 children and young adults ranging in age from 4 to 18 years. They found that the volume of both the hippocampus and amygdala increased in both sexes, but the amygdala increasing significantly more in males and the hippocampus in females. Suzuki et al. (2005), also using MRI, investigated volume expansion of the hippocampus during adolescence. Their results showed that the volume of the hippocampus was significantly larger in the older males suggesting a maturational process was taking place in the structure during adolescence. No significant age-related changes were observed in the hippocampus of female subjects.

1.14.5 Vomeronasal Organ

Sexual differentiation of the vomeronasal system is very widespread in mammals (see Cooke et al., 1998) and it plays a major role in reproduction. During development gonadotropin-releasing hormone (GnRH) neurones arise in the region of the medial olfactory placode and migrate along branches of the terminal and vomeronasal nerves to reach the basal forebrain (Schwanzel-Fukuda and Pfaff, 1989; Wray et al., 1989). In humans the vomeronasal organ is generally regarded as a vestigial non-functional organ. However, Berliner et al. (1996) produced evidence for functional connections between the vomeronasal organ and a variety of hypothalamic regions in men and women. Furthermore, a clear sex difference was found in the responses of the vomeronasal organ to the vomeropherin pregna-4,didiene-3,6-dione. This was administered as pulses to the lumen of the vomeronasal organ or to the olfactory and respiratory epithelia of the nasal septum. Berliner et al. found that vomeropherin significantly decreased luteinising hormone (LH) and follicle-stimulating hormone (FSH) pulsatility in men but did not alter this in women. Moreover, prolactin secretion was unaffected in both sexes. The vomeropherin also brought about concurrent reflex autonomic changes.

1.14.6 Bed Nucleus of the Stria Terminalis

Certain parts of the BNST are larger in men (Allen et al., 1989b) and the structure may be related to gender identity. Zhou and his colleagues (1995b) reported that in those wishing to become male to female transsexuals, the central subdivision of the BNST was 44% larger than in other men and similar in size to that found in women. Interestingly Swaab and Hofman (1995) reported that in the Netherlands the prevalence of male-to-female transsexuals was two and a half times greater than of female to male transsexuals. However, it is known that, although the BNST plays an essential part in rodent sexual behaviour and possesses large numbers of both oestradiol and testosterone receptors, it does not appear to be related to sexual orientation in humans. The size of the BNST is just as large in homosexual as in heterosexual men. A little later (Kruijver et al., 2000) examined the number of somatostatin-expressing neurones in the BNST. They found men had almost twice as many as did women and moreover that the number present in male to female transsexuals was similar to those in women. These findings have reinforced the belief that gender identity largely develops as a result of an interaction between the developing brain and the sex hormones, and indicates that sexual differentiation of the CNS and the genitalia can go in opposite directions.

1.14.7 Anterior Commissure

Allen and Gorski (1991) showed that the anterior commissure, the axons of which in primates primarily connect the temporal lobes, was around 12% larger in women than in men. They also found that the massa intermedia which crosses the third ventricle between the two thalami was present in 78% of women and 68% of men; it was 53% larger in women. Allen and Gorski suggested that anatomical sexual dimorphism in those structures connecting the two cerebral hemispheres may, at least in part, reflect functional sex differences in cognitive function and cerebral lateralisation.

1.14.8 Corpus Callosum

As previously mentioned, sexual dimorphism exists in the corpus callosum, it being larger in women (see Holloway et al., 2005). Pinkston et al. (1997), using MRI, reported that males who did not consistently favour the use of one particular hand had a larger mid-sagittal corpus callosum than males who used one hand preferentially. The exact opposite was true in women; those who used one hand preferentially had a smaller midsaggital corpus callosum than did those who showed no preference for using one particular hand. A little earlier Witelson (1989) had shown that hand preference was related to the size of the isthmus of the callosum in men, though not in women. Witelson suggested that the larger corpus callosum might be due to various histological changes such as more myelination, thicker fibres and different fibre packing density with a larger number of fibres. Overall, Witelson believed these features could allow for a greater degree of function between the two hemispheres and consequently less functional asymmetry. Witelson and Goldsmith (1991) proposed that hand preference in men and isthmus size co-varied as a result of the pruning by testosterone of corpus callosum axons originating in outlying temperoparietal regions. Reduced levels of testosterone during fetal life in males would result in less axon elimination and as a consequence a larger isthmus and an increased incidence of left-handedness. More recently Tuncer et al. (2005) utilised midsagittal MRI to investigate whether the morphology of the corpus callosum and its components might be related to gender and hand preference in young adults. In a comprehensive study they reported that although the overall size of the corpus callosum was larger in males it was not significantly so. However they did find significant gender differences within various parts of this brain area. Left-handedness in males was associated with a larger anterior and posterior body and isthmus. In contrast, no significant differences in size were found in these areas between female left- and right-handed individuals. Right- but not left-handed males were also found to have a significantly larger rostrum and isthmus than did right-handed females. Irrespective of handedness, Tuncer et al. (2005) noted that the rostrum and posterior midbody were larger in men than in women.

1.14.9 Sexual Dimorphism in Handedness

Sexual dimorphism also exists in handedness itself. Around 90% of the population is naturally right-handed and almost all of these have the control of speech localised in the left hemisphere, as also do 70% of left-handed individuals (see Loring and Meador, 1990). Geschwind and Galaburda (1985) suggested that intrauterine exposure to testosterone slows the growth of parts of the left hemisphere leading to corresponding regions of the right hemisphere growing more rapidly in the fetus. As a result, because of the exposure of male fetuses to higher levels of testosterone than females they show greater degree of a shift to right hemisphere participation in handedness and language control. The incidence of left-handedness is also significantly higher in males than in females. In the United Kingdom a survey by Bradley (1992) found 11% of men and 9% of women to be left-handed with 1% of both sexes being ambidextrous. Amongst undergraduate students at the University of Glasgow the occurrence of left-handedness has been recorded as being almost three times higher in males than in females (Gilmore, unpublished observations). In a study comprising 298 twins and 1192 age- and sex-matched singletons, Coren (1994) found the incidence of left-handedness and inverted handwriting postures to be significantly higher in twins, being 14.5% and 16.4% respectively, compared to 9.9% and 11.4% in the controls. This was true for both monozygotic and dizygotic twins. Coren suggested that his results were consistent with twinning being a risk factor for possible neurological damage leading to changes in adult handedness and motor coordination patterns. No sex differences were mentioned in this study. Left-handedness has also been associated with immune system problems and it appears that there is a higher incidence of homosexuality amongst left-handed men and women (see McCormick et al., 1990). However, the latter has been disputed by Marchant-Haycox et al. (1991). Dyslexia is a condition more common in males and in left-handed individuals, and an association has been found between it and with planum temporale asymmetry (Beaton, 1997), abnormalities in the corpus callosum (Robichon and Habib, 1998) and neural ectopias primarily in the left hemisphere (Galaburda, 1995).

1.15 Sexual Orientation and the CNS

It is now generally accepted that programmed gender identity is irreversible. Until the 1980s, a few researchers still argued that in the human, sexual orientation was not innate, but learnt, and that homosexuality and transsexuality were the result of formative or disordering sexual experiences during childhood and adolescence. If this were true, then up to a certain age a child could quite easily adjust to being either male or female. However, a great deal of evidence has now been provided to refute this argument (see Wilson and Rahman, 2005). Observations on female patients inadvertently exposed to androgens in utero as well as the studies by Imperato-McGinley et al. (1979, 1991) have indicated the importance of brain sexual differentiation on an individual's sexual orientation. As a consequence of the males in the Dominican Republic having had their brains masculinised in utero, they apparently had less problems than might be expected in adjusting to their male sex reassignment at puberty. Furthermore, those genetic males whose tissues are insensitive to the effects of both testosterone and 5α-DHT (the testicular feminisation syndrome) are female in both appearance and sexual orientation. On the other hand, girls born with ambiguous genitalia because of exposure to androgens in utero, for example due to congenital adrenal hyperplasia, appear to have psychological problems as adolescents in adjusting to a feminine lifestyle. Because their brains have been programmed to be male, yet the individuals are forced to live in a female environment and encouraged to show feminine behaviour, as well as being exposed to high levels of female gonadal hormones following menarche, they frequently are unable to cope with their situation. Girls exhibiting congenital adrenal hyperplasia syndrome are likely to be more tomboyish than their female peers. As children they are often keen athletes and highly energetic, preferring boys as playmates. In later life, according to Money et al. (1984), they apparently show an increased incidence of bisexuality and lesbianism.

Dörner and his colleagues, working at the Humbolt University in Berlin, investigated the effects of maternal stress on the sexual orientation of men (Dörner et al., 1980, 1983). They concluded that maternal stress during pregnancy led to an increase in homosexuality in the sons. Dörner's findings were in agreement with his earlier work on rodents (1977, 1980, 1981) which indicated that the depressed testosterone levels caused by stress at around the time that brain sexual differentiation takes place permanently alter the way in which noradrenaline, dopamine and 5-HT mediate the effects of the sex hormones on brain cells and on behaviour throughout life. Although this extrapolation of the findings in rodents to that in humans has not been universally accepted, more recent studies have suggested that stress suffered by a mother during pregnancy does increase the chances of her son being homosexual (see Ellis and Cole-Harding, 2001). Furthermore, as recently pointed out by Bao and Swaab (2011), sufficient evidence has now accumulated to reinforce the belief that prenatal androgen exposure during the second half of human fetal life interacts with multiple genes to influence sexual differentiation of the brain in a manner that affects one's gender identification.

In 1993, Hamer and his colleagues claimed to have pinpointed a region on the X chromosome that predisposed some men to homosexuality. Hamer's group studied family histories of 114 homosexuals and found that 13.5% of the gay men's brothers were homosexual compared to 2% in the general population. Furthermore, they found that maternal uncles and maternal male cousins were also more likely to be homosexual. Hamer and his colleagues (1993) then analysed the X chromosomes of 40 pairs of gay brothers and found that 33 pairs had co-inherited markers on the X_{q28} region. Although this region represents only 0.02% of the human genome, it may carry several hundred genes. Subsequent studies failed to replicate the finding by Hamer et al. until very recently. Sanders et al. (2014) have just published the results from research on a genome-wide linkage scan carried out on 409 pairs of homosexual brothers. They identified two regions of linkage, the X_{q28} region on the X chromosome as previously mentioned and the pericentromeric region on chromosome 8. How the genes might operate to predispose individuals to a homosexual orientation is as yet unclear. They may act directly or indirectly on a sex specific region of the brain, for example the $INAH_3$ nucleus, or indirectly through affecting a personality trait.

1.16 Conclusions

As is evident from the material included in this chapter, there are many issues still unresolved, despite the enormous amount of work that has been undertaken in the field of sexual dimorphism in the CNS. This is particularly true when attempting to determine how much of the sexual dimorphism in brain structure can be correlated with function. Obvious structural sex differences such as those in the spinal nuclei can easily be related to the action of the gonadal steroids during early development. However, when one attempts to correlate structure with gender differences in behaviour (especially in the human) the situation becomes highly complex. In particular much still remains to be discovered about how exactly the numerous chemical, hormonal and psychosocial factors acting postnatally affect both the anatomical structure of the CNS and influence sexual orientation and consequent reproductive and other behaviour.

Acknowledgements

I would like to express my sincere gratitude to Ms. Caroline Morris for her valued assistance with the artwork and thank Drs. Sarah Mackay and the late Suzanne Ullmann for their very helpful comments on the manuscript.

References

Allen, L.S. and Gorski, R.A. 1991. Sexual dimorphism of the anterior commissure and massa intermedia of the human brain. *Journal of Comparative Neurology* **312**, 97–104.

Allen, L.S., Hines, J.E., Shryne, J.F. and Gorski, R.A. 1989a. Two sexually dimorphic cell groups in the human brain. *Journal of Neuroscience* **9**, 497–506.

Allen, L.S., Hines, J.E., Shryne, J.F. and Gorski, R.A. 1989b. Sex differences in the bed nucleus of the stria terminalis of the human brain. *Journal of Comparative Neurology* **9**, 467–502.

Arendash, G.W. and Gorski, R.A. 1982. Enhancement of sexual behavior in female rats by neonatal transplantation of brain tissue from males. *Science* **217**, 1276–1278.

Arendash, G.W. and Gorski, R.A. 1983. Effects of discrete lesions of the sexually dimorphic nucleus of the preoptic area or other medial preoptic regions on the sexual behavior of male rats. *Brain Research Bulletin* **10**, 147–154.

Bakker, J., De Mees, C., Douhard, Q., Balthazart, J., Gabant, P., Szpirer, J. and Szpiper, C. 2006. Alpha-fetoprotein protects the developing female mouse brain from masculinization and defeminization by estrogens. *Nature Neuroscience* **9**, 220–226.

Bao, A.-M. and Swaab, D.F. 2011. Sexual differentiation of the human brain: Relation to gender identity, sexual orientation and neuropsychiatric disorders. *Frontiers in Neuroendocrinology* **32**, 214–226.
Barthold, J.S., Mahler, H.R. and Newton, B.W. 1994. Lack of feminization of the cremaster nucleus in cryptorchid androgen insensitive rats. *Journal of Urology* **152**, 2280–2286.
Beaton, A.A. 1997. The relation of planum temporale asymmetry and morphology of the corpus callosum to handedness, gender and dyslexia: A review of the evidence. *Brain and Language* **60**, 255–322.
Benno, R.W. and Williams, T.H. 1978. Evidence for intracellular localization of alpha-fetoprotein in the developing rat brain. *Brain Research* **142**, 182–186.
Berliner, D.L., Monti-Bloch, L., Jennings-White, C. and Diaz-Sanchez, V. 1996. The functionality of the human vomeronasal organ (VNO): Evidence for steroid receptors. *Journal of Steroid Biochemistry* **58**, 259–265.
Beyer, C., Green, S.J., Barker, P.J. and Hutchison, J.B. 1994. Aromatase–immunoreactivity is localized specifically in neurones in the developing mouse hypothalamus and cortex. *Brain Research* **638**, 203–210.
Bowles, J. and Koopman, P. 2007. Retinoic acid, meiosis and germ cell fate in mammals. *Development* **134**, 3401–3411.
Bradley, N. 1992. British survey of left-handedness. *The Graphologist* **10**, 176–182.
Breedlove, S.M. and Arnold, A.P. 1980. Hormone accumulation in a sexually dimorphic motor nucleus in the rat spinal cord. *Science* **210**, 564–566.
Byne, W. 1998. The media preoptic and anterior hypothalamic regions of the rhesus monkey: Cytoarchitectonic comparison with the human and evidence of sexual dimorphism. *Brain Research* **793**, 346–350.
Byne, W., Tobet, S. Mattiace, L.A., Lasco, M.S., Kemether, E., Edgar, M.A., Morgello, S. Buchsbaum, M.S. and Jones, L.B. 2001. The interstitial nuclei of the human anteriorhypothalamus: An investigation of variation with sex, sexual orientation and HIV status. *Hormones and Behavior* **40**, 86–92.
Chassot, A.A., Ranc, F., Gregoire, E.P., Roepers-Gajadien, H.L., Taketo, M.M., Camerino, G., de Rooij, D.G., Schedl, A. and Chaboissier, M.C. 2008. Activation of β-catenin signalling by Rspo1 controls differentiation of the mammalian ovary. *Human Molecular Genetics* **17**, 1264–1277.
Christensen, L.W. and Gorski, L.A. 1978. Independent masculinization of neuroendocrine systems by intracerebral implants of testosterone or estradiol in the neonatal female rat. *Brain Research* **146**, 325–340.
Cooke, B.M., Hegstrom, C.D., Villeneuve, L.S. and Breedlove, S.M. 1998. Sexual differentiation of the vertebrate brain: Principles and mechanisms. *Frontiers in Neuroendocrinology* **19**, 323–362.
Cooke, B.M., Tabibnia, G. and Breedlove, S.M. 1999. A brain sexual dimorphism controlled by adult circulating androgens. *Proceedings of the National Academy of Sciences USA* **96**, 7538–7540.
Coren, C. 1994. Twinning is associated with an increased risk of left-handedness and inverted hand writing posture. *Early Human Development* **40**, 23–27.
Cowburn, P.J. and Payne, A.P. 1992. The effects of serotonin manipulation during the postnatal period on development of sexually dimorphic and non-dimorphic lumbosacral motoneuron groups in the Albino Swiss rat. *Developmental Brain Research* **66**, 59–62.
Crowley, W.R., O'Connor, L.H. and Feder, H.H. 1989. Neurotransmitter systems and social behavior. In: J. Balthazart (Ed.), *Advances in Comparative and Environmental*

Physiology. 3. Molecular and Cellular Basis of Social Behavior in Vertebrates. Springer-Verlag, Berlin, pp. 162–237.

De Jonge, F.H., Louwerse, A.L., Ooms, M.P., Evers, P., Endert, E. and Van de Poll, N.E. 1989. Lesions of the SDN-POA inhibit sexual behavior of male Wistar rats. *Brain Research Bulletin* **23**, 483–492.

Dewing, P., Shi, T., Horvath, S. and Vilain, E. 2003. Sexually dimorphic gene expression in mouse brain precedes gonadal differentiation. *Molecular Brain Research* **118**, 82–90.

Dörner, G. 1977. Hormone dependent differentiation, maturation and function of the brain and sexual behaviour. *Endokrinologie* **69**, 306–320.

Dörner, G. 1980. Sexual differentiation of the brain. *Vitamins and Hormones* **38**, 325–381.

Dörner, G. 1981. Sex hormones and neurotransmitters as mediators for sexual differentiation of the brain. *Endokrinologie* **78**, 129–139.

Dörner, G., Geier, T., Ahrens, L., Krell, L., Műnx, G., Sieler, H., Kittener, E. and Műller, H. 1980. Prenatal stress as possible aetiogenetic factor of homosexuality in human males. *Endokrinologie* **75**, 365–368.

Dörner, G., Schenk, B., Schmiedel, B. and Ahrens, L. 1983. Stressful events in prenatal life of bi-sexual and homosexual men. *Experimental and Clinical Endocrinology* **81**, 83–87.

Ellis, L. and Cole-Harding, S. 2001. The effect of prenatal stress and of prenatal alcohol and nicotine exposure on human sexual orientation. *Physiology and Behavior* **74**, 213–226.

Evans, R.M. 1988. The steroid and thyroid hormone receptor super family. *Science* **240**, 889–895.

Fabre-Nys, C. 1998. Steroid control of monoamines in relation to sexual behavior. *Reviews of Reproduction* **3**, 31–41.

Fadem, B.H. and Barfield, R.J. 1981. Neonatal hormonal influences on the development of proceptive and receptive feminine sexual behavior in rats. *Hormones and Behavior* **15**, 282–288.

Fernández-Guasti, A., Kruijver, P.M., Fodor, M. and Swaab, D.F. 2000. Sex differences in the distribution of androgen receptors in the human hypothalamus. *The Journal of Comparative Neurology* **425**, 422–435.

Galaburda, A.M. 1995. Developmental dyslexia: Four consecutive cases with cortical anomalies. *Neurocase* **1**, 179–181.

Geschwind, N. and Galaburda, A.M. 1985. Cerebral lateralization. Parts I and II. *Archives of Neurology* **42**, 428–459 and 521–552.

Giedd, J.M., Castellanos, F.X., Rajapakse, J.C., Vaituzis, A.C. and Rapoport, J.L. 1997. Sexual dimorphism of the developing human brain. *Progress in Neuro-Psychopharmacology and Biological Psychiatry* **21**, 1185–1201.

Gilmore, D.P. 2002. Sexual dimorphism in the central nervous system of marsupials. *International Reviews of Cytology* **214**, 193–224.

Goldstein, J.M., Seidman, L.K.J., Horton, N.J., Makris, N., Kennedy, D.N., Caviness, V.S., Farone, S.V. and Tsuang, M.T. 2001. Normal sexual dimorphism of the adult human brain assessed by in vivo magnetic resonance imaging. *Cerebral Cortex* **11**, 490–497.

Gorski, R.A. 1963. Modification of ovulatory mechanisms by postnatal administration of estrogen to the rat. *American Journal of Physiology* **205**, 842–844.

Gorski, R.A. 1988. Sexual differentiation of the brain: Mechanisms and implications for neuroscience. In: S.S. Easter, K.F. Barald and B.M. Carlson (Eds.), *From Message to Mind.* Sinaeur, Sutherland, Massachusetts, pp. 256–271.

Gorski, R.A. Gordon, J.H., Shryne, J.E. and Southam, A.M. 1978. Evidence for a morphological sex difference within the medial preoptic area of the rat brain. *Brain Research* **148**, 333–346.

Gorski, R.A., Gordon, J.H., Shryne, J.E. and Southam, A.M. 1980. Evidence for a morphological sex difference within the medial preoptic area of the rat. *Journal of Comparative Neurology* **193**, 529–539.

Goy, R.W. 1966. Role of androgens in the establishment and regulation of behavioral sex differences in mammals. *Journal of Animal Science* **25**, 21–31.

Hamer, D.H., Hu, S., Nagnuson, V.L., Hu, N. and Pattatucci, A.M. 1993. A linkage between DNA markers on the X chromosome and male sexual orientation. *Science* **261**, 321–327.

Hammer, R.P., Ricalde, A.A. and Seatriz, J.V. 1989. Effects of opiates on brain development. *Neurotoxicology* **10**, 475–484.

Hiscock, M., Inch, R., Jacek, C., Hiscock-Kalil, C. and Kalil, C.M. 1994. Is there a sex difference in human laterality? I. An exhaustive survey of auditory laterality studies from six neuropsychology journals. *Journal of Clinical and Experimental Neuropsychology* **16**, 423–435.

Hiscock, M., Israelian, M. Inch, R., Jacek, C. and Kalil-Hiscock, C. 1995. Is there a sex difference in human laterality? II. An exhaustive survey of visual laterality studies from six neuropsychology journals. *Journal of Clinical and Experimental Neuropsychology* **17**, 590–610.

Holloway, R.L., Anderson, P.J., Defendini, R. and Harper, C. 2005. Sexual dimorphism of the human corpus callosum from three independent samples: Relative size of the corpus callosum. *American Journal of Physical Anthropology* **92**, 481–498.

Houtsmuller, E.J., Brand, T., De Jonge, F.H., Joosten, R.N., Van de Poll, N.E. and Slob, A.K. 1994. SDN-POA volume, sexual behavior, and partner preference in male rats affected by perinatal treatment with ATD. *Physiology and Behavior* **56**, 535–541.

Huston, A.C. 1983. Sex-typing. In: E.M. Hetherington (Ed.), *Handbook of Child Psychology Socialization, Personality and Social Development,* Vol. 4, 4th Edition. Wiley, New York, pp. 387–467.

Imperato-McGinley, J., Miller, M., Wilson, J.D., Peterson, R.E., Shackleton, C. and Gajdusek, D.C. 1991. A cluster of male pseudohermaphrodites with 5α-reductase deficiency in Papua New Guinea. *Clinical Endocrinology* **34**, 293–298.

Imperato-McGinley, J., Peterson, R.E., Gautier, T. and Sturla, E. 1979. Androgens and the evolution of male gender identity among male pseudohermaphrodites with 5α-reductase deficiency. *New England Journal of Medicine* **300**, 1233–1237.

Ishunina, T.A., Salehi, A., Hofman, M.A. and Swaab, D.F. 1999. Activity of vasopressinergic neurons of the human supraoptic nucleus is age- and sex-dependent. *Journal of Neuroendocrinology* **11**, 251–258.

Johansen, J.A., Jordan, C.L. and Breedlove, S.M. 2004. Steroid hormone masculinization of neural structures in rats: A tale of two nuclei. *Physiology and Behavior* **83**, 271–277.

Johnston, H.M., Payne, A.P. and Gilmore, D.P. 1992. Perinatal exposure to morphine affects adult sexual behaviour of the male golden hamster. *Pharmacology Physiology and Behavior* **42**, 41–44.

Johnston, H.M., Payne, A.P. and Gilmore, D.P. 1994. Effect of exposure to morphine throughout gestation on feminine and masculine adult sexual behaviour in golden hamsters. *Journal of Reproduction and Fertility* **100**, 173–176.

Johnston, H.M., Payne, A.P., Gilmore, D.P. and Wilson, C.A. 1990. Neonatal serotonin manipulation alters the adult sexual behavior of golden hamsters. *Pharmacology Biochemistry and Behavior* **35**, 571–575.

Jordan, C.L., Breedlove, M. and Arnold, A.P. 1982. Sexual dimorphism and the influence of neonatal androgen in the dorsolateral nucleus of the rat spinal cord. *Brain Research* **259**, 309–314.

Kent, J., Pert, C.B. and Herkenham, M. 1982. Ontogeny of opiate receptors in rat forebrain: Visualization by in vitro autoradiography. *Developmental Brain Research* **2**, 487–504.

Kornblum, H.I., Hurlbut, D.E. and Leslie, F.M. 1987. Postnatal development of multiple opioid receptors in rat brain. *Developmental Brain Research* **37**, 21–41.

Kruijver, F.P.M., Zhou, J.-N., Pool, C.W., Hofman, M.A., Gooren, L.J.G. and Swaab, D.F. 2000. Male-to-female transsexuals have female neuron numbers in a limbic nucleus. *Journal of Clinical Endocrinology and Metabolism* **85**, 2034–2041.

Kudwa, A.E., Michopoulos, V., Gatewood, J.D. and Rissman, E.F. 2006. Roles of estrogen receptors alpha and beta in differentiation of mouse sexual behavior. *Neuroscience* **138**, 921–928.

Lauder, J.M. 1990. Ontogeny of the serotonergic system in the rat: Serotonin as a developmental signal. *Annals of the New York Academy of Sciences* **600**, 297–314.

Lephart, E.D.1996. A review of brain aromatase cytochrome P450. *Brain Research Reviews* **22**, 1–26.

Le Vay, S. 1991. A difference in hypothalamic structure between heterosexual and homosexual men. *Science* **235**, 1034–1037.

Lombardo, M.V., Ashwin, E., Auyeung, B., Chakrabarti, B., Taylor, K., Hackett, G., Bullmore, E.T. and Baron-Cohen, S. 2012. Fetal testosterone influences sexually dimorphic gray matter in the human brain. *Journal of Neuroscience* **32**, 674–689.

Loring, D. and Meador, K. 1990. Cerebral language lateralization. *Neuropsychologia* **28**, 831–838.

Lowrie, M.B. and Vrbová, G. 1992. Dependence of postnatal motoneurons on their targets: Review and hypothesis. *Trends in Neuroscience* **15**, 80–84.

Maatouk, D.M., DiNapoli, L., Alvers, A., Parker, K.L., Taketo, M.M. and Capel, B. 2008. Stabilization of β-catenin in XY gonads, causes male-to-female sex-reversal. *Human Molecular Genetics* **17**, 2949–2955.

Maccoby, E.E. and Jacklin, C.N. 1974. *The Psychology of Sex Differences*. Stanford University Press, Stanford, California.

Mackay, S. 2000. Gonadal development in mammals at the cellular and molecular levels. *International Reviews of Cytology* **200**, 47–99.

MacLusky, N.J., Philip, A., Hurlburt, C. and Naftolin, F. 1985. Estrogen formation in the developing rat brain: Sex differences in aromatase activity during early post-natal life. *Psychoneuroendocrinology* **10**, 355–361.

Madeira, M.D. and Lieberman, A.R. 1995. Sexual dimorphism in the mammalian limbic system. *Progress in Neurobiology* **45**, 275–333.

Magnan, J. and Tiberi, M. 1989. Evidence for the presence of μ- and κ- but not δ-opioid sites in the human fetal brain. *Developmental Brain Research* **45**, 275–281.

Maguire, E.A., Burgess, N. and O'Keefe, J. 1999. Human spatial navigation: Cognitive maps, sexual dimorphism and neural substrates. *Current Opinion in Neurobiology* **9**, 171–177.

Marchant-Haycox, S.E., McManus, I.C. and Wilson, G.D. 1991. Left-handedness, homosexuality, HIV infection and AIDS. *Cortex* **27**, 49–56.

Matsumoto, A. and Arai, Y. 1986. Male-female differences in synaptic organization of the ventromedial nucleus of the hypothalamus in the rat. *Neuroendocrinology* **42**, 232–236.

Mazur, A. and Booth, A. 1998. Testosterone and dominance in men. *Behavioral Brain Science* **21**, 353–363.

McCormick, C.M., Witelson, S.F. and Kingstone E. 1990. Left-handedness in homosexual men and women: Neuroendocrine implications. *Psychoneuroendocrinology* **15**, 69–76.

McEwen, B.S. 1991. Non-genomic and genomic effects of steroids on neural activity. *Trends in Pharmacological Sciences* **12**, 141–147.

McEwen, B.S., Lieberburg, L., Chaptal, C. and Krey, L.C. 1977. Aromatization: Important for sexual differentiation of the neonatal rat brain. *Hormones and Behavior* **9**, 249–263.

McLaughlin, P.J., Tobias, S.W., Lang, C.M. and Zagon, I.S. 1996. Chronic exposure to the opioid antagonist naltrexone during pregnancy: Maternal and offspring effects. *Physiology and Behavior* **62**, 501–508.

Melis, M.R. and Argiolas, A. 1995. Dopamine and sexual behavior. *Neuroscience and Biobehavioral Reviews* **19**, 19–38.

Micevych, P.E., Park, S.S., Akesson, T.R. and Elde, R. 1987. Distribution of cholecystokinin-immunoreactive cell bodies in male and female rats. *Journal of Comparative Neurology* **225**, 124–136.

Miesfield, R.M. 1989. The structure and function of steroid receptor proteins. *Critical Reviews in Biochemistry and Molecular Biology* **24**, 101–117.

Money, J., Schwartz, M. and Lewis, V.G. 1984. Adult heterosexual status and fetal hormonal masculinization and demasculinization: 46,XX congenital virilizing adrenal hyperplasia and 46,XY androgen insensitivity syndrome compared. *Psychoneuroendocrinology* **9**, 405–414.

Morris, J.A., Jordan, C.L., King, Z.A., Northcutt, K.V. and Breedlove, S.M. 2008. Sexual dimorphism and steroid responsiveness of the posterodorsal medial amygdala in adult mice. *Brain Research* **1190**, 115–121.

Nagy, J.I. and Senba, E. 1985. Neural relations of cremaster motoneurons, spinal cord systems and the genitofemoral nerve in the rat. *Brain Research Bulletin* **15**, 609–627.

Nordeen, E.J., Nordeen, K.W., Sengelaub, D.R. and Arnold, A.P. 1985. Androgens prevent normally occurring cell death in a sexually dimorphic spinal nucleus. *Science* **229**, 671–673.

Parker, M.G. 1990. Mechanisms of action of steroid receptors in the regulation of gene transcription. *Journal of Reproduction and Fertility* **88**, 717–720.

Payne, A.P. 1996. Gonadal hormones and the sexual differentiation of the nervous system: Mechanisms and interactions. In: W. Stone (Ed.), *CNS Neurotransmitters and Neuromodulators Neuroactive Steroids*. CRC Press, Boca Raton, FL, pp. 153–175.

Petrillo, P., Tavani, A., Verotta, D., Robson, L.E. and Kosterlitz, H.W. 1987. Differential postnatal development of μ-, δ-, and κ-opioid binding sites in rat brain. *Developmental Brain Research* **31**, 53–58.

Pfaus, J.G. and Everitt, B.J. 1995. Psychopharmacology of sexual behavior. In: F.E. Blood and D.J. Kupfer (Eds.), *Psychopharmacology: The Fourth Generation of Progress*. Raven Press, New York, pp. 743–758.

Pinkston, J.B., Johnson, S.C., Bigler, E.D. and Blatter, D.D. 1997. Effects of handedness and gender on the surface area of the human corpus callosum: A preliminary study using magnetic resonance imaging. *Archives of Clinical Neuropsychology* **12**, 386.

Pomerantz, R.W., Goy, R.W. and Roy, M.M. 1986. Expression of male-typical behavior in adult female pseudohermaphroditic rhesus: Comparisons with normal males and neonatally gonadectomized males and neonatally gonadectomized males and females. *Hormones and Behavior* **20**, 483–500.

Raz, N., Gunning-Dixon, F., Head, D., Rodrigue, K.M., Williamson, A. and Acker, J.D. 2004. Aging, sexual dimorphism, and hemispheric asymmetry of the cerebral cortex: Replicability of regional differences in volume. *Neurobiology of Aging* **25**, 377–396.

Ricalde, A.A. and Hammer, R.P. 1990. Perinatal opiate treatment delays growth of cortical dendrites. *Neuroscience Letters* **115**, 137–143.

Robichon, F. and Habib, M. 1998. Abnormal callosal morphology in male adult dyslexics: Relationship to handedness and phonological abilities. *Brain and Language* **62**, 127–146.

Roselli, C.E., Larkin, K., Resko, J.A., Stellflug, J.N. and Stormshak, F. 2004. The volume of a sexually dimorphic nucleus in the ovine medial preoptic area/anterior hypothalamus varies with sexual partner preference. *Endocrinology* **145**, 478–483.

Sanders, A.R., Martin, E.R., Beecham, G.W., Guo, S., Dawood, K., Rieger, G., Badner, J.A. et al. 2014. Genome-wide scan demonstrates significant linkage for male sexual orientation. *Psychological Medicine* **17**, 1–10.

Scallet, A.C. 1982. Effects of conditioned fear and environmental novelty on plasma β-endorphin in the rat. *Peptides* **3**, 203–206.

Schwanzel-Fukuda, M. and Pfaff, D.W. 1989. Origin of luteinizing hormone-releasing hormone neurons. *Nature* **338**, 161–164.

Sengelaub, D.R. and Arnold, A.P. 1986. Development and loss of early projections in a sexually dimorphic rat spinal nucleus. *Journal of Neuroscience* **6**, 1613–1620.

Sholl, S.A. and Kim, K.L. 1990. Aromatase, 5-alpha-reductase and androgen receptor levels in the fetal monkey brain. *Neuroendocrinology* **52**, 94–98.

Siddiqui, A. and Gilmore, D.P. 1988. Regional differences in the catecholamine content of the rat brain: Effects of neonatal castration and androgenization. *Acta Endocrinologica (Copenh)* **118**, 483–494.

Siddiqui, A., Gilmore, D.P. and Clark, J. 1989. Regional differences in the indoleamine content of the rat brain: Effects of neonatal castration and androgenization. *Biogenic Amines* **6**, 105–114.

Simerly, R.B. and Swanson, L.W. 1987. Castration reversibly alters levels of cholecystokinin-immunoreactivity within cells of three interconnected sexually dimorphic forebrain nuclei of the rat. *Proceedings of the National Academy of Sciences USA* **84**, 2087–2091.

Singh, H.H., Purohit, V. and Ahluwalia, B.S. 1980. Effect of methadone treatment on the fetal testes and hypothalamus in rats. *Biology of Reproduction* **22**, 480–485.

Suzuki, M., Hagino, H., Nohara, S., Zhou, S.-Y., Kawasaki, Y., Takahashi, T., Matsui, M., Seto, H., Ono, T. and Kurachi, M. 2005. Male-specific volume expansion of the human hippocampus during adolescence. *Cerebral Cortex* **15**, 187–193.

Swaab, D.F. and Fliers, E. 1985. A sexually dimorphic nucleus in the human brain. *Science* **228**, 1112–1115.

Swaab, D.F. and Hofman, M.A. 1990. An enlarged suprachiasmatic nucleus in homosexual men. *Brain Research* **537**, 141–148.

Swaab, D.F. and Hofman M.A. 1995. Sexual differentiation of the human hypothalamus in relation to gender and sexual orientation. *Trends in Neuroscience* **18**, 264–270.

Swaab, D.F., Zhou, J.N., Ehlhart, T. and Hofman, M.A. 1994. Development of vasoactive intestinal polypeptide neurons in the human suprachiasmatic nucleus (SCN) in relation to birth and sex. *Brain Research: Developmental Brain Research* **79**, 249–259.

Tempel, A., Habas, J., Paredes, W. and Barr, G.A. 1988. Morphine-induced down regulation of μ-opioid receptors in neonatal rat brain. *Developmental Brain Research* **41**, 129–133.

Tobet, S.A. 2003. Genes controlling hypothalamic development and sexual differentiation. *European Journal of Neuroscience* **16**, 373–376.

Tobet, S.A., Dellovade, T.L., Parker, K. and Homanics, G. 2002. Positioning estrogen receptor alpha-containing cells during hypothalamic development. In: R.J. Handa, S. Hayashi, E. Teresawa and M. Kawata (Eds.), *Development and Steroid Hormone Action*. CRC Press, Florida, pp. 59–75.

Tobin, A.M. and Payne, A.P. 1991. Perinatal androgen administration and the maintenance of sexually dimorphic and nondimorphic lumbosacral motor neurone groups in female Albino Swiss rats. *Journal of Anatomy* **177**, 47–53.

Toran-Allerand, C.D. 1985. On the genesis of sexual differentiation of the central nervous system: Morphogenetic consequences of steroid exposure and possible role of α-fetoprotein. *Progress in Brain Research* **61**, 63–98.

Toscano, J.E., Bauman, M.D., Mason, W.A. and Amaral, D.G. 2009. Interest in infants by female rhesus monkeys with neonatal lesions of the amygdala or hippocampus. *Neuroscience* **162**, 881–891.

Tuncer, M.C., Hatipoğlu, E.S. and Őzateş, M. 2005. Sexual dimorphism and handedness in the human corpus callosum based on magnetic resonance imaging. *Surgical and Radiologic Anatomy* **27**, 254–259.

Vainio, S., Kispert, M., Chin, N. and McMahon, A.P. 1999. Female development in mammals is regulated by Wnt-4 signalling. *Nature* **397**, 405–409.

Van der Schoot, P., Payne, A.P. and Kersten, W. 1999. Sex difference in target seeking behavior of developing cremaster muscles and the resulting first sign of somatic sex differentiation in marsupial mammals. *Anatomical Record* **255**, 130–141.

Walters, M.R. 1985. Steroid hormone receptors and the nucleus. *Endocrine Reviews* **6**, 515–543.

Ward, I.L. 1983. Effects of maternal stress on the sexual behavior of male offspring. *Monographs in Neural Science* **9**, 169–175.

Ward, Jr., O.B., Orth, J.M. and Weisz, J. 1983. A possible role of opiates in modifying sexual differentiation. *Monographs in Neural Science* **9**, 194–200.

Wilson, G. and Rahman, Q. 2005. *Born Gay – The Psychobiology of Sex Orientation*. Peter Owen Publishers, London.

Witelson, S.F. 1989. Hand and sex differences in the isthmus and genu of the human corpus callosum: A postmortem morphological study. *Brain* **112**, 799–835.

Witelson, S.F. and Goldsmith, C. 1991. The relationship of hand preference to the anatomy of the corpus callosum in men. *Brain Research* **545**, 175–182.

Wray, S., Grant, P. and Gainer, H. 1989. Evidence that cells expressing luteinizing hormone mRNA in the mouse are derived from progenitor cells in the olfactory placode. *Proceedings of the National Academy of Sciences USA* **86**, 8132–8136.

Zagon, I.S. and McLaughlin, P.J. 1987. Endogenous opioid systems regulating cell proliferation in the developing rat brain. *Brain Research* **412**, 68–72.

Zhou, J.N., Hofman, M.A., Gooren, L.J.G. and Swaab, D.F. 1995a. VIP neurons in the human SCN in relation to sex, age and Alzheimer's disease. *Neurobiology of Aging* **16**, 571–576.

Zhou, J.N., Hofman, M.A., Gooren, L.J.G. and Swaab, D.F. 1995b. A sex difference in the human brain and its relation to transsexuality. *Nature* **378**, 68–70.

Chapter 2

Mammalian Gonadotropin-Releasing Hormone

Edathil Vijayan

Contents

2.1 Introduction ... 31
2.2 Structure ... 32
2.3 Control of GnRH Release ... 32
2.4 GnRH Action in the Gonadotrope ... 33
2.5 GnRH Variants and Receptor Genes ... 34
2.6 Potential Functions of GnRH I and GnRH II ... 36
2.7 Functional GnRH Receptor in Humans ... 37
Acknowledgements ... 38
References ... 38

2.1 Introduction

Luteinising hormone (LH)–releasing activity was first found in hypothalamic extracts utilising the ovarian ascorbic acid depletion assay of Parlow to measure LH released following injection of acid extracts of stalk-median eminence (McCann et al. 1960). Similarly prepared extracts were shown to evoke ovulation following intrapituitary injection (Campbell et al. 1964). The LH-releasing activity from the extract was rapidly purified, shown to be a peptide present in large amounts in the hypothalamus, and stimulates the release of LH and follicle-stimulating hormone

(FSH); its output into hypophyseal portal vessel blood under experimental and physiological conditions correlates well with the output of LH from the anterior pituitary gland. The decapeptide, now known as GnRH (gonadotropin-releasing hormone), is the primary hypothalamic signal necessary for the structural and functional integrity of the pituitary gonadotropes; regulates the synthesis, glycosylation, and secretion of LH and FSH; and has the apparently unique quality of being able to increase the responsiveness of the pituitary to itself (the priming effect of GnRH) which, in turn, stimulates gonadal steroidogenesis and gametogenesis (Fink 1988). The purification, sequencing and synthesis of the decapeptide by Schally in 1971 is considered a milestone in discovering the role of neuroendocrine system in reproduction (Matsuo et al. 1971).

GnRH is considered a neurohormone, a hormone produced in a specific neural cell and released at its terminal. A key area for production of GnRH is the preoptic area of the hypothalamus, which contains most of the GnRH-secreting neurons. GnRH is secreted into the hypophyseal portal bloodstream at the median eminence. The portal blood carries the GnRH to the pituitary gland, which contains the gonadotrope cells, where GnRH activates its own receptor, gonadotropin-releasing hormone receptor (GnRHR) (Stojilkovic et al. 1994); this is a seven-transmembrane G protein-coupled receptor (Kakar et al. 1992) that stimulates the beta isoform of phosphoinositide phospholipase C, which goes on to mobilise calcium and protein kinase C. This cascade results in the activation of proteins involved in the synthesis and secretion of the gonadotropins, LH and FSH leading to steroidogenic and gametogenic functions of the gonads of both sexes (Fink 1988). GnRH is degraded by proteolysis within a few minutes. Because of GnRH rate-limiting role in reproductive processes, a very large number of GnRH analogs (thousands) have been studied. These have found several clinical uses including important roles in treatment for prostate cancer, and another as a component of the treatment regimen for ovulation induction in women undergoing in vitro fertilisation (Vijayan 1985; Conn and Crowley 1994; Lipman 1998; Fauser et al. 1999; Hong et al. 2008; Schneider and Rissman 2008; Montagnani et al. 2009; Wu et al. 2009).

2.2 Structure

pyroGlu–His–Trp–Ser–Tyr–Gly–Leu–Arg–Pro–Gly–$CONH_2$

2.3 Control of GnRH Release

Pulsatile LH release in mammals including the human appears to be a prerequisite for normal gonadal function. Basal serum LH levels are not the result of continuous GnRH release into portal vessels, but rather due to GnRH pulses which are being released at relatively regular intervals. The frequency of occurrence and the regularity of these GnRH pulses are quite similar in all species (Wuttke et al. 1983).

The pulsatile release of gonadotropins is the consequence of rhythmic discharges of GnRH into the pituitary portal circulation. Inactivation of this hypothalamic decapeptide by the administration of antisera against GnRH causes an abrupt reduction in plasma LH and FSH concentrations (McCormack et al. 1977). The intermittent delivery of endogenous GnRH to the pituitary gland must be the consequence of a synchronous discharge of GnRH-containing neurons effected by some neuronal signal generator or oscillator (Knobil 1981; Wuttke et al. 1983). Since complete neural disconnection of the medial basal hypothalamus (MBH) from the remainder of the central nervous system does not interfere with the circhoral pattern of gonadotropin secretion in ovariectomised monkeys (Krey et al. 1975), the circhoral oscillator as well as the cell bodies of GnRH-producing neurons must reside within the MBH. The physiologic significance of the rhythmic, pulsatile pattern of gonadotropin secretion remained obscure until recently when it was recognised that the functioning of the hypophysiotropic control system that directs gonadotropin secretion is obligatorily intermittent. This conclusion was derived from the observation that continuous infusion of GnRH into female monkeys with lesions of the arcuate nucleus, that abolished endogenous GnRH secretion, was incapable of reinitiating sustained gonadotropin secretion, whereas infusion of the decapeptide as 6 min pulses once every hour reestablished normal pituitary function (Nakai et al. 1978). Furthermore, continuous infusion of GnRH profoundly inhibited gonadotropin secretion previously established by the pulsatile administration of the decapeptide to these animals (Belchetz et al. 1978; Babu and Vijayan 1983). A variety of neuroleptic drugs and alpha-adrenergic blocking agents, acting at the neural level rather than directly on the pituitary, promptly interrupts the pulsatile discharge of gonadotropins (Vijayan and McCann 1978, 1980). The secretory pattern of gonadotropins during menstrual cycle is the consequence of negative and positive feedback effects of ovarian estradiol. Hypothalamic GnRH secretion plays a permissive, and albeit obligatory, role in the feedback effects of estradiol at the pituitary gland.

2.4 GnRH Action in the Gonadotrope

The ability of GnRH to regulate gonadotropin release is important for human and veterinary medicine. Consequently there has been a great deal of interest in understanding the molecular basis of GnRH action. The availability of large number of analogs, some of which are metabolically stable and can be attributed to a high specific activity, has been useful in understanding the subcellular locus, physiological regulation, and chemical nature of the receptor. The interaction of GnRH with its plasma membrane receptor results in receptor regulation, altered cellular responsiveness, and release of gonadotropins, LH and FSH. A model for GnRH action proposed by Conn et al. (1987) is presented in Figure 2.1.

The GnRH receptor is coupled to a calcium ion channel, and structure-activity studies with ion channel antagonists indicate that this channel is similar, but not

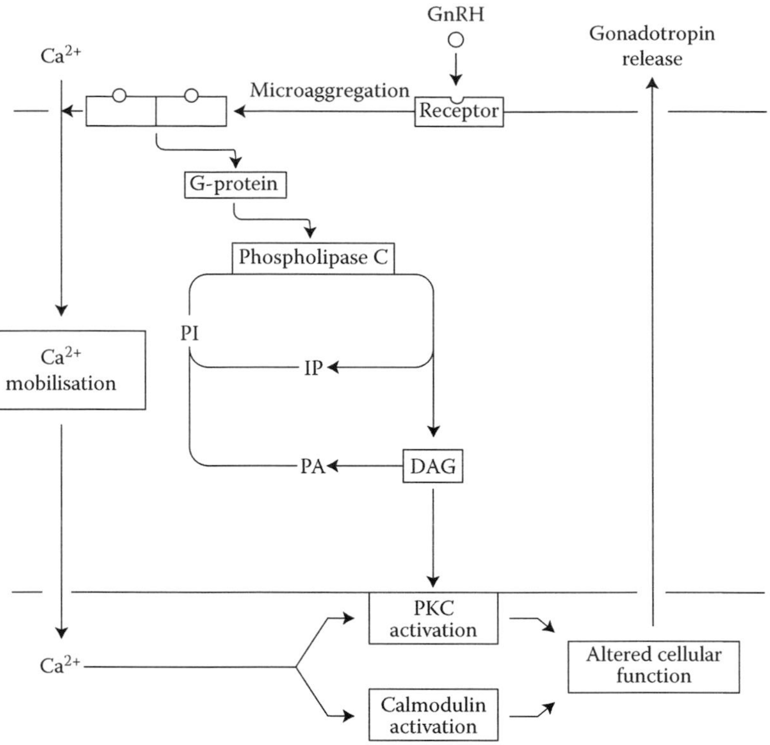

Figure 2.1 Molecular events mediating GnRH-stimulate gonadotropin release. (Adapted from Conn PM et al. 1987. *Biology of Reproduction* 36:17–35.)

identical, to that observed in nerve and muscle tissue. Calcium mobilised from the extracellular space fulfills the requirement of a second messenger, and calcium is mobilised in response to GnRH, and can activate LH release. Calmodulin appears to behave as an intracellular messenger for calcium. The receptor is also intimately coupled to a GTP-binding protein (G-protein) that can activate phospholipase C, which leads to the production of diacylglycerols, and activate protein kinase C redistribution and its activation can synergise with calcium ionophores to provoke gonadotropin release. Agents which provoke inositol phospholipids hydrolysis (GTP and its stable analogs) or activate PKC (DAGs and phorbol esters) can cause the release of LH from gonadotrops (Conn et al. 1987).

2.5 GnRH Variants and Receptor Genes

The traditional view is that mammals express only a single form of GnRH known as mammalian GnRH I. However, at least 16 novel decapeptides have now been identified as GnRH structural variants in lower vertebrates (Grove-Strawser et al. 2002), and more than one form of GnRH is expressed in most species. Among these, a form of GnRH originally isolated from chicken brain (Chicken GnRH II: pGlu-His-Ser-His-Gly-Trp- Tyr-Pro-Gly-NH_2) is universally conserved from fish to *Homo sapiens* (Miyamoto et al. 1984; Sherwood et al. 1993; Sealfon et al. 1997; Chen et al. 1998; White et al. 1998). GnRH II has been cloned from monkey brain (Lescheid et al. 1997; Urbanski et al. 1999) and is expressed in a concentrated fashion in at least three distinct regions of the male rhesus monkey hypothalamus: the supraoptic and paraventricular nuclei and the medial basal hypothalamus. However, GnRH II is not co-expressed with GnRH I peptide (Latimer et al. 2000), suggesting its unique regulation and function. Recently, a seven-transmembrane, G protein-coupled receptor that is highly selective for GnRH II was cloned and characterised from the monkey (Miller et al. 2001; Neill et al. 2001) as well as from a number of fish and amphibians (Tensen et al. 1997; Wang et al. 2001). Miller et al. (2001) identified mRNA receptor by polymerase chain reaction (PCR) in the marmoset pituitary and Neill et al. (2001) by Northern blot analysis in the human pituitary using a cDNA to rhesus monkey pituitary GnRH II receptor. In both human and non-human primates GnRH II receptor expression was found at similar levels in a wide variety of extrapituitary tissues. Unlike the type I GnRH receptor (Kakar et al. 1992), the type II receptor has a C-terminal cytoplasmic tail that is phosphorylated when the receptor binds GnRH II, leading to receptor internalisation and desensitisation (Blomenrohr et al. 1999). The type II GnRH receptor was also identified by immunocytochemistry in the majority of ovine gonadotrops (Miller et al. 2001) but so far has not been identified in gonadotrops of primates. In rams, 10 μg of GnRH II iv stimulated LH and FSH secretion, although less effectively than GnRH I. Moreover, the ratio of FSH to LH secretion was higher following GnRH II administration than following GnRH I, suggesting that GnRH II could play a role in differential secretion of FSH and LH (Miller et al. 2001). In monkeys a robust rise in LH secretion occurred after GnRH II administration, but FSH was not measured (Lescheid et al. 1997). Expression of GnRH II receptors in the monkey hypothalamus (Miyamoto et al. 1984; Sherwood et al. 1993) and identification of a putative GnRH II receptor in human and marmoset pituitary by Neill et al. (2001) and Miller et al. (2001) raise the question whether GnRH II activates this receptor to regulate synthesis and secretion of gonadotropin in primates (Terasawa 2003). Okada et al. (2003) using pituitary cell cultures from adult male monkeys showed GnRH II stimulation of LH and FSH release which was less effective than GnRH I. Further gonadotropin stimulatory actions of GnRH II, like those of GnRH I, were blocked by a GnRH I antagonist Antide. Addition of GnRH II failed to stimulate LH or FSH beyond the maximal stimulation by GnRH I alone. Both GnRH II and GnRH I stimulated FSH and LH release within 6 h in plated cultures. These data imply that GnRH II stimulates LH and FSH release from monkey pituitary cells by activating GnRH I receptors (Okada et al. 2003). Densmore and Urbanski (2003) also reported blockade of GnRH II effects by Antide and GnRH I and GnRH II produced similar elevations of circulating LH and FSH in female monkeys. They concluded that gonadotropin stimulation by GnRH II was possibly mediated via GnRH I receptor. The finding that GnRH II is less effective than GnRH I to stimulate LH and FSH in monkey pituitary cell cultures is also consistent with earlier results for sheep pituitary cell cultures (Miller et al. 1986). GnRH II was approximately one-third as potent as GnRH I in pituitary cell cultures from rats where GnRH II receptors appear to be absent (Neill 2002). GnRH II was 5- to 6-fold more potent than chicken GnRH I in releasing LH from pituitary cell cultures from male chickens (King et al. 1988) and 2.5-fold more potent using perifused pituitary cells from male turkeys (Guemene and Williams 1992). Thus the GnRH II system appears to be more active in non-mammalian species.

2.6 Potential Functions of GnRH I and GnRH II

Differential regulation of FSH and LH release by hypothalamic GnRH in rats has been studied in great detail (Lumpkin et al. 1980). Chicken GnRH II was found to have twofold greater potency to release FSH versus LH when compared to stimulation by chicken GnRH I from pituitary cells of mature hens (Miller et al. 1986). Chicken GnRH II was no more potent than mammalian GnRH to stimulate FSH when compared to LH from rat hemipituitaries (Yu et al. 1997). Mean ratio of circulating FSH to LH was twofold higher in rams following 10 μg i.v. doses of GnRH II compared with GnRH I (Miller et al. 2001). Monkey pituitary cell culture studies also did not support the notion that GnRH II plays any major role in the selective secretion of FSH in male primates (Okada et al. 2003). Miller et al. (2001) identified GnRH II receptor mRNA in the marmoset pituitary by PCR using a cDNA from marmoset brain, and in the preoptic area, amygdala and periventricular area of the hypothalamus in the monkey by immunocytochemistry using an antiserum to a synthetic peptide from the human (Miller et al. 2001). GnRH I and GnRH II mRNA and peptide are not co-expressed in the monkey brain and that there is no evidence that GnRH II is found at high concentration in hypothalamic portal blood, and functions of these similar decapeptides may be quite different (Latimer et al. 2000). GnRH II has been shown to regulate sexual behaviour (Sherwood et al. 1993). Administration of GnRH II, but not GnRH I, stimulated courtship behaviour in female sparrows (Maney et al. 1997). Administration of GnRH II into the cerebral ventricle of female musk shrews, a nocturnal mammal, stimulated sexual behaviour if the animals were food restricted for 48 h (Temple et al. 2003). GnRH II mRNA expression was found to be substantially higher in the mediobasal hypothalamus in adult compared with juvenile monkeys (Latimer et al. 2001). Though our understanding of the neuroendocrine control of behaviour in male primates is limited, greater presence of GnRH II mRNA expression in the hypothalamus may influence this in primates. Overall it is possible that the GnRH II receptor may have a unique function in the monkey brain and pituitary other than as a regulator of gonadotropin secretion. Human GnRH II receptor sequences are located on chromosomes 1 and 14. The chromosome 14 sequence is comprised of only exons 2 and 3, contains an intron, and is transcribed in the antisense orientation (Miller et al. 1999). A second human type II receptor based on exon 1 expression was found by dot blotting in human sperm and faintly in central nervous system and peripheral tissues, but not in human pituitary. This gene contains a stop codon and a frame shift, and it may be a pseudogene. Thus, whether a full-length functional GnRH II receptor is expressed in humans remains uncertain (Kakar and Winters 2002; Van Biljon et al. 2002).

2.7 Functional GnRH Receptor in Humans

Despite GnRH II receptor transcripts being detected in a series of human tissues, no full length appropriately processed transcript representing GnRH II receptor in human tissues was found using reverse transcription (RT)-PCR. Most transcripts detected differ from the monkey GnRH II receptor cDNAs in one or more respects leading to the conclusion that if a functional GnRH II receptor is expressed in humans, it may be a 5TM domain receptor lacking TM regions I and II. A definitive test of the existence of a functional GnRH II receptor in humans will be the description of a GnRH II–specific response in human cells or tissues. GnRH II response specificity will be a response that is more sensitive to GnRH II than to GnRH I and/or a GnRH II response that is not inhibited by GnRH I antagonists. A functional GnRH II receptor transcript should be demonstrable under such conditions. Thus, GnRH II gene is expressed in all tissues and cells studied so far as incompletely processed transcripts, and under the appropriate physiological conditions a properly processed GnRH II receptor will be found (Neill 2002).

GnRH has been used extensively as a tool in neuroendocrine research. Early attempts to use this decapeptide clinically for the treatment of reproductive disorders supposed to be due to an inadequate secretion of endogenous GnRH were, however, of only limited success. Effective therapeutic use had to await further progress in the understanding of the physiologic mechanisms that control gonadotropin secretion and gonadal function. The pattern of hypophysiotropic stimulation is of critical importance and the elucidation of the physiologic significance of pulsatile gonadotropin secretion have provided the rational basis for the efficient use of synthetic GnRH in the treatment of GnRH deficiency. These findings have also furthered the understanding of seemingly paradoxical antifertility effects of long-acting GnRH analogues initially designed to compensate for the short action of the parent decapeptide and thus to simplify the treatment of infertility (Knobil 1981; Vijayan 1985). GnRH is found in organs outside the hypothalamus and pituitary, and its role in other processes is poorly understood. The nucleotide sequencing of the human genome provided the opportunity to determine which of these peptides and receptors might be expressed in primates. Of the four GnRHs reportedly expressed in mammals, only GnRH I (mammalian GnRH) and GnRH II (chicken GnRH II) genes were identified in the human genome. The experimentally expressed GnRH II receptor was functional with and specific for GnRH II, and was, unlike the GnRH I receptor, desensitised to continuous GnRH treatment. GnRH II receptor mRNA is expressed ubiquitously in human tissues.

Questions remain about the potential functions of the primate GnRH II receptor such as regulation of gonadotropin secretion, female sexual behaviour and tumor cell growth, and also about whether it is expressed as a full-length, functional gene transcript in humans.

Acknowledgements

This chapter is dedicated to my mentors: Dr. AG Sathyanesan, FNA, who introduced me to neuroendocrinology, and Dr. SM McCann, with whom I had the privilege of working for several years, for inspiring my career.

I wish to thank Dr. CS Paulose, professor and head, Department of Biotechnology and Director, Centre for Neuroscience, CUSAT, and the Indian Council of Medical Research (ICMR), New Delhi for the Emeritus Medical Scientist Award.

References

Babu N, Vijayan E. 1983. Antifertility actions of luteinizing hormone-releasing hormone (LHRH) in rat. *Ind J Exp Biol* 21: 408–409.

Belchetz PE, Plant TM, Nakai Y, Keogh EJ, Knobil E. 1978. Hypophysial responses to continuous and intermittent delivery of hypothalamic gonadotropin releasing hormone. *Science* 202: 163.

Blomenrohr M, Heding A, Sellar R, Leurs R, Bogerd J, Eidne KA, Willars GB. 1999. Pivotal role for the cytoplasmic carboxyl-terminal tail of a nonmammalian gonadotropin-releasing hormone receptor in cell surface expression, ligand binding, and receptor phosphorylation and internalization. *Mole Pharmacol* 56: 1229–1237.

Campbell HJ, Feuer G, Harris GW. 1964. The effect of intrapituitary infusion of median eminence and other brain extracts on anterior pituitary gonadotropin secretion. *J Physiol (London)* 170: 474–478.

Chen A, Yahalom D, Ben-Aroya N, Kaganovsky E, Okon E, Koch YA. 1998. A second isoform of gonadotropin-releasing hormone is present in the brain of human and rodents. *FEBS Lett* 435: 199–203.

Conn PM, Crowley WJ Jr. 1994. Gonadotropin-releasing hormone and its antagonists. *Annu Rev Med* 45: 391–405.

Conn PM, McArdale CA, Andrews WV, Huckle WR. 1987. The molecular basis of gonadotropin-releasing hormone (GnRH) action in the pituitary gonadotrope. *Biology of Reproduction* 36: 17–35.

Densmore VS, Urbanski HF. 2003. Relative effect of gonadotropin-releasing hormone (GnRH) I and GnRH II on gonadotropin release. *J Clin Endocrinol Metab* 88: 2126–2134.

Fauser BC, Devroey P, Yenn SC, Gosden R, Crowley WF Jr, Baird DT, Bouchard P. 1999. Minimal ovarian stimulation for IVF: Appraisal of potential benefits and drawbacks. *Human Reprod* 14: 2681–2686.

Fink G. 1988. Gonadotropin secretion and its control. In: Knobil E, Neill JD, eds., *The Physiology of Reproduction,* 1349–1377. New York: Raven Press.

Grove-Strawser D, Sower SA, Ronsheim PM, Connolly JB, Boun CG, Rubin BS. 2002. Guinea pig GnRH: Localization and physiological activity reveal that it, not mammalian GnRH, is the major neuroendocrine form in guinea pigs. *Endocrinology* 143: 1602.

Guemene D, Williams JB. 1992. In vitro and in vivo responses to chicken LHRH-I and chicken LHRH-II in male turkeys (*Meleagris gallopavo*). *J Endocrinol* 132: 387–393.

Hong S, Cheung AP, Leung PCK. 2008. Gonadotropin-releasing hormones I and II induce apoptosis in human granulose cells. *J Clin Endocrinol Metab* 93: 3179–3185.

Kakar SS, Musgrove LC, Devor DC, Sellers JC, Neill JD. 1992. Cloning, sequencing, and expression of human gonadotropin releasing hormone (GnRH) receptor. *Biochem Biophys Res Commun* 189: 289–295.

Kakar SS, Winters SJ. 2002. Gonadotropin-releasing hormone receptor: Cloning, expression and transcriptional regulation. In: Parhar IS, ed., *Progress in Brain Research*, vol. 141, 129–147. Elsevier Science, Orlando, FL.

King JA, Davidson JS, Miller RP. 1988. Interaction of endogenous chicken gonadotropin-releasing hormone I and II on chicken pituitary cells. *Endocrinology* 117: 43–49.

Knobil E. 1981. Patterns of hypophysiotropic signals and gonadotropin secretion in the rhesus monkey. *Biol Reprod* 24: 44.

Krey LC, Butler WR, Knobil E. 1975. Surgical disconnection of the medial basal hypothalamus and pituitary function the rhesus monkey. I. Gonadotropin secretion. *Endocrinology* 96: 1073.

Latimer VS, Kohama SG, Garyfallou VT, Urbanski HF. 2001. A developmental increase in the expression of messenger ribonucleic acid encoding a second form of gonadotropin releasing hormone in the rhesus macaque hypothalamus. *J Clin Endocrinol Metab* 86: 324–329.

Latimer VS, Rodrigues SM, Gryfallou VT, Kohama SG, White RB, Fernald RD, Urbanski HF. 2000. Two molecular forms of gonadotropin-releasing hormone (GnRH I and GnRH II) are expressed by two separate populations of cells in the rhesus macaque hypothalamus. *Brain Res Mole Brain Res* 75: 287–292.

Lescheid DW, Terasawa E, Abler LA, Urbanski HF, Warby CM, Miller RP, Sherwood NM. 1997. A second form of gonadotropin-releasing hormone (GnRH) with characteristics of chicken GnRH II is present in the primate brain. *Endocrinology* 138: 5618–5629.

Lipman ME. 1998. Endocrine responsive cancer. In: Wilson JD, Foster DW, Kronenber HM, Larsen PR, eds., *Williams Textbook of Endocrinology*, 1675–1692. Philadelphia: WB Saunders Co.

Lumpkin MD, Vijayan E, Ojeda SR, McCann SM. 1980. Does the hypothalamus of infantile female rats contain a separate follicle stimulating hormone-releasing factor? *Neuroendocrinology* 30: 23–32.

Maney DL, Richardson RD, Wingfield JC. 1997. Central administration of chicken gonadotropin-releasing hormone II enhances courtship behavior in a female sparrow. *Horm Behavior* 32: 11–18.

Matsuo H, Baba Y, Nair RMG, Arimura A, Schally AV. 1971. Structure of the porcine LH-and FSH-releasing hormone. I. The proposed amino acid sequences. *Biochem Biophys Res Commun* 43: 1344–1350.

McCann SM, Taleisnik S, Friedman HM. 1960. LH-releasing activity in hypothalamic extracts. *Proc Soc Exp Biol Med* 104: 432–438.

McCormack JT, Plant TM, Hess DL, Knobil E. 1977. The effect of luteinizing hormone releasing (LHRH) antiserum administration on gonadotropin secretion in the rhesus monkey. *Endocrinology* 100: 663.

Miller R, Conklin D, Lofton-Day C, Hutchinson E, Troskie B, Illing N, Sealfon SC, Hapgppd JA. 1999. A novel GnRH receptor homologue gene: Abundant and wide tissue distribution of the antisense transcript. *J Endocrinol* 162: 117–126.

Miller R, Lowe S, Conklin D, Pawson A, Maudsley S, Troskie B, Ott T, et al. 2001. A novel mammalian receptor for the evolutionally conserved type II GnRH. *Proc Natl Acad Sci USA* 98: 9636–9641.

Miller RP, Milton RC, Follet BK, King JA. 1986. Receptor binding and gonadotropin-releasing activity of a novel chicken gonadotropin-releasing hormone (His5,Trp7, Tyr8 GnRH) and a d-Arg6 analog. *Endocrinology* 119: 224–231.

Miyamoto K, Hasegawa Y, Nomura M, Igarashi M, Kangawa K, Matsuo H. 1984. Identification of the second gonadotropin-releasing hormone in chicken hypothalamus: Evidence that gonadotropin secretion is probably controlled by two distinct gonadotropin releasing hormones in avian species. *Proc Natl Acad Sci USA* 81: 3874–3878.

Montagnani M, Moretti RM, Mai S, J-Caulier J, Motta M, Limonta P. 2009. Type I gonadotropin-releasing hormone receptor mediates the antiproliferative effects of GnRH II on prostate cancer cells. *J Clin Endocrinol Metb* 94: 1761–1767.

Nakai Y, Plant TM, Hess DL, Keogh EJ, Knobil E. 1978. On the sites of the negative and positive feedback actions of estradiol in the control of gonadotropin secretion in the rhesus monkey. *Endocrinology* 102: 1008.

Neill JD. 2002. Minireview: GnRH and GnRH receptor genes in the human genome. *Endocrinology* 143: 737–743.

Neill JD, Duck LW, Sellers JC, Musgrove LC. 2001. A gonadotropin-releasing hormone (GnRH) receptor specific for GnRH II in primates. *Biochem Biophys Res Commun* 282: 1012–1018.

Okada Y, Kawanao AM, Kakar SS, Witers SJ. 2003. Evidence that gonadotropin-releasing hormone (GnRH) II stimulates LH and FSH secretion from monkey pituitary cultures by activating GnRH I receptor. *Biol Reprod* 69: 1356–1368.

Schneider JS, Rissman EF. 2008. Gonadotropin-releasing hormone II: A multipurpose neuropeptide. *Integr Comp Biol* 48: 588–595.

Sealfon SC, Weinstein H, Miller RP. 1997. Molecular mechanisms of ligand interaction with the gonadotropin-releasing hormone receptor. *Endocr Rev* 18: 180–205.

Sherwood NM, Lovejoy DA, Coe IR. 1993. Origin of mammalian gonadotropin releasing hormone. *Endocr Rev* 14: 241–254.

Stojilkovic SS, Reinhart J, Catt KJ. 1994. Gonadotropin releasing hormone receptors: Structure and signal transduction pathways. *Endocrine Rev* 15: 462–499.

Temple JL, Miller RP, Rissman EF. 2003. An evolutionarily conserved form of gonadotropin-releasing hormone coordinates energy and reproductive behavior. *Endocrinology* 144: 13–19.

Tensen C, Okuzawa K, Blomenrohr M, Rebers F, Leurs R, Bogerd J, Schulz R, Goos H. 1997. Distinct efficacies for two endogenous ligands on a single cognate gonadoliberin receptor. *Eur J Biochem* 243: 134–140.

Terasawa E. 2003. Gonadotropin-releasing hormone II: Is this neuropeptide important for mammalian reproduction? *Endocrinology* 144: 3–4.

Urbanski HF, White RB, Fernald RD, Kohama SG, Gryfallou VT, Desmore VS. 1999. Regional expression of mRNA encoding a second form of gonadotropin-releasing hormone in the macaque brain. *Endocrinology* 140: 1945–1948.

Van Biljon W, Wykes S, Scherer S, Krawetz SA, Hapgood J. 2002. Type II gonadotropin releasing hormone receptor transcripts in human sperm. *Biol Reprod* 67: 1741–1749.

Vijayan E. 1985. Recent developments in female contraception: LHRH. *J Steroid Biochem* 23: 827–831.

Vijayan E, McCann SM. 1978. Re-evaluation of the role of catecholamines in control of gonadotropin and prolactin release. *Neuroendocrinology* 25: 150–165.

Vijayan E, McCann SM. 1980. Effect of blockade of DA receptors on acetylcholine (Ach) induced alterations of plasma gonadotropin and prolactin in conscious ovariectomized rats. *Brain Res Bull* 5: 23–30.

Wang L, Bogerd J, Choi HS, Seong JY, Soh JM, Chun SY, Blomenrohr M et al. 2001. Three distinct types of GnRH receptor characterized in the bullfrog. *Proc Natl Acad Sci USA* 98: 361–366.

White RB, Eisen JA, Kasten TL, Fernald RD. 1998. Second gene for gonadotropin-releasing hormone in humans. *Proc Natl Acad Sci USA* 95: 305–309.

Wu HM, Cheng JC, Wang HS, Huang HY, MacCalman CD, Leug PCK. 2009. Gonadotropin-releasing hormone type II induces apoptosis of human endometrial cancer cells by activating GADD45(alpha). *Cancer Res* 69: 4202–4208.

Wuttke W, Roosen-Runge G, Demlin J, Stock KW, Vijayan E. 1983. Neuroendocrine control of pulsatile luteinizing hormone release. In: Leyendecker G, Stock H, Wildt L, eds., *Brain and Pituitary Peptides II*, 1–10. Basel: S Karger.

Yu WH, Karanth S, Walczewska A, Sower SA, McCann SM. 1997. A hypothalamic follicle-stimulating hormone-releasing decapeptide in the rat. *Proc Natl Acad Sci USA* 94: 9499–9503.

Chapter 3

Gonadotropins: Past, Present and Future

Kambadur Muralidhar, Rajesh Chaudhuri, Shah Saddad Hussain and Nikki Kumari

Contents

3.1 Introduction 43
3.2 Isolation and Structural Features 44
3.3 Biosynthesis and Regulation of Secretion 51
3.4 Bioassay and Immunoassay for Gonadotropins and Prolactin 56
3.5 Hormone Action with Special Reference to Gonadotropins and Prolactin 60
3.5.1 General Aspects of Hormone Action..61
3.5.2 Gonadotropin Action...63
3.6 Non-Gonadal Actions of Gonadotropins 69
3.7 Clinical Use 71
Acknowledgements..73
References 73

3.1 Introduction

Reproductive physiology of vertebrates especially mammals has been the focus of attention of a number of biologists. Different aspects of this complex process have been studied by a variety of biologists, anatomists, physiologists, biochemists, embryologists and so forth (Gorbman, 1979; Bolander, 1989; Knobil and Neill, 2006). The perfection of hypophysectomy by Smith in 1912 (referred to in Smith, 1963) led to an enormous body of scientific information on the role of hypophysis in regulating the physiological processes underlying reproduction. Pituitary control of reproduction has been studied by three approaches developed in three successive periods in the last seven decades. The first approach was surgical hypophysectomy and replacement therapy. An offshoot of this approach was biochemical endocrinology dealing with (1) purification and characterisation of hormonal principles; (2) biosynthesis and secretion of hormones; (3) mechanism of action of hormones; (4) structure and regulation of expression of genes for hormones and their cognate receptors; and (5) study of hormone–receptor interaction using physico-chemical techniques. The second approach was immuno-hypophysectomy where a hormone in the extra-cellular fluid and that bound to cell surface receptors was neutralised by specific antibody and the deleterious effect was noted. In this approach replacement with pure hormone should restore normal condition (Moudgal, 1989). An off-shoot of this approach was studies on immuno-reproduction leading to the development of immuno-contraceptive vaccine candidates like follicle-stimulating hormone (FSH) and human chorionic gonadotropin (hCG) (Gupta, 1991). The third approach is of recent origin and involves knocking out a desired gene of interest at the embryonic stem cell level and studying the consequences in the future adult individual (Kumar et al., 1995; Kumar, 2007). A parallel activity here is the study and use of transgenic animals with foreign genes (or modified normal genes) in their body cells (Matzuk et al., 2001).

There has been an enormous body of information generated by the three approaches in experimental endocrinology of reproduction. Our understanding of the process of reproduction has benefited from information arising from all three approaches. One should not underestimate of course the non-endocrinological knowledge about reproduction as a phenomenon.

This chapter is an attempt to present a coherent view of gonadotropins of pituitary origin and their role in the regulation of gonadal functions. The discussion is restricted to mammalian systems and essentially to biochemical information. Selected topics concerning pituitary protein hormones and their influence on gonadal processes will be discussed. These include their purification, structural highlights, microheterogeneity, *in vivo* and *in vitro* bioassay, regulation of biosynthesis and secretion, their receptors in gonadal tissue and some unsolved problems.

3.2 Isolation and Structural Features

Pituitary secretes protein hormones like lutenising hormone (LH), follicle-stimulating hormone (FSH) and thyroid-stimulating hormone (TSH) in all vertebrates, whereas chorionic gonadotropins are secreted by placental cells of primates and equids (Bousfield et al., 1994). These pituitary hormones including prolactin (PRL) from a number of mammalian species have been purified (Ellis, 1961; Muralidhar et al., 1998). They have been extensively characterised with regard to their chemical structure and immunological properties. Nevertheless, our understanding of their three-dimensional structure, structural requirement for hormone assembly *in vitro*, antigenic epitopes, structural basis of recognition by their cognate receptors and so forth is nebulous. Purification of these hormones is desirable not only to undertake biochemical and physiological studies but also to meet the therapeutic demands. Many endocrine disorders affecting reproduction are documented and to restore health in such individuals, hormone replacement therapy is the only course of action and hence pure hormones or characterised preparations are essential. For obvious reason, human pituitary hormones were the first to be produced in bulk quantities from urine of post-menopausal women. Urine was a cheap source material to isolate menopausal pituitary FSH and LH (as a mixture called hMG-human menopausal gonadotropins). Similarly, human chorionic gonadotropin (hCG) was isolated from urine of pregnant women. Later pituitaries from farm animals and laboratory experimental animals served as source material (cattle, pig, horse, rat, rabbit, mouse and buffalo) for bulk preparations (Cohen, 2003; Lunenfeld, 2004; Macklon et al., 2006).

Meanwhile, comparative endocrinologists added non-mammalian vertebrates as a source for pituitary gonadotropins. In the case of human pituitaries, the U.S. government organises national pituitary agencies for collection and distribution to selected laboratories. Commercial companies collect glands from farm animals at abattoirs. India does not have such an agency. The procedures for purifications still today are mostly biochemical in nature. Hormones are initially extracted with salt or organic solvents, and the extract is then subjected to bulk separation methods like salt or organic solvent fractionations. In the final step, the fraction enriched in the given hormone is subjected to fine separation methods like chromatography and electrophoresis. A wide range of variations in the scheme at the fine separation stage is available from different laboratories around the world. The major problem with protocols using the aforementioned schemes is the low final yield of the hormones. This is due to two main reasons: (1) low initial abundance of hormones in the source material and (2) losses during purification through multistep protocols. While the initial abundance of hormones cannot be increased in the natural sources of the hormones, their losses during isolation can be decreased. One way to do this is development of protocols having fewer but efficient purification steps (Jia et al., 1993). Pseudo-affinity chromatography using triazine textile dyes (e.g. Cibacron blue F3GA) immobilised to various matrices can be used for this purpose. Flow chart 1 shows two schemes for the use of Cibacron blue F3GA for isolation of bubaline LH from pituitary glands. The structure of these dyes is known to mimic nucleotide cofactors (NADP) and interact with many proteins specifically at their nucleotide fold region. Nucleotide fold has not been observed in the crystal structure of gonadotropins and investigations have revealed that affinity of bubaline LH to Cibacron blue arises due to complex interaction constituting both electrostatic and hydrophobic forces (Vijayalakshmi, 2002; Muralidhar and Chaudhary, 2008).

The pituitary hormones which are directly involved in the regulation of gonadal activity are heterodimeric glycoproteins FSH and LH. The gametogenic and steroidogenic actions of gonadotropins are not observed in hypo-prolactinemic

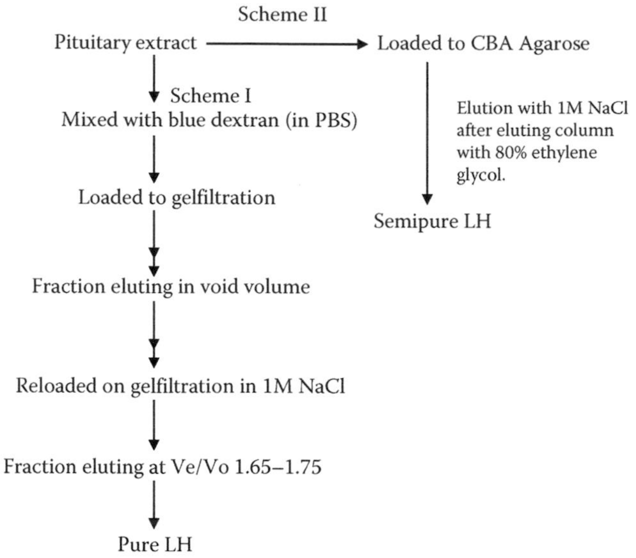

Outline of steps in the protocol for the isolation of bubaline LH from whole pituitaries. In Scheme I pituitary extract was mixed with blue dextran and the mixture loaded onto a gel-filtration column. When the LH-Dye complex from the void volume was chromatographed on the gel-filtration column in the presence of 1M NaCl, pure LH eluted at a Ve/Vo 1.65-1.75. In Scheme II pituitary extract was loaded on the Cibacron Blue Agarose (CBA) column after washing unbound proteins the semipure LH was eluted with 1 M NaCl.

condition as well as hyper-prolactinemic condition as in lactational amenorrhea. In addition, growth hormone and PRL have direct effects on gonadal tissue (Sinha, 1995). The physico-chemical properties of pituitary hormones from water buffaloes are listed in Table 3.1.

The amino acid sequence of more than 30 gonadotropins isolated from a number of mammalian species has been determined. A large number of workers in the United States and Europe have contributed to this information. In general for FSH and LH the α-subunit has 92 to 96 amino acids while the β-subunit has 114 to 121 amino acids depending on species. As they are glycoproteins they have oligosaccharide chains linked to either specific asparagine (N–linked oligosaccharides) or specific serine (O-linked oligosaccharides) residues. The number of oligosaccharide chains is two in α-subunit, both of N type. The β-subunit could have either one (e.g. oLH, buLH) or two N-linked chains (e.g. hFSH). In addition hCG, equine CG (eCG) and eLH have O-linked oligosaccharides in their β-subunits.

In humans the pituitary LH (hLH) and placental gonadotropins (hCG) differ in many structural properties. The amino acid sequence of hCG β-subunit (145 amino acids long) is different from that of hLH. A major difference is the presence

Table 3.1 Properties of Buffalo Pituitary Hormones Involved in the Regulation of Gonadal Activity

Hormones	*FSH*	*LH*	*GH*	*TSH*	*PRL*
Yield (mg/kG)	3–4	150–250	800–1000	9–10	250–300
Polypeptides	2	2	1	2	1
Mol mass (kDa)	30	30	22.2	30	24
Amino acids	210	210	190	–	199
Tryptophane	Absent	Absent	Present	Absent	Present
Nature	Glycoprotein	Glycoprotein	–	Glycoprotein	Glycoprotein?
Carbohydrate (%)	–	15	–	–	4
N-ter	–	F/S	F/M	F/G	T
Sulphate at sugar	–	Sulphated at sugar	–	–	Sulphated and tyrosine

of a C-terminal extra peptide in hCG-β while it is absent in hLH-β. Further hCG-β as mentioned earlier has four extra O-linked carbohydrate chains in the C-terminal peptide region. However in equids, eLH and eCG appear to differ only in carbohydrate structure. The amino acid sequence of the two subunits of eLH and eCG are identical. hCG from which either the C-terminal peptide or O-linked oligosaccharides have been removed is fully active *in vitro*. But in eCG when C-terminal peptide is removed or O-linked carbohydrate is removed, affinity to receptor was found to decrease (Bousfield et al., 1994).

The subunit nature of gonadotropins and thyrotropin was discovered in the 1960s (Li and Starman, 1964). Very soon it was realised that the β-subunit is hormone specific and α-subunit is species specific. Hence inter-specific and inter-hormone hybrids can be made where the β-subunit contributes to specificity in biological activity. For example ovine LH-β and human FSH-α when combined would be LH like in hormonal activity. In these studies it was always observed that hCG-β when combined with any α-subunit gives not only LH-like activity but gives higher potency than the homologous LH-β. This was attributed to the long plasma half-life of hCG-β. However in the case of equids, observations were different. It was the equine pituitary α-subunit when combined with any β-subunit that would give higher potency than the homologous α-subunit. That means oLH-β and oLH-α when combined would be less potent than hybrid oLH-β and eLH-α subunit (Roser et al., 1986). Another interesting property is that equine FSH is more potent than oFSH or porcine FSH or bovine FSH (Bousfield et al., 1996).

Extensive chemical and enzymatic analyses have provided us with detailed information about the structural variations in these carbohydrates (Manjunath et al., 1982; Sairam and Jiang, 1992; Edge, 2003). The sugar composition (in terms of mannose, galactose, glucosamine, galactosamine, sialic acid, fucose etc.) and the arrangement of these sugars differ from hormone to hormone and from species to species. An example for types of oligosaccharide attached to buLH, hCG and eCG is given in Figure 3.1.

Prolactin and growth hormone are simple proteins. They have varying degrees of homology in primary structure between them and with placental lactogens from human and ovine species. The homology in the structure is reflected in overlap in function including recognition by receptors. Until 1970s it was believed that human prolactin did not exist as human GH exhibits all the known properties of prolactin (Sinha, 1995). We now believe that all species have separate PRL and GH in their pituitaries. The glycosylated PRL is not as active as non-glycosylated PRL (Table 3.2). One of the unique features of buffalo and sheep pituitary PRL is the presence of tyrosine-O-sulphate. Our laboratory discovered this and reported it in 1988 (Kohli et al., 1988). The sulphated tyrosine may play a role in exportable glycoproteins. But this has not been experimentally proven (Moore, 2003).

One of the major structural features of the gonadotropic and lactogenic protein hormones is microheterogeneity. By this we mean that variants of the given hormone exist in the source material. These structural differences could be in the

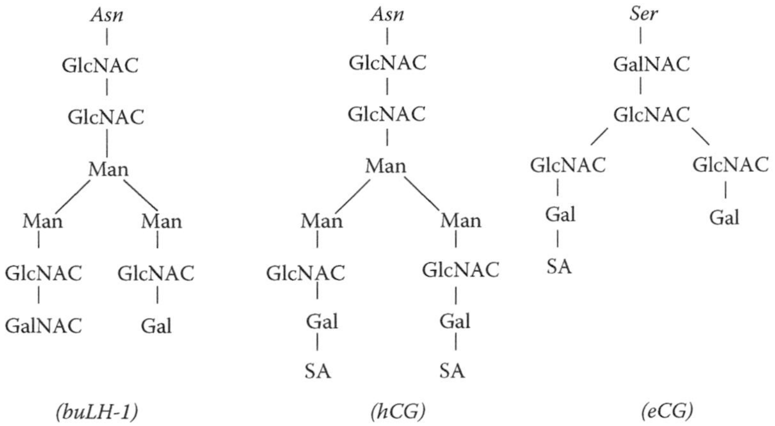

Figure 3.1 Representative oligosaccharide structures in pituitary and placental gonadotropins. Asn, asparagines; Ser, serine; GlcNAC, N-acetylglucosamine; GalNAC, N-acetylgalactosamine; Man, mannose; Gal, galactose; SA, sialic acid.

N-terminal amino acid, content of acidic amino acids like aspartic/glutamic acids compared to content of their amidated forms like glutamine/asparagines; glycosylation pattern including numbers of chains, sugar composition and sugar linkages; and in certain post-translational modification like sulphation, phosphorylation, sialylation etc. There have been numerous chemical and biological studies indicating that these structural variants differ in immunological and/or biological potencies. This would greatly affect potency estimates depending upon types of assay utilised. What is the physiological significance of the microheterogeneity? They are not allelic differences. There have been suggestions that surge level gonadotropins and tonic level gonadotropins are structural variants. There has been suggestion

Table 3.2 Biopotency of Different Prolactins in Nb2 Rat Lymphoma Cell Line Proliferation Assay

Hormone	*Biopotency (ED50) (ng/well)*
Ovine prolactin	1.6
Glycosylated buffalo prolactin	16.8
Buffalo prolactin	1.9
Buffalo prolactin (lower pI)	2.1
Buffalo prolactin (higher pI)	1.6
Porcine prolactin	16.0
Rat prolactin	0.9

that there are different FSH in male and female. Analysis of secretory dynamics of these variants and analysis of blood plasma for these variants are very difficult at the present level of our endocrine technology (Hartree and Renwick, 1992; Ulloa-Aguirre et al., 1995, 2003).

Sulphation of LH at their oligosaccharide portion has been hypothesised as having a role in its intracellular trafficking and episodic release from pituitaries. However in a recent study it was demonstrated that inhibition of sulphation in pituitaries did not change the pattern of release of LH (Pearl and Boime, 2009). This indicated that sulphation at sugar residues did not play role in the regulated release of LH. Sulphation of gonadotropins has major influence on their plasma clearance rate. Heavily sulphated hormones are cleared from circulation faster (Fiete et al., 1991). This is due to the presence of a receptor on the Kupffer cells which specifically recognises sulphated oligosaccharides and removes them from the blood thus decreasing the plasma half-life suggesting that it may be the basic isoforms which imparts hormone activity *in vivo*. Also, basic LH isoforms are more potent *in vitro* than the acidic LH isoforms (Sairam et al., 1994).

As in the case of hCG there has been a number of reports about the presence of prolactin in extra pituitary tissue. Human decidua appears to produce prolactin but its secretion is not affected by dopamine, TRH and VIP. Only progesterone was found to stimulate decidual prolactin production. But this could be an indirect effect and not a specific direct effect. Other tissues suspected of synthesising PRL are uterine myometrium, mammary gland, lymphocytes, hypothalamus and skin. Interesting functions have been attributed to these extra pituitary forms of prolactin (Ben-Jonathan et al., 1996).

The glycoprotein hormone biosynthesis represents a selected example of the assembly of functional multi-subunit proteins, in which the common α-subunit associates with the hormone-specific β-subunit to give rise to a unique biologically active entity. The two subunits combine non-covalently early in the secretory pathway, and formation of this heterodimer is necessary for specific binding to the cognate receptor (Pierce and Parsons, 1981). Some critical mutations in either of the two subunits have been reported to give rise to aberrant annealing of the subunits with loss of biological activity (Lin et al., 1999). Moreover, only the functional heterodimer and not the monomeric subunits can efficiently bind to and activate the cognate heptahelical receptor. Thus, the correct inter-subunit interactions between the two subunits in the heterodimer determine the functionality of the glycoprotein hormones. Hence, the process of heterodimerisation is highly reliant on the 'correct' folding of both the subunits and their secretion in 'adequate' quantities (Bedows et al., 1992; Muyan et al., 1998). The folding process of both hormone subunits is dictated by the disulphide bridge formation and glycosylation (Darling et al., 2000; Xing et al., 2001). These structure–function experiments showed that several determinants in both glycoprotein hormone subunits have remarkable effects on the recovery of the heterodimer (Boime and Ben-Menahem, 1999).

In the past, recombinant expression of the glycoprotein hormones has been attempted using a variety of expression systems such as mammalian and insect cell expression systems (Garcia-Campayo and Boime, 2001). However, due to the inherent technical complexities of these hormones, they were not expressed in adequate quantities required for structure–function studies. Further, the requirement of glycosylation for correct folding of the hormone subunits limits the use of bacterial expression system. The yeast expression system using the methylotropic yeast, *Pichia pastoris*, is an ideal system for hyper-expression of glycoprotein hormones as it blends the advantages of both prokaryotic as well as eukaryotic expression systems. Using this expression system, all the members of glycoprotein hormone family and their analogs have been expressed. Expression, characterisation and purification (Gadkari et al., 2003) of the hormones allowed determination of the glycan moiety of the recombinant hormones (Blanchard et al., 2007). Using recombinant DNA techniques as novel approaches for designing therapeutic hormone analogs have been made. Expressing them in *Pichia pastoris* have yielded valuable tools for structure–function analyses. However, possible mutagenesis induced defects in subunit association are major concerns in structure–function studies of any multimeric proteins, particularly in case of glycoprotein hormones where subunit association is a prerequisite for hormone binding and eliciting response. There are many examples of loss of subunit association in case of these hormones. Disruption of disulphide bond formation and lack of glycosylation results in absence of subunit association. As a result, assessing the exact role of the mutated residues on biological activities of hormones is very difficult. And therefore, a new strategy which bypasses the limitations of subunit assembly needed to be evolved. Biologically active single chain analogs of these hormones have been made and this is one of the approaches to overcome limitations imposed on subunit association. Thus a number of reports allude to biological activity of single chain gonadotropins wherein the α- and the β-subunits are joined at the genetic level in different orders (Garcia-Campayo *et al.*, 2005; Jablonka-Shariff et al., 2006; Setlur and Dighe, 2007). In addition it has been unequivocally demonstrated that a construct containing two tandemly linked hCG β-subunits cDNA yielded upon translation a receptor blocker of LH/CG R (Figure 3.2; Roy et al., 2007).

3.3 Biosynthesis and Regulation of Secretion

A major research activity in endocrine physiology is to understand the secretory dynamics of hormones and the various factors which affect hormone synthesis and secretion. Until recently the approach taken to study this problem was to expose pituitaries *in vivo* or *in vitro* to various factors and quantitate the level of gonadotropins or PRL in circulation or in the medium. Very few studies could distinguish release of presynthesised hormone from *de novo* synthesis and release of hormone. Immuno-cytochemical analysis using antisera to the hormone-specific β-subunit

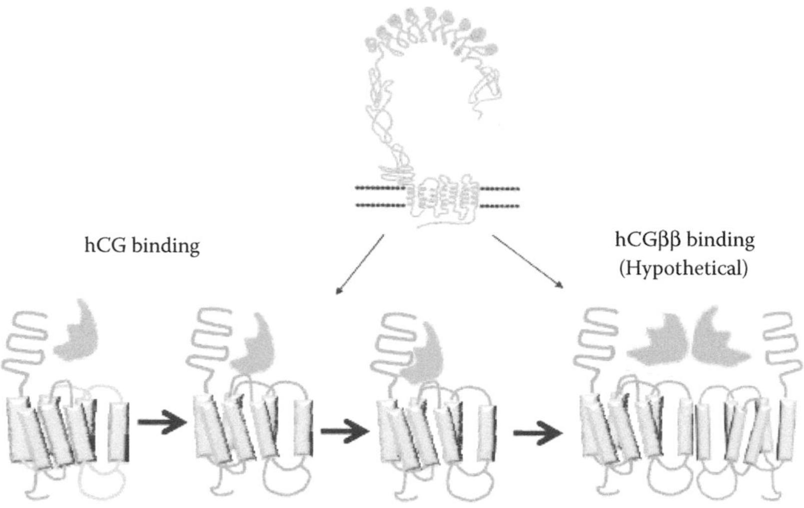

Figure 3.2 (See colour insert.) A cartoon suggesting the way hCGββ could bind gonadal LH receptor.

of FSH and LH have shown that among the gonadotropes, a certain percentage can make both hormones. While this would explain the regulation of LH and FSH release by a single GnRH, it would not explain many other problems such as differential regulation of FSH and LH secretion by GnRH.

It was only in late 1970s and early 1980s that the molecular details of protein hormone biosynthesis were unraveled. This part of the work had to await development of certain techniques in molecular biology. The guiding principle for the work was Steiner's pioneering research on anglerfish insulin. Using *in vitro* translation systems and isolated messenger RNAs it was realised very soon that gonadotropins subunits are synthesised on separate polysomes and are coded by separate messenger RNAs (Pierce and Parsons, 1981). This was completely different from the story of insulin where a single polypeptide precursor is made and this precursor with intra- and interchain disulphide bonds gets cleaved at appropriate places to yield insulin. Once this idea of separate messenger RNAs for α- and β-subunits was confirmed the next step was to characterise the nascent polypeptide. Two features were noteworthy: (1) the nascent polypeptide was larger than the size of a mature subunit by 20 to 25 amino acids and (2) there were no disulphide bonds as antiserum to denatured α- and β-subunits could recognise and immuno-precipitate the *de novo* synthesised precursors protein, while antiserum to native α and β could not. This means that the *de novo* synthesised pre-hormone was without the tertiary structure. When the precursor proteins were sequenced, the reading frame for the sequence of protein hormone in question could be recognised. The remaining extra portion of the hormone was the signal peptide. Subsequently the pre-protein was cleaved inside the endoplasmic reticulum so that the signal peptide comes off.

In the case of gonadotropin subunits, another event, also a co-translational event like signal peptide cleavage, is the N-glycosylation of asparagine residues (Lodish et al., 2007).

At the same time many studies have been done on the mode of biosynthesis of the carbohydrate portion of the glycoprotein. These studies found that independent of final structure, the oligosaccharide is assembled in two steps. This is true for gonadotropins or for that matter for any glycoprotein. In the first step a common lipid-linked oligosaccharide is first synthesised. This is then transferred to an acceptor protein. In the second step the sugar structure then undergoes peripheral sugar processing. Thus, three major types of oligosaccharides can be seen in mature proteins. One is the high mannose type, the second is the complex type and the third is a hybrid structure. The sugar types and lipid-linked oligosaccharide structure is given in Figure 3.3. Detailed information is available on aspects of this process like the enzymes involved in individual sugar activation (e.g. ADP-glucose, GDP mannose etc.), the assembly of the lipid-linked oligosaccharide, the en bloc transfer of the lipid-linked oligosaccharide onto the nascent incoming protein and the subsequent post-translational changes in the terminal sugars. All these occur during the transit of the protein through the endoplasmic reticulum and the Golgi body in the pituitary cell. In the case of hCG, the secretion automatically follows assembly. Hence secretion is unregulated. In the case of pituitary gonadotropins, the *de novo* synthesised subunits of the hormone combine and are stored in secretory granules. Under specific stimuli only, the pituitary releases the stored hormone (e.g. GnRH for LH etc.). Some molecular details of the assembly

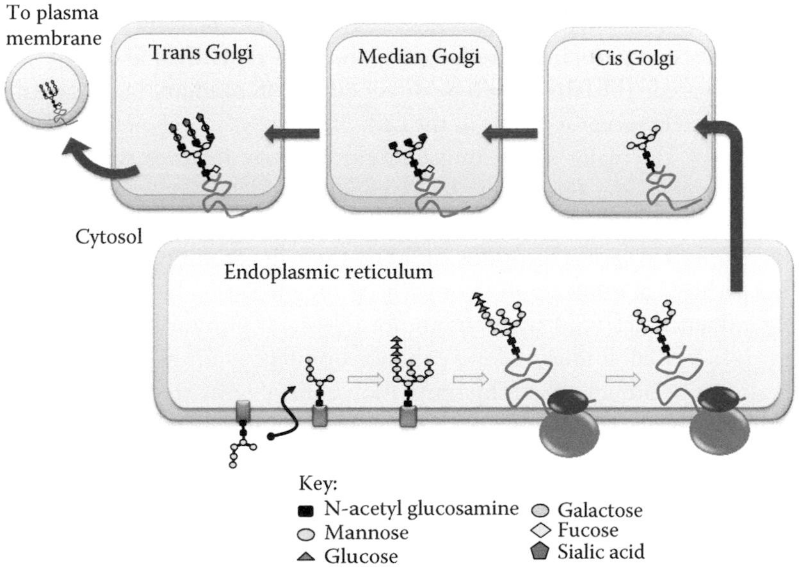

Figure 3.3 (See colour insert.) Dolichol linked glycosylation of proteins.

of gonadotropins are known largely due to the work of Ruddon et al. (1996) and Boime and Ben-Menahem (1999).

Assembly of the hormone dimer apparently occurs after glycosylation. However altered processing of the N-glycosidically linked oligosaccharide chain or O-glycosylation of α-subunit or a post-translational modification like phosphorylation results in a non-combination of α-subunit with β-subunit. Free subunits uncombined are not only found in the pituitaries but secreted as well (Bloomfield et al., 1978; Arora et al., 1999). hCG is an exception as in this case the O-glycosylation occurs after the assembly of the dimer.

The genes for gonadotropin subunit are located on different chromosomes in all the animal species thus far investigated (mouse, rat, cattle, pig, sheep and human pituitaries and certain tumor cell lines). In the case of humans, the α-subunit gene is located on chromosome 6 and the genes for LH β and FSH β are located on chromosome 19 and 11, respectively. Although there is a single copy for the α-subunit, there are multiple copies of the β-subunit gene. In humans, the LH β and hCG β-subunit genes are separate. However in equids there is a single gene for both eLH β and eCG β-subunits. The α-subunit gene has 4 exons and 3 introns. The common α-subunit gene is much larger (8016.5 kb) than the multiple β-subunit genes (1.5 kb). The FSH β-subunit mRNA is considered longer (1.7 kb) than the LH-β (0.7 kb). The placental β-subunit mRNAs are around 1 kb in length. The hLH β-subunit gene is not expressed in the placenta, while the hCG gene is probably expressed in the pituitary also to a small extent (Bolander, 1989; Konig, 1993).

Many factors are known to influence gonadotropin secretion rate and pattern. These factors are gonadal steroids, inhibin, hypothalamic GnRH or environmental factors like temperature, photoperiod etc. There could be only two mechanisms for their effects on pituitary secretion. One must be via GnRH and the other non-GnRH mediated. The latter could be direct effects on pituitary but may still influence post-GnRH receptor events in the cell. How does GnRH elicit response from the pituitary? The major signal transduction pathway involves the inositol phosphate turnover. Some researchers had provided evidence earlier that the cAMP pathway is involved. However, the majority of researchers in the field do not believe this any more. However exceptions could be there in certain phylogenic groups.

The pathway of influence with or without the mediation by GnRH should ultimately influence LH/FSH synthesis and/or release. If pituitary content of gonadotropins is increased, it must involve enhanced protein synthesis. Enhanced protein synthesis can be brought about by translation control or by transcriptional control. By transcriptional control we refer to changes in the mRNA levels by affecting transcription process. By post-transcriptional control we refer to the changes in the stability of mRNA and by translational control we refer to the altered translational rate of mRNA or stabilisation of mRNA.

The levels of mRNAs for gonadotropins α- and β-subunit change during the estrous cycle. The time course of change while paralleling the change in gene product level in circulation is not temporally synchronised. It was also found that both

FSH-β and LH-β mRNA levels increase once again in metestrus and diestrus in rats without parallel changes in gene product. In sheep, however, the temporal changes in both LH-α and β-subunit mRNAs coincide with gene product levels. This was not true for FSH-β. The kinetic picture of post-gonadectomy rise in mRNA is complex. Even in response to GnRH the increase in FSH-β and α-subunit mRNA was linear with increase in GnRH pulse amplitude. However for LH-β mRNA levels it was a biphasic response. With regard to GnRH frequency changes, lower frequency stimulates FSH-β mRNA levels while higher frequency of GnRH stimulates LH-β mRNA levels. Very interestingly the effects of inhibin on mRNA levels are as equivocal as its effects on basal- and GnRH-stimulated FSH and LH release.

The cAMP responsive element has been found in sequences upstream of hLH-α gene but not in bovine LH-α gene flanking sequences. This has made some people speculate that this is the reason for absence of placental gonadotropins in species other than human and equids. However this cannot be true as in the equine LH-α gene flanking sequences, there is no cAMP responsive element. Probably other factors are involved in the expression of gonadotropin genes in placenta.

Transgenic mice having hFSH-β gene have been shown to consistently express it in pituitaries. The heterodimer that is mouse FSH-α subunit and human FSH-β subunit was found secreted higher in males than in female mice showing sexual dimorphism in mice. It is also interesting to note that castrated mice when given androgen did not affect the content of normal mouse pituitary FSH content. However, in the transgenic mouse, the hFSH-β gene expression was completely suppressed showing species differences between mouse and man regarding gonadal feedback mechanism. It is also interesting to note that this androgen effect on hFSH-β expression in a mouse environment was not due to aromatisation to estrogen.

As far as PRL is concerned, the gene has been localised both in humans and rats. In rats it is on chromosome 17 while in humans it is on chromosome 6. Five exons and four introns comprise the rat gene. PRL is made on 1 kb mature mRNA corresponding to a 227-amino acid polypeptide. Hence the signal peptide must be about 28 amino acids long. The regulatory elements in the rat PRL gene (5′-flanking sequence) have been studied. Estrogen is known to enhance PRL production, as estrogen response element (ERE) has been found in the regulatory sequences of flanking PRL gene. The human PRL gene has six exons. The release of PRL is inhibited by dopamine. Thyrotropin-releasing hormone (TRH) stimulates PRL release. Both Ca^{++} influx pathway and voltage-dependent calcium channels-mediated Ca^{++} influx are involved. TRH acts through a G-protein binding membrane receptor type (Bolander, 1989).

Thus in the near future, we should uncover in great detail the remaining part of the gonadotropin assembly in the pituitary gonadotroph that is the site of α and β hybrid formation, the mechanism of native folding in terms of existence of separate chaperones for the α and β subunits, the intra-cellular sorting of two β-subunits, that of FSH-β and LH-β, in the same cell etc. The role of non-target tissue in the metabolism of gonadotropins and PRL/GH has also to be understood.

3.4 Bioassay and Immunoassay for Gonadotropins and Prolactin

Assay of hormonal activity is a prerequisite to the discovery of a hormone. Quantitative measurements of hormonal levels in circulation or other putative sources become mandatory in the design of endocrine investigations. Hence, development and use of reliable assays for gonadotropins and prolactin have always been a challenging task. Hormonal assay measures biological effects of the hormone on the target tissue or an overt property like antigenicity. The nature of these bioassays and immunoassays has undergone many changes in configuration and, over a period of time, has become more precise, sensitive, accurate and specific. Bioassay can be performed either in whole animal where the *in vivo* response to the hormone is measured or using tissue or cells where the response is measured *in vitro*. In some cases radioreceptor binding assays are being used where no response to the hormone is measured other than binding. Pharmacological antagonists would be biologically active in this assay system but are not actually active when administered *in vivo* or *in vitro* as they do not elicit any response. The response measured could be a change in the size or morphology of the target tissue or change in a biochemical parameter like enzyme activity, output of metabolite or biosynthesis of a macromolecule etc.

Immunoassay measures the antigenicity of the given hormone. Epitopic basis of antigenicity tells us that immunoassay potency units need not always correlate with bioassay potency estimates. Antigenic sites and receptors binding/activation sites need not be the same on the topography of the hormone. Further, microheterogeneity in these protein hormones has resulted in presence of a number of variants including fragments in circulation, in the pituitary or in urine, and these variants are obviously biologically inactive in conventional assays but would answer immunoassay procedures. Thus a number of reports have alluded to this paradoxical situation of having an immunologically active but biologically inactive protein hormone. Also in this area rDNA techniques have opened up newer ways of performing *in vitro* bioassays using cloned receptors and reporter genes (Jaffe and Behrman, 1978).

For gonadotropins more than a dozen bioassays have been developed and applied (Van Damme et al., 1979). FSH is assayed by Steelman-Pohley assay (Figure 3.4) and Igarashi-McCann assay in rats and mice, respectively. The assay is also called hCG augmentation assay in which increase in ovarian weight as a function of dose of FSH injected is measured. The rat version is not sensitive. Radioreceptor assay (RRA) using ^{125}I-FSH and rat testicular tubular preparations or bovine testes membrane is also being practiced for FSH. In general, interference from serum in receptor binding assay is heavy and hence these assays do not give correct estimates. In addition, Sertoli cells or granulosa cells can be used *in vitro* to measure either estrogen formation from testosterone (Jia and Hsueh, 1986) or increase in cAMP levels in response to dose of FSH. If 1-^{3}H-testosterone is used, release of ^3H_2O can be directly measured instead of estrogen. Improvisations in this assay system permit

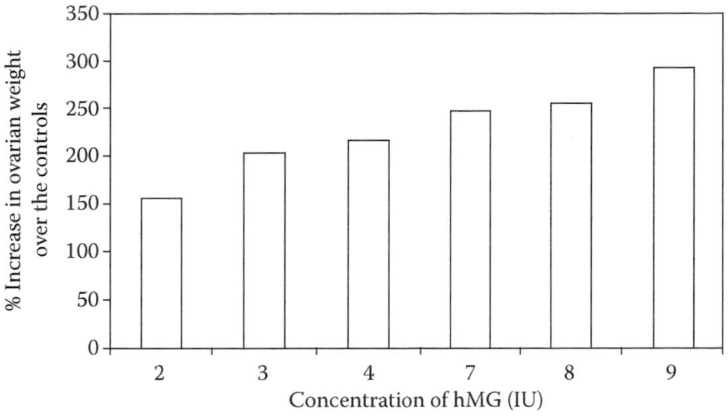

Figure 3.4 Percentage increase in ovarian weights over the control in the Steelman–Pohley bioassay model over a range of hMG concentration. The values are the average of only two measurements.

one to measure picograms of human FSH. Aromatase can also be measured in granulose cells obtained from rats given estrogens. Plasminogen activator activity is another end point of FSH bioassay but LH poses interference in this assay.

For LH, Parlow (1961) initially devised the ovarian ascorbic acid depletion assay where a dose-dependent depletion of ovarian ascorbic acid content (Figure 3.5) in the super-ovulated immature rat was demonstrated. The super ovulation response in rats and mice has been compared (Wilson and Zarrow, 1962). In addition to

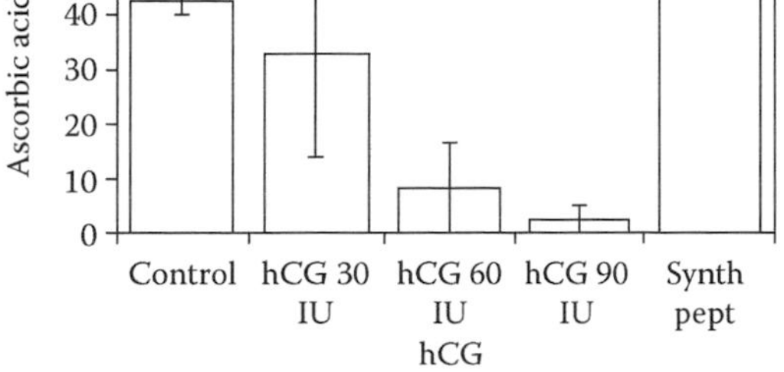

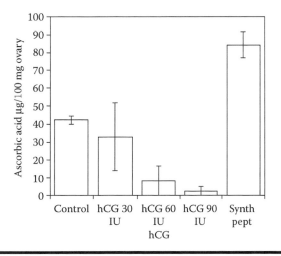

Figure 3.5 Effect of hCG and a PRL-derived synthetic peptide on ovarian ascorbic acid content in super-ovulated immature rat (Parlow model). Values are mean ± SEM.

LH some other factors also influence ascorbic acid content in ovaries. Observation by another group demonstrated the changes in the same followed by the injection of prostaglandin (Engelbart et al., 1973; Igarashi et al., 1974). It is also shown that vasopressin also possesses appreciable activity like that of LH (McCann and Taleisnik, 1960). Purified preparation of other anterior pituitary hormones failed to affect the levels of ascorbic acid in ovary (McCann and Taleisnik, 1960; Schemidt-Elmendotff and Loraine, 1962). This depletion in ovarian ascorbic acid can also be observed with cAMP derivatives and was found to be cycloheximide insensitive (Table 3.3). The classical ventral prostate/seminal vesicle weight increase assay in immature male rats is still a valid assay for LH. *In vitro* assay for LH includes testosterone production and release response measurement. A mouse Leydig cell tumor derived cell culture is a popular version of such a bioassay for LH (Table 3.4).

When receptors for FSH and LH were cloned, a number of assays using cloned receptors came into vogue. Appropriate host cells (e.g. L-cells, human fetal kidney cells etc.) transfected with rat or human FSH receptor genes respond to FSH. The responses measured are either cAMP levels or a reporter gene activity like luciferase enzyme. Other versions are also in vogue. Similar cloned receptor–reporter gene assays are in trend for LH. Problems of species specificity, mistaken use of truncated cDNA and other disadvantages have to be overcome in these assays.

The biological assay for prolactin has always been difficult to develop. Sensitive immunoassays of course are available for prolactin from a number of species. Although there are more than 100 reported biological effects of PRL from fish to mammals, not a single is dose dependent. The pigeon crop sac assay long promoted as an assay for PRL is not at all specific. One of the interesting effects of PRL is the inhibition of osmotic uptake of water by fish gill pre-incubated in saline and re-incubated in water. We have found that buffalo PRL exhibits a dose-dependent inhibitory effect and is surprisingly more active than fish PRL.

Prolactin and GH are structurally related and also have overlapping functions. Although their receptors have been identified they can occupy each other's receptors

Table 3.3 Inability of Cycloheximide to Block LH Action on Ovarian Ascorbic Acid

Group	*Treatment*	*Dose (µg/rat)*	*Ascorbic Acid (mg%)*	*Depletion (%)*
I	Saline	–	89.84 ± 11.2	0.00
II	LH	3.2	50.59 ± 5.02	43.68#
III	LH + Cycloheximide	3.2 + 250	53.36 ± 11.83	40.67*

Note: Values are mean ± S.D.

# $p < 0.005$; d.f. = 9 (I: II)

* $0.1 > p > 0.025$; d.f. = 9 (II:III)

Table 3.4 List of Possible Bioassays for Gonadotropins

		FSH Bioassays
1.	Steelman and Pohley (1953)	Since 1937 quantitative measurement of FSH was achieved by measuring the increase in ovarian weight in female rats.
2.	Igarashi and McCann (1964)	Intact immature mice are injected with various doses of FSH and the uterine weight is recorded. LH affects at only very high doses.
3.	Reichert and Bhalla (1974)	This is a radio receptor assay in which radioiodinated hormone and increasing doses of unlabelled hormone compete for the limited receptor preparation.
4.	Van Damme et al. (1979)	This assay measures the estrogen production from cultured rat Sertoli cells in response to doses of hormone.
5.	Jia and Hsueh (1986)	This *in vitro* assay is based on measurement of estrogen from granulose cells incubated with androgen and in response to hormone.
6.	Beers and Strickland (1978)	It Is based on the stimulation of plasminogen activator secreted by cultured granulose cells. It responds to both FSH and LH.
		LH Bioassays
1.	Greep et al. (1941)	Increase in seminal vesicle or prostate tissue weight in response to hormone measured in hypophysectomised rats.
2.	Witschi (1942), as reported in Gorbman (1979)	The seasonal breeding and cock plumage of male birds is regulated by gonadotropins. The melanin reaction in feathers of breast of males is measured for the assay.
3.	Wilson and Zarrow (1962)	Ovulation in immature rats is measured in response to doses of hormone injected.
4.	Parlow (1961)	The dose depletion of ovarian ascorbic acid caused by the hormone in super-ovulated immature rats is measured.
5.	Ellis (1961)	The increase in the content of injected radio iodinated serum albumin in immature rat ovaries in hyperaemic response to hormone is measured.

(*Continued*)

Table 3.4 (*Continued*) List of Possible Bioassays for Gonadotropins

		LH Bioassays
6.	Bell et al. (1964)	Similar to OAAD but measures cholesterol depletion.
7.	Moudgal et al. (1971)	Radioreceptor binding assay.
8.	Dufau et al. (1974)	This assay measures testosterone production in rat Leydig cells in response to hormone.
9.	Van Damme et al. (1979)	This measures testosterone produced by mouse Leydig cells *in vitro* in to hormone.
10.	Muralidhar and Moudgal (1976)	This measures receptor-bound hormone by radioimmunoassay.
11.	Jia et al. (1993)	Bioassay using fetal kidney cell line 293 transfected with hLH receptor cDNA and a luciferase gene driven by cAMP-dependent promoter.

also (Ben-Jonathan et al., 1996), and they have multiple biological effects. GH is more involved in general metabolic regulation, while PRL has more of a role in reproduction and immune response. Accordingly, one of the bioassays for GH and PRL uses rat Nb_2 lymphoma cell lines whose mitotic response is dose dependent. This is related to the lactogenic hormone receptors. Hence PRL shows mitogenic effects and among GHs only human GH shares this effect. The assay is sensitive and can measure PRL in the subnanogram range.

Radioreceptor assays for PRL are always in vogue. A competitive binding assay based on rat liver receptor works well for PRL. But, the biological response to this binding, however, is not very clear. Even if it is related to induction to estrogen receptors, the response is not strictly dose dependent. Recently PRL has been shown to cause dose-dependent release of PGE_2 from uterine stromal cells. Will this serve as a new and reliable bioassay for PRL?

3.5 Hormone Action with Special Reference to Gonadotropins and Prolactin

The most important characteristic that distinguishes all living organisms from nonliving objects is the extraordinary ability of living organisms to respond to external signals and regulate their own metabolism to ensure continual survival in the steady-state condition. In this property, the unicellular organisms utilise genetic mechanism more often than higher eukaryotic organisms. The latter use non-genetic mechanisms (process not affecting gene expression). However, under extreme provocation (e.g. during ontogeny adaptation to environmental insult etc.) all organisms higher or lower use genetic mechanism to tide over the given adverse situation.

The tropic nature of gonadotropins on gonadal tissues is more or less accepted as a matter of fact. While in the case of FSH studies have clearly demonstrated such an action on ovaries and Sertoli cells, the tropic action of LH was demonstrated only in the 1970s. In pregnant mare serum gonadotropin (PMSG) primed immature female rats, treatment with human chorionic gonadotropin (hCG) or ovine LH resulted in stimulation of the rate of synthesis of polyadenylic acid (poly A)-rich RNA in the ovaries. The rate of total RNA synthesis was not significantly affected by hormone treatment, whereas protein synthesis was enhanced. The increase in the rate of synthesis of poly A-rich RNA in the ovaries could be inferred as induction of messenger RNA synthesis after the hormone treatment. The level of cyclic AMP in the ovaries of such rats was also elevated after administration of LH, the increase coincided with the increase in the rate of synthesis of poly A-rich RNA (Prasad et al., 1978). Even the so-called compensatory ovarian hypertrophic (COD) action of FSH was shown to be inclusive of hyperplasia (Bhagat et al., 1993).

3.5.1 General Aspects of Hormone Action

Hormone action can be discussed as a special aspect in the regulation of cellular metabolism or as a model to study communication between cells in the multicellular organism. In the first case the emphasis is on mechanism by which a tissue responds to hormonal signal and in the second case the emphasis is on what constitutes the hormonal signal/trigger. Historically speaking the discovery of hormones was by surgical extirpation of a part of the body and observing the alterations in the rest of the body. For example hypophysectomy leads to disturbances in adrenal, growth, reproductive and other physiologies and hence led to the discovery of growth hormones etc. Subsequently, hormonal effects as observed by administration of hormones *in vivo* and *in vitro* were documented. This extensive documentation led to rationalisation of hormonal effects into metabolic and developmental effects. Such a classification served well for many years. The myth that hormonal effects cannot be observed in broken cell preparation was blown when Sutherland (referred to in Robison et al., 1971) discovered cAMP and brought in the concept of second messengers. The enormous details of biochemical changes in the target tissue reinforced the ideas of unifying biochemical mechanism of action for all hormones and the focus shifted to understanding specificity of hormone action and hence receptors were discovered. The two major classes of receptors that is membrane bound and intracellular only brought back the ideas of metabolic and developmental hormones being different than their receptors were, respectively, membrane bound and nuclear in location. These receptors have been purified biochemically and their genes cloned and expressed in heterologous cells. Reconstitution experiments have given us an idea of structure–function relationship.

In the case of membrane receptors, the formation of hormone receptor complex is supposed to lead to activation of a number of enzyme systems like adenylate cyclase, phospholipase C etc. The work of Rodbell, Gilman and others (referred to in Litwack, 1972) brought in a third component of the receptor complex: G protein. Subsequent to the discovery of cAMP, studies on the mechanism of action of cAMP led to the unraveling of the role of protein kinase A. Krebs and Fischer (referred to in Litwack, 1972) largely contributed to this development. The link between cAMP and hormone receptor was G protein. All the membrane proteins interacting with G proteins have similar protein structure that is they have an extra cellulardomain (ECD), a transmembrane domain (TMD) and an intracellular domain (ICD). The TMD typically spans the membrane seven times. One of the cytoplasmic loops interacts with G proteins (Figure 3.6). The detail of this process is not known. Notwithstanding all that have been said, there are cases (e.g. insulin, epidermal growth factors etc.) where the membrane receptor itself acquires an enzymatic activity like tyrosine kinase. Taken together, this would mean, extracellular signals (e.g. ligands) can either get internalised through receptors if they are bringing in nutrients (LDL receptors) or cause opening of ion channels or activate tyrosine kinase or interact with G protein leading to the activation of secondary enzyme systems. A battery of protein kinases are involved in the conversion of signals into altered

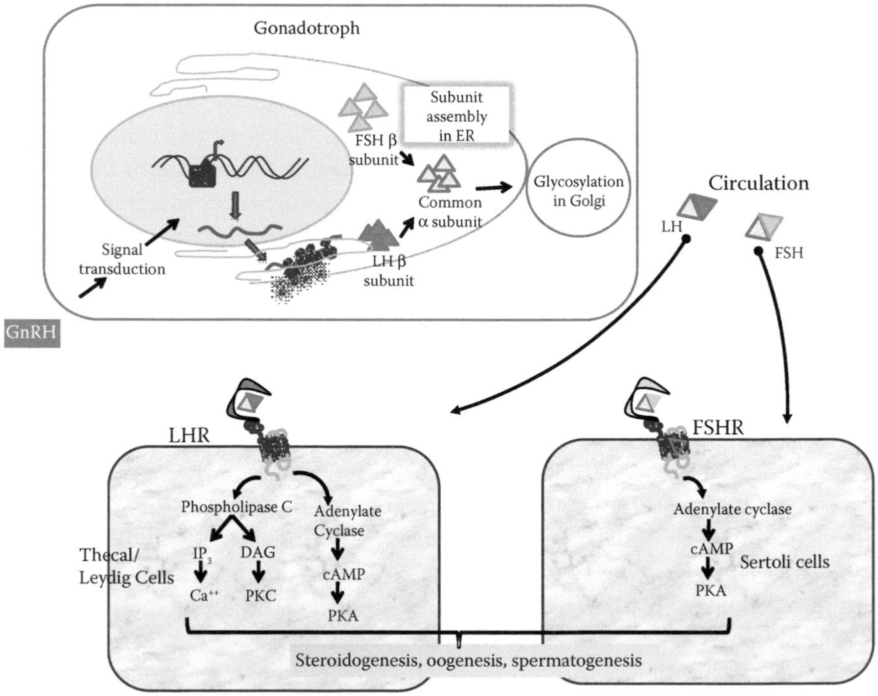

Figure 3.6 (See colour insert.) Signal transduction pathway of gonadotropins.

cellular metabolism. Metabolic pathways are altered with regard to rate of flux or direction and this can include at a later time period alteration in gene expression. On the other hand, lipophilic hormones like steroids enter the target cells and activate receptors in the nucleus. These nuclear receptors have been shown to possess supra secondary features just like those of DNA binding proteins and binding to cis-acting elements, that is specific sequences on the upstream side of structural genes, and cause either activation or suppression of specific gene expression. Many of these cis-acting enhancers and silencers have been mapped and sequenced (Trist et al., 1997).

The studies on structure–function relationships in hormones (e.g. protein hormones) and receptors have been greatly stimulated by recombinant DNA techniques. Site-directed mutagenesis as well as cDNA cloning and expression of truncated receptors and hormones as well as fusion proteins in both cases have given us an idea of the topological surfaces/portions involved in hormone receptor interaction on one hand and receptor effectors system interaction on the other hand.

3.5.2 Gonadotropin Action

Study of hormone action in the recent years has gotten a boost with gene knockout experiments. Using embryonic stem cell technology and transgenic animals the expression or absence of expression of a specified gene has been studied. This has led to new and unexpected discoveries with regard to the physiological role of already existing hormones. There are still unsolved problems in this field. For example stoichiometry of binding between hormones and their putative receptors is not known with certainty. What really constitutes the hormonal signal? Whether it is change in concentration of hormone or change in response is also not very clear. We are in a new and perhaps closing era of hormone action study where the final picture of mechanism of hormone action will emerge.

Advances in the last two decades in manipulating the mouse genome by random and site-specific mutagenesis have been invaluable to understanding the biology of gonadotropins. Using transgenic and gene targeting approaches, both transgenic (gain of function) as well as knockout mice lacking the hormone-specific gonadotropin subunits, have been produced and are used as models to study the biological roles of FSH and LH (Table 3.5; Kumar, 2007).

From the studies of Greep et al. (1941, 1942) it is almost a paradigm that FSH is a gametogenic hormone and LH is a steroidogenic hormone. Subsequent studies have only added biochemical details and unraveled signal transduction mechanisms in the gonads. The finer details of the role of FSH in follicular maturation and growth, the role of LH in ovulation and maintenance of corpus luteum, and the role of prolactin in permitting the action of both are being worked out. In males the relative roles of FSH and LH are very clear. There appears to be certain species differences in the absolute requirement for maintenance of spermatogenesis.

Gonadotropin action or gonadal response to pituitary and placental gonadotropins has been studied at molecular level. In the 1960s and early 1970s, in line

Table 3.5 Summary of Phenotypes of Transgenic and Gene Knockout Mouse Models for Gonadotropins

Model	*Male Phenotype*	*Female Phenotype*
Transgenic		
MT-hFSH	Normal size testes, apparently normal spermatogenesis, infertility, elevated serum testosterone, enlarged seminal vesicles	Infertility, hemorrhagic and cystic ovaries, elevated serum estradiol, progesterone, androgens, kidney defects, bladder obstruction
MT-hLHβ	Small testes, infertility, elevated testosterone, enlarged seminal vesicles, Leydig cell hyperplasia, vacuolated tubules, abnormal tubules containing only Sertoli cells	Infertility, less hemorrhagic but mostly cystic ovaries, elevated serum estradiol and progesterone, disrupted folliculogenesis with no pre-ovulatory follicles and corpora lutea, abnormal thecal cells, enlarged and cystic kidneys
Knockout		
FSHβ	Small testes, normal fertility, decreased Sertoli cell number, Leydig cell normal, no change in serum testosterone levels, reduced sperm number and motility	Infertility, small ovaries, pre-antral stage block in folliculo-genesis, defects in granulosa cell, proliferation, suppressed aromatase, respond to PMSG/hCG treatment
LHβ	Small testes, infertility, reduced serum testosterone and intra-testicular testosterone levels, suppressed steroid pathway enzymes, aberrant Sertoli cell gene expression, spermato-genesis blocked at the round spermatid stage	Small ovaries, thin uteri, reduced serum estradiol and progesterone, no pre-ovulatory follicles and corpora lutea, PMSG/hCG treatment rescues LH deficiency

with such studies on other hormones, the specificity of gonadotropins action on gonadal tissue was explained by the presence of specific cognate receptors only on the gonadal tissues. The nature of receptor and receptor-mediated gonadotropins action was then investigated in great detail. The location of receptors both at the organ and subcellular level was achieved by the use of radiolabel gonadotropins

(Moudgal et al., 1971). Scatchard analysis of the binding data gave the values of receptor number and affinity. Radioreceptor assay performed in the same way as radioimmunoassay soon replaced *in vitro* and *in vivo* bioassays. It helped also in quick evaluation of structural requirements to be recognised by gonadal receptors. Such structure–function relationship studies revealed that deglycosylated gonadotropins could bind to receptors but were unable to elicit signal transduction *in vitro*. Although this led to the euphoric feeling that a classical pharmacological receptor antagonist for gonadotropins in the form of chemically or enzymatically deglycosylated FSH and LH are available, *in vivo* studies quickly demonstrated that they could not compete with endogenous or exogenous gonadotropins in blocking their action simply because the rate at which they come off the receptor is the same as that of native gonadotropins. A hepatic receptor which recognises asialo-glycoproteins has been well described. Desialylated gonadotropins are cleared from circulation by this receptor and hence are biologically much less active (Morell et al., 1971).

For some years in the 1980s, research workers reported the purification and physicochemical characterisation of gonadotropin receptors. A lot of confusing data on the size of the receptors and the number of polypeptides comprising the LH receptor appeared in literature. The situation became clear only when genetic engineering methods were employed. Both LH (hCG) and FSH receptors have been cloned and expressed. The FSH receptor consists of 10 exons and 9 introns while the LH receptor comprises 11 exons and 10 introns. The receptors have three portions: an extracellular domain, the region spanning the target cell membrane and an intracellular portion. LH receptor has been cloned from rodent, pig and human gonadal tissue. The portion of the transcript from the first 10 exons codes for the extracellular portion of the receptor. The receptor structure is similar to other G-protein-interacting receptors like β-adrenergic receptors. The LH receptor is about 669 to 742 amino acid residues long. Once again, recombinant DNA techniques have established that the extracellular N-terminal region of LH receptor recognises and binds the hormone. Similar studies should also bring out the structural basis of FSH, LH and TSH recognition by their cognate receptors which otherwise exhibit a lot of homology. The cytoplasmic domain of the receptor is a site of regulation. More than 80% homology exists among murine, porcine and human LH receptors. Nevertheless, the potency *in vivo*, of animal hormones in sub-human primates and humans is very negligible. One of the observations with these receptors is that truncated forms of LH and FSH receptors have been detected or artificially created (by cloning and transfection experiments), which have interesting structural and functional properties.

The pharmacological receptor comprises three parts: (1) the biochemical receptor described earlier; (2) the G protein; and (3) the adenylate cyclase or other effector systems. The β_2-adrenergic receptor system which is similar to the gonadotropin receptor system except that its gene has only one exon, has been extensively studied with regard to genetics, physiochemical properties, molecular events during signal transduction and termination in terms of the second messenger that is cAMP. The cellular biochemical events following elevation of cAMP levels have been extensively studied in many systems and a fairly large body of information exists. Both the FSH receptor in Sertoli and granulosa cells and the LH receptor in luteal, thecal and Leydig cells are linked to G protein and function using cAMP as a second messenger (Dufau et al., 1974; Dorrington and Armstrong, 1975). The post-cAMP rise events have also been described for gonadotropin action. Cholesterol ester hydrolase could be the enzyme whose function is regulated by cAMP-dependent protein kinase (Bousfield et al., 1994).

In addition to the major endocrine pathway involving pituitary gonadotropins and gonadal cAMP, other intra-gonadal cell types including immune competent cells like local macrophages could influence the function of Leydig cells and granulose cells through paracrine pathways. An increasing body of literature points to this immuno- endocrine and paracrine system of regulation of gonadal function.

As outlined in the introduction, transgenic animal models are being increasingly used to explore problems of hormone action. While the primary role of FSH LH in male reproduction appears to have been well accepted ever since Roy Greep's observation in the 1930s, the role of FSH in the initiation/maintenance of spermatogenesis has again come under investigation. In recent years whether FSH is required for spermatogenesis in primates has been realised largely by those who are working on immuno-contraception vaccines. A recent publication further adds to this confusion. Using gene knockout experiments with embryonic stem cells in a model mouse, FSH-deficient mice were created by deleting most of the mouse FSH-β subunit gene. Both heterozygous (FSH-β null/+ animals) and homozygous (FSH-β null/FSH-β null) animals generated from heterozygous animals by intercrossing were analyzed. Male mice homozygous for mutated alleles were fertile at 6 weeks of age though testes size was small. Total sperm count came down by 75%. The female homozygous FSH-deficient mice were however sterile. In the case of females, there were no corpora lutea and no follicles beyond pre-antral stage. This data has raised the fundamental question of what is the role of FSH in spermatogenesis (Kumar and Low, 1993). Thus in the field of gonadotropin action, there are many unsolved areas like stoichiometry of hormone receptor which are involved in binding and activating the receptor and others. A peptide receptor antagonist to gonadotropin would be the best contraceptive agent. But it is yet to be developed.

The metabolic effects of gonadotropins have been documented for over five decades. In the post-genomics era, this has been documented by the use of microarray platforms, metabolomics and proteomics approaches. Notwithstanding this, the detail about signalling pathways through which such regulation occurs is not known. For example it was reported in 1985 that FSH administration to female rats can result in an increase in ovarian glycogen content (Chadha et al., 1985). FSH normally acts through cAMP pathway, say in causing stimulating estrogen production in Sertoli cells. In the action of glucagon or epinephrine on liver glycogen, cAMP pathway again is utilised thereby activating gycogenolysis. How does cAMP therefore bring about increased glycogen synthesis as in the case reported by Chadha et al. (1985)? Another example of gonadotropin action is the contrasting effects of FSH and LH on ovarian ascorbic acid content. In the typical Parlow bioassay for LH-like activity (otherwise called ovarian ascorbic acid depletion assay or OAAD assay), super-ovulated immature rats are used and in the model administration of LH results in a dose-dependent decrease in ascorbic acid content in the gonad (Parlow, 1961). It was demonstrated earlier that cAMP pathway is involved and that no new protein synthesis is required for this effect to be observed (Table 3.3). In the same model, the animals are to be given PMSG (FSH-like) to increase follicular growth before the formation of corpus luteum induced by hCG injection. If one were to measure ascorbic acid in ovaries following PMSG injection one notices increase in vitamin C content (Arora et al., 2012). Two paradoxes in this observation are (1) PMSG acts like LH in horses and mares but works like FSH in heterologous animals under certain circumstances (Matteri et al., 1986; Kumari et al., 2013) and (2) FSH and LH have opposite effects on ovarian ascorbic acid levels. In order to know more details, it was hypothesised that LH and FSH might be regulating L-gulonate dehydrogenase activity and through that regulate the metabolite traffic of L-gulonic acid either towards ascorbic acid formation or towards xylitol formation (Hussain et al., 2012). It was observed that both LH (hCG) and FSH (PMSG) cause increase in specific activity of the enzyme L-gulonate dehydrogenase (Figures 3.7 and 3.8) while having opposite effects on ascorbic acid content. As *in vitro* studies showed inhibition of the enzyme activity by ascorbic acid (Sharma et al., 2013), it was thought that increase in activity of L-GuDH was due

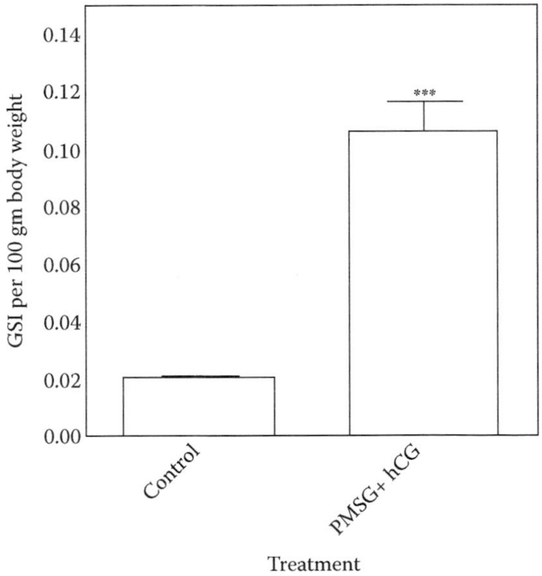

Figure 3.7 GSI in animals without and after injection of PMSG plus hCG.

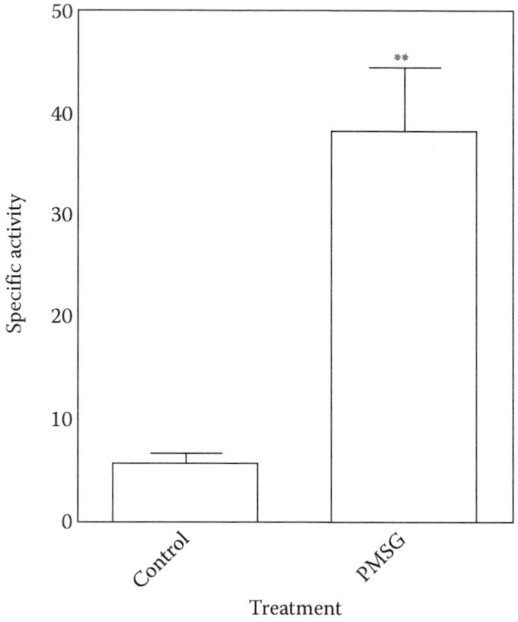

Figure 3.8 Specific activity of enzyme in ovary after PMSG (50 IU) injection.

to release from inhibition by ascorbic acid during LH action. However simultaneous increase in both ascorbate levels and L-gulonate dehydrogenase activity of the gonad as observed during FSH (PMSG) action could not be explained by the same logic. Further, LH action was an acute effect while FSH action was a chronic effect as increased protein synthesis follows FSH action. It was tentatively concluded that under *in vivo* conditions, increase or decrease in specific activity of L-gulonate dehydrogenase has no effect on ascorbate concentration while a decrease in ascorbate could result in an increase in activity of L-gulonate dehydrogenase.

The actions of prolactin have been studied for over six decades. For long time the signal transduction mechanism for PRL action was not clear as no cAMP or IP3 fluctuations were observed in the target tissue. Later it was recognised that PRL and cytokines belong to the same group of protein factors called helix bundle peptide hormones. Their effects are essentially mediated by receptors which have protein tyrosine kinase activity. The PRL receptor has been cloned from rat liver and ovary. There are two forms of it one which is 591 amino acids long and another short form which is 291 amino acids long. Yet another 391 amino acid long PRL receptor has been recognised in the rat Nb2 cell line. Alternate splicing appears to lead to these isoforms. The receptor gene also has been located on chromosome 5 in human. The structural features of the receptor required for ligand binding have been worked out. Receptor dimerisation during hormone binding appears to be necessary for signal transduction. The JAK tyrosine kinase, especially JAK-2 appears to mediate the action of PRL. Further changes in the target cells involve phosphorylation of specific substrate proteins called signal transducer and activators of transcription (STAT) proteins (Kelly et al., 2001).

In the recent years evidence has accumulated that some actions of PRL are more metabolic in nature and not mitogenic or developmental. Chronic and transient hyper-prolactinemia has been associated with luteal phase dysfunction. Evidence suggests that elevated PRL may exert its anti-gonadal effects through reducing available ovarian LH receptors. Previous *in vivo* studies have also shown that in male rabbits, prolactin inhibits the testosterone production stimulated by LH or hCG. Prolactin is also shown to decrease responsiveness to LH-RH acting at the level of the hypothalamus in rats. We have also observed a similar kind of effect where PRL derived anti-angiogenic 14-mer peptide increases the ovarian ascorbic acid content contrary to LH in Parlow assay (Figure 3.5). Other actions of PRL may be permissive whose molecular correlates are not easy to spell out. The mitogenic actions of PRL can be seen in its effects on lymphocytes or in amphibian growth during the tadpole stage. Its action on the prostate is more metabolic in nature. The cytokine-like effects of PRL on lymphocytes do not involve protein kinase C (PKC) action while its metabolic effects like increase in citrate level of lateral prostate in rats or transaminase activity in ventral prostate involves increase in PKC activity. Prolactin has both synergistic effects with androgens and also androgen independent effects. *In vitro* effects of PRL on cells would clearly distinguish the two modes of metabolic effects.

Many of the oncogenes appear to resemble and hence stimulate the hormone receptor protein of G protein or some other protein in the pathway of hormone action. It has led to the integration of hormone action at two levels. One, at the level of signal, that is hormone is only one among the many signals that target cells process. Two, at the level of response, that is response to hormones, drugs, neurotransmitters, growth factors and environmental factors appear to be similar suggesting evolution of receptor mechanism to take care of signals which disturb steady state.

3.6 Non-Gonadal Actions of Gonadotropins

For over one century since their discovery, gonadal action of gonadotropins has been studied in considerable detail. In addition to the direct action on gonads, gonadotropins influence the activity of non-gonadal tissues as well, whether directly or indirectly. On the basis of the expression of full length receptor mRNA, radioreceptor assay, immunocytochemistry and other techniques it has been demonstrated that the receptors for gonadotropic hormones are present in brain, adrenal gland, seminal vesicle, bone, epididymis, placenta, skin, breast, uterus, oviducts, fetal membranes, sperm, early embryos, cervix, retina of humans and other vertebrates. Except for a few cases, the functional relevance of these receptors is yet not clear. It needs much more time to conceptualise the physiological significance of presence of these receptors in non-gonadal tissues (Kumar et al., 2010).

In various species including humans these receptors have been demonstrated to be present in the two layers of uterus, endometrial and myometrial vascular smooth muscles and endothelium (Fields and Shemesh, 2004). These receptors have been indicated to cause vasodilatation (mediated by gonadotropins through induction of prostaglandin synthesis) and induction of synthesis of progesterone by this tissue. Keeping this in mind it has been therefore suggested that gonadotropins are directly involved in the process of implantation, regulation of blood flow, growth and proliferation of uterus, and maintenance of early pregnancy.

FSH has been found to directly regulate the growth of epididymis and bone (Dahiya and Rao, 2006; Iqbal et al., 2006). It has been demonstrated that animals deprived of FSH show atrophied epididymis (decreased size of lumen and uneven surface of luminal epithelium) and the effect is more pronounced in the caudal region even though androgen levels were normal. hCG has been found to bind to the caudal region of epididymis (possibly through LH receptor) and cause induction of aromatase activity resulting in the synthesis of E_2. This observation provides an alternate site other than testis for the synthesis of estrogen in males.

Normal adrenal gland (zona reticularis region) has been shown to express LH receptor, and where secretion of androgens by fetal adrenal gland under the influence of hCG has been reported. Adenomas of adrenal gland are known to express LH/hCG receptor causing these cells to secrete androgens. In these cases except for the diseased states, a clear evidence of presence of LH receptor and function is however not yet forthcoming.

Studies have shown that there is a correlation between elevated gonadotropin levels and incidence of Alzheimer's disease. It is also indicated that elevated LH (as seen during menopause/andropause) levels cause processing of amyloid precursor protein through an amyloidogenic pathway resulting in the production in Aβ peptide. The accumulation of Aβ peptides may lead to the development of disease (promoting ischemia, impaired nutrition transport). This lead provides a new drug target for treatment of Alzheimer's disease.

Osteoporosis in older females after menopause (a condition where serum gonadotropins are elevated and estrogens level is decreased) has been believed to be due to the decrease in the levels of estrogens in the blood. However recent studies have indicated that FSH is directly responsible for the bone loss in elderly females. In *in vivo* studies it was found that mice nullizygous for FSH-β and FSH receptor have preserved bone mass despite elevated estrogen levels in the blood. It has been suggested that FSH stimulates osteoclast formation, function and survival to enhance bone resorption via the activation of a pertussis toxin-sensitive G_i-coupled FSH receptor present on these cells. In addition to this, FSH was also found to enhance the production of osteoclastogenic cytokines, TNFα within the bone marrow (Kumar et al., 2010).

hCG and LH share structural similarities with TSH and the definitive role of hCG and LH as thyroid stimulators has been established. This effect was first demonstrated in 1967 (Burger, 1967). Work by Pekary et al. (1988) stated that hCG causes a more delayed and prolonged thyroidal iodine release (TIR) than TSH. Cohn et al. (1990) first claimed the *in vivo* effect of hCG on thyroid function in pregnant women. This group made the observation both *in vivo* and *in vitro*. In the *in vitro* system, that is the FRTL-5 cells in culture, a partially purified preparation of hCG stimulated both growth and function as indicated by parameters like iodide uptake and tritiated thymidine incorporation respectively. The *in vivo* effect of serum hCG on thyroid during pregnancy in both the iodide uptake and ^{3}H-thymidine incorporation assays has been observed. Biochemically and clinically overt hyperthyroidism has been observed in patients with trophoblastic tumors that secrete large amounts of hCG. Subsequently thyrotrophic activity has been found in several highly purified hCG preparations from molar tissue and normal pregnancy urine. These results indicate that one unit of hCG had thyrotrophic activity equivalent to that in 0.5 μU of TSH in mouse thyroid bioassay (Nisula et al., 1974; Kraiem, 1994). There are some speculations that a non-hCG thyroid stimulating molecule may also be secreted by trophoblastic cells (Shemesh et al., 1982). Another report has suggested that human thyroid tissue may contain specific mRNA for the LH/hCG receptors (Frazier et al., 1990). Highly purified hCG (CR127) stimulates Chinese hamster ovary (CHO) cells which have been transfected with human TSH receptor (Davies et al., 1992) Moreover, numerous studies have demonstrated that purified hCG stimulates cAMP generation and iodide uptake by animal thyroid cells. However, several workers have judged hCG to be a negligible human thyroid stimulator because it has little or no measurable stimulating effect on adenylate cyclase in human thyroid membranes or human monolayer cell culture system. The receptor involved in this effect still needs to be isolated. Speculations are made that hCG acts through conventional TSH-R on the thyroid membranes and that the TSH-R is ambiguous in its ligand binding. In this context it is interesting to note that both buLH and hCG were found to stimulate increase in cAMP levels in buffalo and rat thyroid tissue minces, incubated *in vitro* (Figure 3.9). A similar research problem is to understand how PMSG acts as LH in equids but as FSH in other heterologous species. The molecular details of interaction between PMSG and these receptors are not yet known.

3.7 Clinical Use

Gonadotropins are also pharmaceuticals. Clinical handling of infertility cases among women requires use of gonadotropins. Assisted reproduction technologies (ART) utilise gonadotropins. hCG and hMG are the major gonadotropins employed in clinical practices. Similarly veterinary clinical practice employs FSH and PMSG to induce super-ovulation as part of the Multiple Ovulation and Embryo Transfer (MOET) programme. In addition the majority of pregnancy diagnostic methods

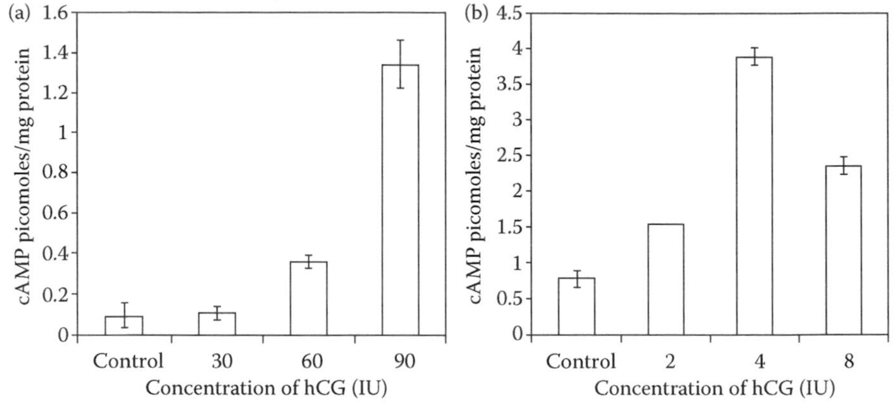

Figure 3.9 Thyrotropic effect of HCG on cAMP of rat thyroid follicles content. (a) *In vivo* effects and (b) *in vitro* effects.

in humans are based on detection of hCG in urine. A spectrum of diagnostic kits based on immunological techniques (agglutination, dipstick, ELISA, RIA etc.) is available in the market and research laboratories for this purpose.

Until recently preparations of gonadotropins and PRL from human and farm animals were made by use of conventional biochemical techniques. Reference preparations as bioassay standards, as radioiodination standards and as immunoassay standards are available for human, bovine, ovine, porcine and murine species. For buffalo, of course, no such standard preparation is available from any commercial or government organisation. Our laboratory has made these reference standards for water buffalo. These are yet to be validated by multicentre analysis and field testing.

In India, buffalo constitutes a major milk-producing animal. The major reason for failure of the MOET programme using buffalo in India could be unresponsiveness of buffalo to heterologous gonadotropins (PMSG, which is used to obtain FSH activity in farm animals) and production of antibody against heterologous gonadotropins when introduced in the animal for the induction of super-ovulation (European Study Group, 1998; Muralidhar and Chaudhary, 2008).

Of late, recombinant DNA technology has made heavy inroads into reproductive biology research concerning gonadotropins and prolactin and also into the logistics of gonadotropins supply. Recombinant DNA technique–based human LH, FSH PRL, and hCG are available from research laboratories and commercial sources (Blanchard et al., 2007). Biologically active recombinant buffalo TSH and FSH have been reported to have been made (Nidhi Vashistha et al., 2014). A more exciting field of activity has been using such techniques to throw new light on problems of gonadotropins actions and structure–function relationships. More thermostable artificial variants, long-acting analogues (by increasing plasma half life) made by genetically fusing C-terminal peptides to FSH/LH, single-chain gonadotropins (where α- and β-subunits are fused in tandem and expressed as single chain) and single-chain multi-domain gonadotropins (made by fusing α-subunits with β-subunits of two or three different gonadotropins in tandem) have been made in laboratories. These variants have been found to be biologically active *in vitro* systems and are waiting to get commercialised after trials. These analogues also pose a new problem in front of workers engaged to elucidate structure–function relationships of gonadotropins.

Some scientists have even extended these and are projecting whole transgenic animals (cows, buffalo) as new biopharmaceutical factories. The idea is to obtain transgenic farm animals expressing the desired foreign protein only in the mammary gland so that the animal can keep producing the protein in bulk and one can simply obtain it in the milk. Subsequent purification from milk then, of course, requires judicious use of biochemical techniques.

Acknowledgements

Research work from our laboratory reported in this article was made possible by generous financial support from CSIR, ICAR, HLRF, DST (in the form of JC Bose National fellowship to KM), DBT and University of Delhi. Financial assistance to NV and NK from CSIR in the form of SRF is gratefully acknowledged.

References

Arora, T., Arundhati, G. and Muralidhar, K. 1999. *J. Endocrinol. Reprod.*, 3(2), 60–69.

Arora, T., Nath, R., Vashistha, N., Kumari, N., Hussain, S.S. and Muralidhar, K. 2012. *World J. Life Sci. Med Research*, 2(4), 150–158.

Bedows, E., Huth, J.R. and Ruddon, R.W. 1992. *J Biol Chem.*, 267(13), 8880–8886.

Beers, W.H. and Strickland, S. 1978. *J. Biol. Chem.*, 253, 3877–3881.

Bell, E.T., Mukherji, S. and Loraine, J.A. 1964. *J Endocrinol.*, 28, 321.

Ben-Jonathan, N., Meharson, J.L., Allen, D.L. and Steinmetry, R.W. 1996. *Endocr. Rev.*, 17, 639–669.

Bhagat, L., Duraiswami, S. and Kumari, G.L. 1993. *J. Biosci.*, 18, 59–72.

Blanchard, V., Gadkari, R.A., Gerwig, G.J., Dighe, R.R., Leeflang, B.R. and Kamerling, J.P. 2007. *Glycoconj. J.*, 24, 33–47.

Bloomfield, G.A., Faith, M.R. and Pierce, J.G. 1978. *Biochim. Biophys. Acta.*, 533, 371–382.

Boime I. and Ben-Menahem, D. 1999. *Recent. Prog. Horm. Res.*, 54, 271–288.

Bolander, F.F. 1989. *Molecular Endocrinology*, Academic Press, New York.

Bousfield, G.R., Butnev, V.Y., Gotschall, R.R., Baker, V.L. and Moore, W.T. 1996. *Mol. Cell. Endocrinol.*, 125(1–2), 3–19.

Bousfield, G.R., Li, J. and Ward, D.N. 1994. Gonadotropins: Chemistry and biosynthesis. In *Knobil, and Neill's the Physiology of Reproduction*, 2nd edition, Neill, J.D., Ed., Chapter 30, Vol. 1, 1581–1634. Elsevier, New York.

Burger, A. 1967. *Acta Endocrinologica*, 55, 600–610.

Chadha, N., Kohli, R. and Muralidhar, K. 1985. *J. Reprod. Biol. Comp. Endocrinol.*, 5, 19–24.
Cohen, J. 2003. *Reprod. Biomed. Online*, 6(3), 361–366.
Cohn, M., Kennedy, R.L., Darne, J., Griffiths, H., Price, A. and Davies, R. 1990. *Horm. Res.*, 33, 177–183.
Dahiya, C.L. and Rao, A.J. 2006. *Biol. Reprod.*, 75, 98–106.
Darling, R.J., Ruddon, R.W., Perini, F. and Bedows, E. 2000. *J. Biol. Chem.*, 275(20), 15413–15421.
Davies, T.F., Tomer, Y. and Huber, G.K. 1992. *J. Clin Endocrinol. Metab.*, 74(6), 1477–1479.
Dorrington, J.H. and Armstrong, D.T. 1975. *Proc. Natl. Acad. Sci. USA*, 72, 2677–2681.
Dufau, M.L., Mendelson, C.R. and Catt, K.J. 1974. *J. Clin. Endocrinol.*, 39, 610–613.
Edge, A.S.B. 2003. *Biochem J.*, 376(2), 339–350.
Ellis, S. 1961. *Endocrinology* 69, 554–570.
European Recombinant Human LH Study Group. 1998. *J. Clin. Endocrinol. Metab.*, 83, 1507–1514.
Fields, M.J. and Shemesh, M. 2004. *Biol. Reprod.*, 71, 1412–1418.
Fiete, D., Srivastava, V., Hindagaul, O. and Baenziger, J.U. 1991. *Cell*, 67, 1103–1110.
Frazier, A.L., Robbins, L.S., Stork, P.J., Sprengel, R., Segaloff, D.L. and Cone, R.D. 1990. *Mol. Endocrinol.*, 4, 1264–1267.
Gadkari, R., Deshpande, R. and Dighe, R.R. 2003. *Protein. Expr. Purif.*, 32(2), 175–184.
Garcia-Campayo, V. and Boime, I. 2001. *Trends Endocrinol. Metab.*, 12(2), 72–77.
Garcia-Campayo, V., Boime, I., Ma, X., Daphna-Iken, D. and Kumar, T.R. 2005. *Biol. Reprod.*, 72, 301–308.
Gorbman, A. 1979. *Amer. Zool.* 19, 1261–1270.
Greep, R.O., van Dyke, H.B. and Chow, B.K. 1941. *Proc. Soc. Exp. Biol. Med.*, 46, 644–649.
Greep, R.O., van Dyke, H.B. and Chow, B.F. 1942. *Endocrinology*, 30, 635–649.
Gupta, S.K. Ed. 1991. *Immunology: Perspective in Reproduction and Infection*, Oxford and IBH Publishing, New Delhi.
Hartree, A.S. and Renwick, A.G.C. 1992. *Biochem. J.*, 287, 665–679.
Hussain, S.S., Sharma, M., Nangia, M. and Muralidhar, K. 2012. *Novus Int. J. Pharmaceutical. Technol.*, 1(4), 19–29.
Igarashi, M. and McCann, S.M. 1964. *Endocrinology*, 74, 446–452.
Igarashi, M., Sato, T., Jyujo, T., Lesaka, T., Taya, K. and Ishikawa, J. 1974. *Endocrinology*, 95, 417–420.
Iqbal, J., Sun, L., Rajendra Kumar, T., Blair, H.C. and Zaidi, M. 2006. *Proc. Natl. Acad. Sci. USA*, 103, 14924–14930.
Jablonka-Shariff, A., Kumar, T.R., Eklund, J., Comstock, A. and Boime, I. 2006. *Mol. Endocrinol.*, 20, 1437–1446.
Jaffe, B.L. and Behrman, H.R. 1978. *Methods of Hormone Radioimmunoassay*, 2nd edition, Academic Press, New York.
Jia, X. and Hsueh, A.J.W. 1986. *Endocrinology*, 119, 1570–1577.
Jia, X.C., Perlas, E., Su, J.G.J., Moran, F., Lasley, B.L., Ny, T. and Hsueh, A.J.W. 1993. *Biol. Reprod.*, 43, 1310–1316.
Kelly, P.A., Benart, N., Freemark, M., Lucas, B., Goffin, V. and Bouchard, B. 2001. *Biochem. Soc. Trans.*, 2, 48–52.
Knobil, E. and Neill, J.D. Eds. 2006. *Knobil and Neill's The Physiology of Reproduction*, 3rd edition, Elsevier, New York.
Kohli, R., Chadha, N. and Muralidhar, K. 1988. *FEBS Letters*, 242, 139–143.
Konig, W. 1993. *Peptide and Protein Hormones*, VCH, New York.

Kraiem, Z., Sadeh, O., Blithe, D.L. and Nisula, B.C. 1994. *J. Clin. Endocrinol Metab.*, 79(2), 595–599.
Kumar, A., Rao, C.V. and Chaturvedi, P.K. 2010. *Gonadal and Nongonadal Actions of Gonadotropins*, Narosa Publishing House, New Delhi.
Kumar, T.R. 2007. *Mol. Cell. Endocrinol.*, 260–262, 249–254.
Kumar, T.R., Donehower, L.A., Bradley, A. and Matzuk, M.M. 1995. *J. Int. Med.*, 238, 233–238.
Kumar, T.R. and Low, M.J. 1993. *Mol. Endocrinol.*, 7, 898–906.
Kumari, N., Vashistha, N., Sharma, P., Godhwal, U. and Muralidhar, K. 2013. *Int. J. Bioassays*, 2(2), 370–375.
Li, C.H. and Starman, B. 1964. *Nature*, 202, 291–292.
Lin, W., Ransom, M.X., Myers, R.V., Bernard, M.P. and Moyle, W.R. 1999. *Mol. Cell Endocrinol.*, 152(1–2), 91–98.
Litwack, G. Ed. 1972. *Biochemical Actions of Hormones*, Vol. 2, Academic Press, New York.
Lodish, H., Berk, A., Kaiser, C.A., Krieger, M., Scott, M.P., Bretscher, A., Ploegh, H. and Matsudaira, P. 2007. *Molecular Cell Biology*, WH Freeman Publishers, San Francisco.
Lunenfeld, B. 2004. *Hum. Reprod. Update*, 6, 653–667.
Macklon, N.S., Stouffer, R.L., Guidice, L.C. and Fauser, B.C.J.M. 2006. *Endocrine Rev.*, 27, 170–207.
Manjunath, P., Sairam, M.R. and Schiller, P.W. 1982. *Biochem. J.*, 207(1), 11–19.
Matteri, R.L., Papkoff, H., Murthy, H.M.S., Roser, J.F. and Chang, Y.S. 1986. *Domestic Animal Endocrinology*, 3(1), 39–48.
Matzuk, M.M., Brown, C.W. and Kumar, T.R. Eds. 2001. *Transgenics in Endocrinology*, Humana Press, New Jersey.
McCann, S.M. and Taleisnik, S. 1960. *Am. J. Physiol.*, 199, 847–850.
Moore, K.L. 2003. *J. Biol. Chem.*, 278, 24243–24246.
Morell, A.G., Gregoriadis, G., Scheinberg, H., Hickman, J. and Ashwell, G. 1971. *J. Biol. Chem.*, 246, 1461–1467.
Moudgal, N.R. 1989. *Current Opinion in Immunology*, 2(5), 736–742.
Moudgal, N.R., Moyle, W.R. and Greep, R.O. 1971. *J. Biol. Chem.* 246(16), 4983–4986.
Muralidhar, K. and Chaudhary, R. 2008. *Proc. Nat. Acad. Sci. (India)*, LXXVII, Section B, Part II, 189–210.
Muralidhar, K. and Moudgal, N.R. 1976. *Biochem.J.* 160, 615–619.
Muralidhar, K., Rajendra Kumar, T. and Usha, V. 1998. *Proc. Nat. Acad. Sci. (India)* LXVIII, Part III and IV, 153–187.
Muyan, M., Ruddon, R.W., Norton, S.E., Boime, I. and Bedows, E. 1998. *Mol. Endocrinol.*, 12(10), 1640–1649.
Nisula, B.C., Morgan, F.J. and Canfield, R.E. 1974. *Biochem. Biophys. Res. Comm.*, 56, 86–91.
Parlow, A.F. 1961. In *Human Pituitary Gonadotropins*, Albert, A., Ed., CC Thomas, Springfield, 300–307.
Pearl, C.A. and Boime, I. 2009. *Mol. Cell. Endocrinol.*, 309, 76–81.
Pekary, A.E., Hershman, R.M., Lee, H.Y., Sugawara, M., Kirell, C.J., Pang, X.P. and Yanagisawa, M. 1988. *J. Clin. Endocrinol. Metab.*, 67, 74–79.
Pierce, J.G. and Parsons, T.F. 1981. *Ann. Rev. Biochem.*, 50, 465–495.
Prasad, M.S.K., Muralidhar, K., Moudgal, N.R. and Adiga, P.R. 1978. *J. Endocr.*, 76, 283–292.
Reichert, L.E. and Bhalla, V.K. 1974. *Endocrinology*, 94, 483–494.

Robison, G.A., Butcher, R.W. and Sutherland, E.W. Eds. 1971. *cAMP*, Academic Press, New York.
Roser, J.F., Carrick, F.N. and Papkoff, H. 1986. *Biol. Reprod.*, 35, 493–500.
Roy, S., Setlur, S., Gadkari, R.A., Krishnamurthy, H.N. and Dighe, R.R. 2007. *Endocrinology*, 148, 3977–3986.
Ruddon, R.W., Sherman, S.A. and Bedows, E. 1996. *Protein Sci.*, 5(8), 1443–1452.
Sairam, M.R. and Jiang, L.G. 1992. *Mol. Cell Endocrinol.*, 85(3), 227–235.
Sairam, M.R., Zaky, A.A.H. and Hassan, A.A. 1994. *J. Endocr.*, 143, 313–323.
Schemidt-Elmendotff, H. and Loraine, J.A. 1962. *J. Endocr.*, 23, 413.
Setlur, S.R. and Dighe, R.R. 2007. *Glycoconj. J.*, 24(1), 97–106.
Sharma, V., Hussain, S.S., Pooja, S., Jangra, S., Nirmala, K. and Muralidhar, K. 2013. *World J. Life Sci. Med Res.*, 3(1), 8–14.
Shemesh, M., Avivi, A. and Schreiber, A.B. 1982. *J. Biol. Chem.*, 257, 11384–11389.
Sinha, Y.N. 1995. *Endocrine Rev.*, 16, 354–369.
Smith, P.E. 1963. *Endocrinol.*, 73, 793–806.
Steelman, S.L. and Pohley, F.M. 1953. *Endocrinology*, 53, 604–616.
Trist, D.G., Humphrey, P.P.A., Leff, P. and Shankley, N.P. 1997. *Receptor Classification: The Integration of Operational, Structural, and Transductional Information* (Annals of New York Academy of Sciences, Vol. 812). New York Academy of Sciences, New York.
Ulloa-Aguirre, A., Midgley, A.R. Jr., Benitins, I.Z. and Padmanabhan, V. 1995. *Endocrine Rev.*, 16, 765–787.
Ulloa-Aguirre, A.,Timossi, C., de Tomasi, B., Maldonado, A. and Nayudu, P. 2003. *Biol. Reprod.*, 69, 379–389.
Van Damme, M.P., Robertson, D.M., Marana, R., Ritzen, E.M. and Diczfaluzy, E. 1979. *Acta Endocrinol.*, 91, 224–237.
Vashistha, N., Dighe, R., Arora, T., Kumari, N. and Muralidhar, K. 2014. *Austin J. Biotechnol. Bioeng.*, 1(4), 1–9.
Vijayalakshmi, M.A., Ed. 2002. *Biochromatography Theory and Practice*, Taylor & Francis, London.
Wilson, E.D. and Zarrow, M.X. 1962. *J. Reprod. Fert.*, 3, 148–158.
Xing, Y., Williams, C., Campbell, R.K., Cook, S., Knoppers, M., Addona, T., Altarocca, V. and Moyle, W.R. 2001. *Protein Sci.*, 10(2), 226–235.

Chapter 4

Endocrine Regulation of Spermatogenesis in Mammals

Hanumanthappa Krishnamurthy and M. Ram Sairam

Contents

4.1 Introduction 78
4.2 Mammalian Testis 79
    4.2.1 Leydig Cells 80
    4.2.2 Peritubular Myoid Cells 80
    4.2.3 Sertoli Cells 80
    4.2.4 Spermatogenic Cells 81
4.3 Hormonal Regulation of Spermatogenesis 81
    4.3.1 Role of FSH in Regulating Spermatogenesis 82
        4.3.1.1 Regulatory Effect of FSH in Rats 83
        4.3.1.2 Regulatory Effect of FSH in Mice 83
        4.3.1.3 Regulatory Effect of FSH in Non-Human Primates 85
        4.3.1.4 Regulatory Effect of FSH on Spermatogenesis in Men 86
    4.3.2 Role of LH in Regulating Spermatogenesis 86
    4.3.3 Role of Testosterone in Regulating Spermatogenesis 87
4.4 Conclusions 88
References 89

4.1 Introduction

Mammalian spermatogenesis is a complex biological process where a spermatogonial stem cell apart from renewing itself, undergoes series of divisions and terminally differentiates into a mature sperm or male gamete. These finely orchestrated and precisely ordered cellular events are regulated by a combination of several endocrine, paracrine and intracrine factors. The data available thus far suggest that a well-defined hypothalmo–pituitary–gonadal axis that operates through both positive and negative feedback loops controls reproductive events in both male and female mammals. In response to the environmental cues and feedback mechanisms, the hypothalamus releases gonadotropin-releasing hormone (GnRH) which signals the pituitary gland to release follitropin or follicle-stimulating hormone (FSH) and lutropin or luteinising hormone (LH) (Hodgson et al. 1983; Fink 1988; Cooke and Saunders 2002). LH acts on Leydig cells in the testicular interstitial compartment whereas FSH acts on Sertoli cells in the tubular compartment through their respective receptors. It appears that FSH-induced Sertoli cell factors, LH-induced testosterone and various paracrine factors are required for quantitative and qualitative spermatogenesis in mammals (Figure 4.1). Both FSH and testosterone acting synergistically are required for the development of germ cells. Although, the regulation of spermatogenesis in primates and rodents is similar, the former depends more

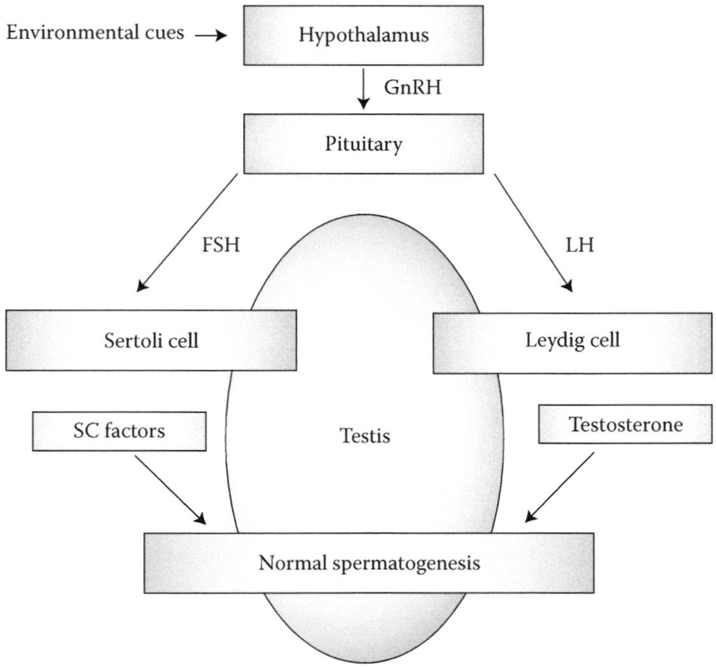

Figure 4.1 A schematic diagram of the hormonal regulation of spermatogenesis in mammals.

on FSH (O'Shaughnessy 2014). FSH has been shown to potentiate the action of testosterone and induce the formation of tight junctions at the Sertoli cell barrier (McCabe et al. 2012). Recently it has also been shown that mice lacking both FSH and androgen receptors can only produce about 3% of normal germ cells on day 20 post-partum (O'Shaughnessy et al. 2012). A clear understanding of these complex processes is required for treatment of infertility and devising novel modes of contraception for humans, and for management of other animals.

4.2 Mammalian Testis

The mammalian testis is encapsulated in tunica albuginea and consists of two distinct compartments: the interstitium and long tubules containing germ cells. While the constituents of the interstitium differ between species, in general it consists of Leydig cells, macrophages, lymph space, blood vessels and endothelial cells (Fawcett 1973). The seminiferous tubules are avascular and consist of different types of germ cells embedded in irregular-shaped Sertoli cells (de Kretser and Kerr 1988). The peritubular myoid cells and a basal lamina are present in between the two compartments, surrounding the seminiferous tubules (Figure 4.2).

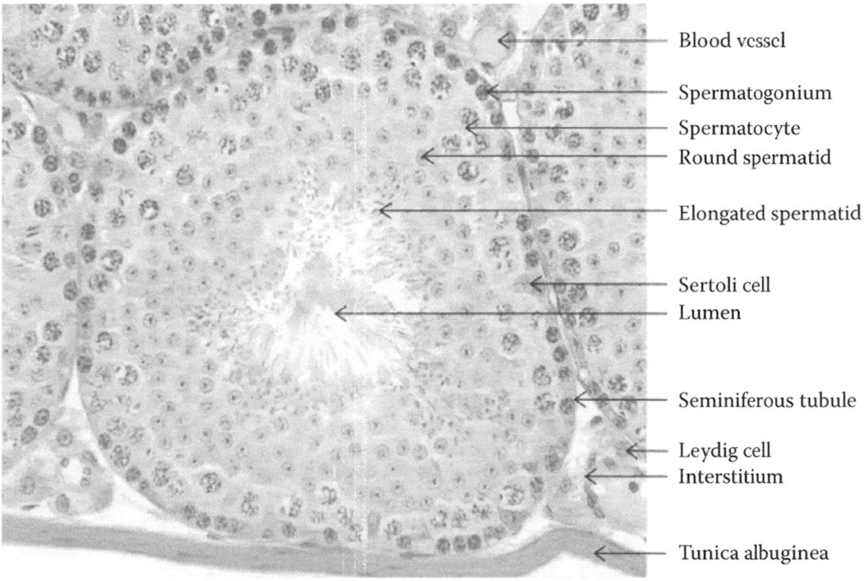

Figure 4.2 (See colour insert.) The cross-section of mouse testis showing interstitial and tubular compartments. The interstitium is comprised of blood vessels, Leydig cells and lymph space. The seminiferous tubule shows various germ cells, Sertoli cells and lumen.

4.2.1 Leydig Cells

Leydig cells (also known as interstitial cells) are present in between the seminiferous tubules. The irregularly shaped Leydig cells generally have more than one nucleoli, fat droplets, pigment granules, and crystalline structures. The number and appearance of Leydig cells vary across species and are entangled between blood and lymphatic vessels. The main function of Leydig cells is the production and secretion of androgens. Testosterone is produced mainly in the testis to supply androgens to various tissues in the body. In most androgen-dependent target tissues, the enzyme 5α-reductase converts testosterone to dihydrotestosterone. Dihydrotestosterone binds to the androgen receptor with a higher affinity than testosterone and is considered to be the most potent of the androgens. It is, however, accepted that testosterone at high concentrations interacts in the same way with the androgen receptor as dihydrotestosterone (Grino et al. 1990). During embryonic development, dihydrotestosterone is necessary for the formation of the male external genitalia and the differentiation of the uro-genital duct into prostate, bulbourethral glands, urethra, periurethral glands and part of the urinary bladder. The Wolffian ducts are unable to convert testosterone into dihydrotestosterone due to the absence of 5α-reductase activity (Siiteri and Wilson 1974). However, testosterone plays an important role in the differentiation of Wolffian duct into epididymis, ductus deferens, seminal vesicle and ejaculatory duct (Wilson and Lasnitzki 1971). Testosterone and dihydrotestosterone are also involved in the manifestation of male secondary sexual characteristics and play an important role in the regulation of spermatogenesis (Hall 1988).

4.2.2 Peritubular Myoid Cells

Peritubular myoid cells (PMC) form a layer of flattened cells between the interstitium and the seminiferous tubules. PMCs together with the Sertoli cells produce an extracellular matrix which provides structural support to the spermatogenic epithelium (Skinner et al. 1985). Furthermore, these cells also seem to be involved in contractile function of the tubules which may play a role in the transport and release of the testicular spermatozoa (Clermont 1958; Fritz and Tung 1986).

4.2.3 Sertoli Cells

Sertoli cells playing the crucial role of 'nursing cells' for developing germ cells occupy a special place within the long and coiled seminiferous tubule of the testis. In the fetal testis, Sertoli cells first differentiate and surround the gonocytes to form seminiferous cords (Magre and Jost 1991). Sertoli cells are the only somatic cell type present in the spermatogenic epithelium and they start proliferating at fetal and neonatal period that ends around day 15 in rats (Orth 1982). Sertoli cells together with peritubular myoid cells and the basal lamina, form the blood–testis barrier (Setchell et al. 1969; Dym and Fawcett 1970; Ploen and Setchell 1992). This protects the germ cells from external elements and provides an immune barrier. The contribution of Sertoli cells to these barriers is important. Tight junctions located between neighbouring Sertoli cells divide the seminiferous tubules into a basal and an adluminal compartment. The passage of growth factors, nutrients and hormones from the basal to the adluminal compartment of the seminiferous tubule is restricted by these junctional contacts between adjacent Sertoli cells. In fact, the tight junctions of Sertoli cells establish the functional blood–testis barrier (Sharpe 1983; Saez et al. 1987; Bardin et al. 1988). Their rather bizarre shape unlike all other epithelial cells in the body allows the attachment and progressive stepwise development of germ cells as they migrate from the basal lamina to the tubular lumen. In many species including man, each healthy Sertoli cell is capable of supporting the development of only a finite number of spermatogonial cells. Recently it has been demonstrated that Sertoli cells can maintain Leydig cell number and peritubular myoid cell activity in the adult mouse testis (Rebourcet et al. 2014). In rodents, the Sertoli cells express androgen receptors soon after birth (Willems et al. 2010) and mice lacking androgen receptors specifically on Sertoli cells (SCARKO) show that androgens act through the Sertoli cells to regulate spermatocyte number and further differentiation thereof (De Gendt et al. 2004; Abel et al. 2008).

4.2.4 Spermatogenic Cells

Spermatogonia are immature germ cells. They divide mitotically and one of the two daughter cells will remain a spermatogonial stem cell and the other will proliferate by mitotic divisions in the seminiferous tubules located at basal lamina (de Rooij 2009). The number of spermatogonial divisions differs from species to species. After various divisions, they differentiate into primary spermatocytes and enter first meiotic division which results in two secondary spermatocytes and these undergo second meiotic division to form four round spermatids. The round spermatids undergo a process called spermiogenesis, where they differentiate morphologically into sperm by nuclear condensation, ejection of the cytoplasm and formation of the acrosome and flagellum. Thus in all species, the process of spermatogenesis includes a number of well-defined stages that help investigators to link molecular changes to specific stages, but the number of stages are not identical in various species.

4.3 Hormonal Regulation of Spermatogenesis

Mammalian spermatogenesis is a well-programmed process where effective interactions between interstitial and tubular compartments of the testis are apparent. The sequence and time taken for the development of spermatogonial stem cell to terminally differentiated spermatozoa is unique for each species. The critical role of the hypothalamo–pituitary–gonadal axis established long time ago has been confirmed and expanded by recent studies. Thus, the two pituitary gonadotropins – FSH and LH – produced in response to hypothalamic GnRH and the feedback loops play important roles in this process (Greep et al. 1936; Sharpe 1994; Cooke and Saunders 2002). LH and FSH are gonadotropins belonging to a family that also includes TSH (Pierce and Parsons 1981). These glycoprotein hormones are heterodimers of α- and β-subunits that are non-covalently bound. The α-subunits of these hormones are identical, however, their β-subunits are specific and unique for each hormone except for hCG and LH (Policastro et al. 1986). LH and hCG are structurally and functionally very similar and these hormones bind to the same receptor. These receptors belong to the G protein-coupled receptor family and typically have an extracellular N-terminal hormone binding domain, a heptahelical transmembrane region and an intracellular C terminal region that couples to G proteins. Abrogation of LH action by any one of several means including the use of LH-RH agonists or antagonists leads to prompt reduction of circulating and testicular androgens and termination of spermatogenesis causing infertility. Therefore, separating the androgenic function of the testis from the process of spermatogenesis is a very difficult but desirable goal for achieving reversible contraception in the male. Although, the suspected problem of cross-contamination of the other hormones during purification of gonadotropins has been addressed by producing the same by recombinant technology, it is still difficult to precisely elucidate the action of one single hormone as they often affect the levels of other hormones (O'Shaughnessy 2014).

4.3.1 Role of FSH in Regulating Spermatogenesis

Understanding the underlying mechanisms of FSH in regulating spermatogenesis has been the focus of investigators working in the area of male reproduction especially on infertility and contraception. This hormone, like LH, acts by coupling to its own receptor in Sertoli cells. The FSH receptor is also a G protein-coupled receptor with a large extracellular domain. FSH receptors are preferentially expressed in the testis but not in many extragonadal tissues unlike the LH receptor (Dankbar et al. 1995). Although, there are some reports of FSH receptor expression in the epididymis (Dahia and Rao 2006), for all practical purposes we will consider expression in Sertoli cells to be the most physiologically relevant. Thus the Sertoli cell is the principal site of FSH action in the testis (Bockers et al. 1994). In immature rats, the FSH receptor is regulated by FSH and not testosterone (Maguire et al. 1997). FSH regulates the proliferation and differentiation of Sertoli cells and the paracrine factors secreted by Sertoli cells also modulate Leydig cell functions (Migrenne et al. 2012). Specific action of FSH on Sertoli cells regulates the sperm output of the seminiferous tubules by controlling the expansion of premeiotic germ cells (Schlatt and Ehmcke 2014). Alternate splicing generates various transcripts of the FSH receptor (Khan et al. 1993; Simoni et al. 1997) coding for structural motifs that could mediate different aspects of hormone signalling (Sairam and Babu 2007).

4.3.1.1 Regulatory Effect of FSH in Rats

FSH receptor expression in the testis begins with the onset of puberty and appearance of spermatocytes (Laborde et al. 1996; Rannikko et al. 1996). The increase in apoptotic cells observed during the first wave of spermatogenesis is stage specific (Rodriguez et al. 1997; Jahnukainen et al. 2004). In adult rats, FSH receptor expression and ligand binding activity were shown to be high in stages XII–II and low in stages VI–VII (Kangasniemi et al. 1990; Heckert and Griswold 1991; Rannikko et al. 1996). The role FSH in mammalian spermatogenesis has been a topic of considerable debate (Moudgal and Sairam 1998) with various interpretations reported based on the results obtained in studies using different species and different modalities of the interruption of hormonal/receptor signalling. Thus it is not surprising that there are different views on hormonal dependence of rodent spermatogenesis especially in adults (Dym et al. 1979; Singh and Handelsman 1996; Kumar et al. 1997). Specific immunoneutralisation of FSH has been shown to induce apoptosis of spermatogonia and pachytene spermatocytes in rats (Shetty et al. 1996). FSH induces germ cell proliferation and development in intact (Matikainen et al. 1994) and hypophysectomised (Vihko et al. 1991; Russell et al. 1993) pre-pubertal rats. Immunisation of rats against FSH receptor reduced the production of sperm by 50% (Graf et al. 1997). The restoration of spermatogenesis by FSH up to round spermatid stage in GnRH immunised adult rats that eliminates gonadotropins substantiates the interplay of hormones (McLachlan et al. 1995).

4.3.1.2 Regulatory Effect of FSH in Mice

Immunisation of mice against the FSH receptor was found to affect fertility in male mice (Remy et al. 1996). The mouse has been a very useful experimental tool to secure various genetic mutants that could help discriminate the contribution of hormones and receptors to the spermatogenic process. For example in the hypogonadotropic (hpg) mouse, a mutation in the GnRH gene deprives the animal of functional GnRH and thus the lack of gonadotropins in circulation leads to failure of gonadal development causing infertility. In hpg male mice FSH has been shown to stimulate spermatogonial proliferation but not spermiogenesis (Singh and Handelsman 1996). Gene disruption is another technique that has been successfully used in recent years to address the differential role of gonadotropins in male reproduction. Thus studies involving both the knockout of the FSH β gene (Kumar et al. 1997) and that of its receptor (Dierich et al. 1998; Sairam et al. 1998) have provided new rodent models useful in elucidating the molecular mechanisms involved in the complex process. Thus, the targeted disruption of the FSH β-subunit and FSH receptor gene have clearly demonstrated the role of FSH receptor signalling, which is essential for the complete development of normal testis and maintaining qualitative and quantitative spermatogenesis. Mouse testis development and function appear to be differently regulated by FSH receptor signalling during fetal and pre-pubertal life (Migrenne et al. 2012). FSH-stimulated glial cell line-derived neurotrophic factor (GDNF) and fibroblast growth factor 2 (FGF2) have been shown to be important for spermatogonial stem cell self-renewal and survival (Mullaney and Skinner 1992; Tadokoro et al. 2002; Goriely et al. 2005; Simoni et al. 1997; Hofmann 2008; Ding et al. 2011).

FSH β mutant male mice are reported to be fully fertile (Kumar et al. 1997) whereas FSH receptor knockout (FORKO) males exhibit reduced fertility (Dierich et al. 1998; Sairam et al. 1998). FORKO males have elevated levels of FSH and decline in testosterone with normal LH levels compared to wild type. The FORKO males show reduced sperm counts and motility and abnormal sperm morphology. However, no such sperm abnormalities, were reported in the FSH β mutants that had normal levels of LH and testosterone (Kumar et al. 1997). Normal testicular size and spermatogenesis was restored in FSH β-subunit mutants by genetic rescue (Kumar et al. 1998).

The onset of spermatogenesis begins with spermatogonial stem cell proliferation and differentiation and completes the first wave by day 40–45 in mice (de Rooij 1998). Recent data show that the mice lacking both FSH and androgen receptors can only produce about 3% of normal germ cells on day 20 post-partum (O'Shaughnessy et al. 2012). In the FSH receptor knockout (FSHRKO) and SCARKO (Sertoli cell only knockout of androgen receptor) mice, there is a significant decrease in germ cell numbers in both groups at 20 days of age, during the first wave of spermatogenesis but before completion of meiosis. However, the loss of germ cells in the FSHRKO mouse is apparently more than in the SCARKO mouse (O'Shaughnessy et al. 2012).

The FSH receptor knockout male mice take 3 weeks longer than wild-type mice to produce the first litter showing that there is a delay in sexual maturity in the absence of FSH receptor (Krishnamurthy et al. 2001a). Additional flow cytometry analysis of testicular cells revealed a compensatory increase in the percentage of cells in 2C compartment which comprises of spermatogonia, spermatocytes and non-germ cells (like Leydig, Sertoli and peritubular myoid cells). Decreases in the absolute number of homogenisation-resistant elongated spermatids were observed in FORKO males (Krishnamurthy et al. 2000a). Interestingly, elongated spermatids of FORKO males exhibited increased propidium iodide binding suggesting reduction in sperm nuclear compaction. The increase in the sperm head size and susceptibility to dithiothreitol-induced decondensation in FORKO males further indicates the inadequate condensation of sperm chromatin. The results of the sperm chromatin structure assay (SCSA) have revealed the susceptibility of FORKO sperm to acid denaturation which was observed in FORKO males (Krishnamurthy et al. 2000a). It has also been reported that the sperm from FORKO males accumulate free radicals that can damage sperm DNA, an effect that can be reduced by preventive treatments suggesting potential clinical applications in handling of sperm from infertile men for in vitro fertilisation or inseminations (Libman et al. 2010).

Leydig cells in FORKO males exhibit less steroidogenic activity compared to wild type male mice indicating that there is impaired intercellular communication between Leydig cells and Sertoli cells (Krishnamurthy et al. 2001b; Baker et al. 2003). Sertoli cells produce a number of proteins including cytokines and other paracrine factors that are also important for maintaining optimal Leydig cell function (Lei et al. 2009). The apoptotic surge observed in adults may be contributing to the sub-optimal fertility seen in the adult FORKO mouse (Krishnamurthy et al. 2000a; O'Shaughnessy 2014).

4.3.1.3 Regulatory Effect of FSH in Non-Human Primates

Currently available data suggest an important requirement of FSH for primate spermatogenesis (Murty et al. 1979; Matsumoto et al. 1986; Van Alphen et al. 1988; Weinbauer et al. 1991). FSH stimulates spermatogonial proliferation and Sertoli cell numbers in immature rhesus monkeys (Arslan et al. 1993; Schlatt et al. 1995). Many studies on passive and active immunisation of FSH in bonnet monkeys have demonstrated that chronic bioneutralisation can produce oligospermia and infertility (Murty et al. 1979; Moudgal et al. 1992). Similarly, immunisation of adult bonnet monkeys with a recombinant receptor protein domain has also been shown to impair testicular function and fertility (Moudgal et al. 1997b). While these monkeys did not show any change in circulating testosterone levels, the poor quality of the voided sperm was apparent as demonstrated by a variety of parameters like decline in motility, viability, gel penetrability, and acrosin and hyaluronidase activities (Moudgal et al. 1992; Moudgal et al. 1997a). The flow cytometric analysis of testicular cells of FSH-immunised rats and monkeys, and FSH receptor-immunised monkeys revealed drastic changes in the percentage of germ cells and their ratios (Moudgal et al. 1997b) indicating disturbances in testicular compartments. In rhesus monkeys, proliferation of type A pale (Ap) spermatogonia occurs in a gonadotropin-independent manner and the differentiation of Ap to B spermatogonia is absolutely gonadotropin dependent and stimulated either by FSH or testosterone (Marshall et al. 2005). Proliferation of Ap spermatogonia leading to increased production of differentiated B spermatogonia is achieved by selectively elevating the level of FSH but not LH (Simorangkir et al. 2009). Thus the aforementioned findings point out some of the drawbacks in the approaches like the elimination of paracrine interactions within pituitary gland by radical surgery such as total hypophysectomy that eliminates all hormones, or repeated injections of GnRH antagonists that suppress both FSH and LH, or the purity of hormones (antigens) used in generating the antibodies in immuno-neutralisation studies (Kumar 2009). Nevertheless, we can clearly draw attention to the fact that such active or passive immunisation studies in rodents or other animals including non-human primates provide the only means of achieving a state equivalent to conditional knockouts of either the gonadotropins or their receptors as the experiments are conducted in the post-pubertal state (Moudgal and Sairam 1998).

4.3.1.4 Regulatory Effect of FSH on Spermatogenesis in Men

Reports on human mutations relevant to gonadotropin action are focusing on the plausible role of FSH and/or its receptor in regulating normal testicular function (Siegel et al. 2013). Autonomous sustenance of spermatogenesis in a man carrying activating FSH receptor mutation in the absence of gonadotropins has been reported (Gromoll et al. 1996). Inactivating mutation of FSH receptor gene in Finnish men led to an incomplete elimination of its signalling function (Tapanainen et al. 1997). In that study, one out of five men were infertile, two were fertile and fertility of the other two was not known but they also had low sperm counts. The azoospermia and hypogonadism condition reported in a man could be due to the inactivating mutation in FSH β-subunit gene (Phillip et al. 1998). However, as serum levels of inhibin B, an indicator of endogenous FSH activity (Anderson et al. 1997) was not completely abolished in four out of the five men studied, the question arises whether this FSH receptor mutation completely abolished all the FSH activity in all the men studied (Weinbauer and Nieschlag 1998). Interestingly, only in the fifth patient, who was infertile, inhibin B was undetectable. Alterations observed in sperm chromatin structure following abrogations in FSH/FSH-R action were similar in man and monkey (Krishnamurthy et al. 2000b). Recent reports also indicate that the presence of anti-FSH antibody in circulation is strongly correlated with sperm quality and quantity in idiopathic male infertility (Wang et al. 2008). Therefore complete suppression of FSH activity could impair/disrupt spermatogenesis in man. Another interesting animal model that shows a clear dependence of FSH requirement for spermatogenesis is the seasonally active male hamster (see Schlatt and Ehmcke 2014).

4.3.2 Role of LH in Regulating Spermatogenesis

Males require LH for Leydig cell proliferation and maturation, and for increasing the synthesis of testosterone, which promotes spermatogenesis and also regulates the function of accessory sex organs (Lei and Rao 2000; Lei et al. 2001). It has long been known that suppression or neutralisation of LH action induces a castration like effect in the adult and is harmful for spermatogenesis (Hayashida 1963). Active immunisation of adult male rats and rabbits with bovine LH results in altered mating behaviour (Talaat and Laurence 1971). The authors have also observed that passive immunisation of rats with anti-LH sera diminished mating capacity to a greater extent than active immunisation. Immunisation of 18-day-old rats with LH receptor results in 50% reduction of testicular sperm counts (Graf et al. 1997). Immunisation of male rabbits against ovine LH reduces serum testosterone and increases circulating FSH concentration and this could be a consequence of the negative feedback effect of the lack of testosterone (Jeyakumar et al. 1995). LH deprivation on spermatogenesis assessed by DNA flow cytometry and histological analyses of testicular biopsy tissue revealed that lack of testosterone primarily results in a rapid reduction and complete absence of round and elongated spermatids. The immediate effect of LH/testosterone deprivation thus appears to be at the step of meiotic transformation of primary spermatocytes to round spermatids (Jeyakumar et al. 1995). Adult male bonnet monkeys were rendered specifically LH/testosterone deficient by immunising them with a heterologous ovine LH (Suresh et al. 1995) or by administration of central inhibitors like Norethisterone alone (Shetty et al. 1997). Both treatments led to azoospermia; however, the level of circulating FSH was unaffected. These studies have clearly established that lack of LH blocks spermatogonial proliferation and meiosis. The overall effect of testosterone deficiency was azoospermia leading to infertility and change in behaviour.

A vast body of recent data also shows that LH may have functional roles in many other tissues (other than the reproductive tract), including breast, adrenals, brain, skin etc. (Rao 1982; Ziecik et al. 1992; Rodway and Rao 1995; Rao 1996, 1998, 1999; Lei and Rao 2000). Earlier studies have used various methods of LH deprivation to understand the importance of its actions in reproduction (Moudgal et al. 1974; Klinefelter et al. 1987; Keeney et al. 1988; Keeney and Ewing 1990; Mendis-Handagama 1997; Sriraman et al. 2000). Although we have appreciable information from this approach, questions have still remained on the presence of circulating residual LH levels (Lei et al. 2001). To understand the precise role of LH on the gonadal and non-gonadal tissues, studies have targeted the disruption of LH receptor in mice (Lei et al. 2001; Zhang et al. 2001). The characterisation of these phenotypes has shown the specific role of reproductive and non-reproductive processes. LH receptor knockout mice had normal gonads except showing hypoplasia indicating that LH receptor signalling is not fully essential for early development. Production of LH independent secretion of testosterone by early fetal Leydig cells and Müllerian inhibiting substance by Sertoli cells may be responsible for their differentiation (George et al. 1978). Serum LH levels were significantly and understandably high in the LH receptor knockout males due to the loss of testosterone and negative feedback of its own secretion through decreased GnRH from hypothalamus (Lei et al. 2001). As one would expect, disruption of LH β-subunit leads to hypogonadism, defective steroidogenesis and infertility in male mice (Ma et al. 2004).

4.3.3 Role of Testosterone in Regulating Spermatogenesis

Testosterone is critical for maintaining normal spermatogenesis. The first spermatogenic wave in rodents begins around birth and in primates there is a quiescent period before the onset of puberty. In rats, a high level of circulating testosterone is recorded around birth and blockade of this surge by GnRH antagonist results in poor ejaculatory behaviour and fertility (Kolho and Huhtaniemi 1989). However, contradicting results were observed in case of marmoset after abolishing the perinatal testosterone surge (Lunn et al. 1997). In the hpg mouse, supplementation of testosterone to weaning pups for eight weeks, results in the production of sperm capable of fertilising the egg in vitro (Singh et al. 1995). Similarly, testosterone supplementation releases GnRH antagonist-induced blockade of spermatid formation in 20-day-old rats (Ganguly et al. 1994). In primates, exogenous testosterone administration alone can induce complete spermatogenesis (Marshall et al. 1984). However, low sperm counts were found in the ejaculates. The effect of androgen deficiency on the testis can be reversed by testosterone supplementation (Kerr et al. 1993).

Testosterone withdrawal has suggested that the initial effect is on stage VII and VIII round spermatids (O'Donnell et al. 1996). Testosterone alone could maintain quantitatively normal sperm counts in GnRH-immunised rats (Awoniyi et al. 1992), but the presence of FSH in such an experimental model precludes a clear interpretation (McLachlan et al. 1994). Immunisation of rhesus monkey with testosterone to reduce circulating levels of the hormone has failed to completely inhibit intratesticular concentrations of testosterone (Weinbauer and Nieschlag 1978). Involution of the germinal epithelium and production of sperm in gonadotropin-deficient monkeys are prevented or delayed by testosterone supplementation and this effect is apparently dose dependent (Weinbauer and Nieschlag 1993). The importance of androgen in maintaining male fertility is also emphasised by studies of the total androgen receptor (Yeh et al. 2002) or the SCARKO (Willems et al. 2011; O'Shaughnessy 2014) knockout mice. Testosterone induces spermatogenesis in rats actively immunised with GnRH to the point of restoring fertility (Awoniyi et al. 1992). Testosterone can also reinitiate spermatogenesis in gonadotropin-deficient monkeys, but the effect is not quantitatively normal (Weinbauer and Nieschlag 1996). Re-initiation of spermatogenesis with testosterone or hCG alone has also been reported (Matsumoto 1994). Interestingly, the Djungarian hamster (another experimental model used in male reproduction) exhibits seasonal changes in spermatogenesis. Animals like the hamster exposed to shorter daylight show testicular involution and disruption of spermatogenesis. Treatment with testosterone or LH does not stimulate sperm production (Milette et al. 1988) as in this animal species the process is entirely dependent on FSH levels (Lerchl et al. 1993; Niklowitz et al. 1997). Mechanisms related to androgenic control of spermatogenesis have been recently reviewed elsewhere (Smith and Walker 2014).

4.4 Conclusions

In this chapter, we considered an overview of many approaches related to the understanding of hormonal control of spermatogenesis in males including several genetically altered mice. Based on the discussion it is clear that this is a very complex process dependent on the interplay of critical hormones from the hypothalamo–pituitary–gonadal axis. An understanding of these processes has been helpful in using hormones for treatment of infertility in men and for attempts in devising approaches to reversible male contraception. Synthetic gonadotropins (including recombinant gonadotropins) are now used for treating infertile men. Experimental animals including non-human primates, genetic models and seasonally breeding animals have been helpful in revealing steps that could be subject to manipulation to achieve desired results in managing fertility. While it is evident that it is relatively easy to abolish spermatogenesis by interfering with LH/GnRH/testosterone secretions or actions and thus render any male infertile, this process also eliminates libido, an effect that is unacceptable for contraception. Therefore, efforts to separate sperm suppression or alter fertilising ability without compromising androgen synthesis or actions are required for acceptable male contraception. Further any modality that would achieve reversal would be highly desirable. Based on studies in monkeys it would appear that FSH inhibition/suppression would be mandatory for effective male contraception (Moudgal and Sairam 1998; Weinbauer et al. 2001).

References

Abel MH, Baker PJ, Charlton HM et al. 2008. Spermatogenesis and Sertoli cell activity in mice lacking Sertoli cell receptors for follicle-stimulating hormone and androgen. *Endocrinology*, 149:3279–3285.

Anderson RA, Wallace EM, Groome NP et al. 1997. Physiological relationship between inhibin B, follicle stimulating hormone secretion and spermatogenesis in normal men and response to gonadotropin suppression by exogenous testosterone. *Human Reproduction*, 12:746–751.

Arslan MA, Weinbauer GF, Schlatt S et al. 1993. FSH and testosterone alone or in combination initiate testicular growth and increase the number of spermatogonia and Sertoli cells in a juvenile non-human primate (*Macaca mulatta*). *Journal of Endocrinology*, 136:235–243.

Awoniyi CA, Zirkin BR, Chandrashekar V et al. 1992. Exogenously administered testosterone maintains spermatogenesis quantitatively in adult rats actively immunized against gonadotropin-releasing hormone. *Endocrinology*, 130:3283–3288.

Baker PJ, Pakarinen P, Huhtaniemi IT et al. 2003. Failure of normal Leydig cell development in follicle-stimulating hormone (FSH) receptor-deficient mice, but not FSH beta-deficient mice: Role for constitutive FSH receptor activity. *Endocrinology*, 144:138–145.

Bardin CW, Cheng CY, Musto NA et al. 1988. The Sertoli cell. In: *The Physiology of Reproduction*, E Knobil and JD Neill, Eds., Raven Press, New York, 933–974.

Bockers TM, Nieschlag E, Kreutz MR et al. 1994. Localization of follicle-stimulating hormone (FSH) immunoreactivity and hormone receptor mRNA in testicular tissue from infertile men. *Cell and Tissue Research*, 278:595–600.

Clermont Y. 1958. Contractile elements in the limiting membranes of the seminiferous tubules in the rat. *Experimental Cell Research*, 15:438–441.

Cooke HJ, Saunders PT. 2002. Mouse models of male infertility. *Nature Reviews Genetics*, 3(10):790–801.

Dahia CL, Rao AJ. 2006. Demonstration of follicle-stimulating hormone receptor in cauda epididymis of rat. *Biology of Reproduction*, 75(1):98–106.

Dankbar B, Brinkworth MH, Schlatt S et al. 1995. Ubiquitous expression of the androgen receptor and testis-specific expression of the FSH receptor in the cynomologus monkey (*Macaca fascicularis*) revealed by ribonuclease protection assay. *Journal of Steroid Biochemistry and Molecular Biology*, 55:35–41.

De Gendt K, Swinnen JV, Saunders PT et al. 2004. A Sertoli cell-selective knockout of the androgen receptor causes spermatogenic arrest in meiosis. *Proceedings of the National Academy of Sciences of the United States of America*, 101:1327–1332.

de Kretser DM, Kerr JB. 1988. The cytology of the testis. In: *The Physiology of Reproduction*, E Knobil and JD Neill, Eds., Raven Press, New York, 837–932.

de Rooij DG. 1998. Stem cells in the testis. *International Journal of Experimental Pathology*, 79:67–80.

de Rooij DG. 2009. The spermatogonial stem cell niche. *Microscopy Research and Technique*, 72:580–585.

Dierich A, Sairam MR, Monaco L et al. 1998. Impairing follicle-stimulating hormone (FSH) signaling in vivo. Targeted disruption of the FSH receptor leads to aberrant gametogenesis and hormonal imbalance. *Proceedings of National Academy of Sciences USA*, 95:13612–13617.

Ding LJ, Yan GJ, Ge QY et al. 2011. FSH acts on the proliferation of type A spermatogonia via Nur77 that increases GDNF expression in the Sertoli cells. *FEBS Letters*, 585:2437–2444.

Dym M, Fawcett DW. 1970. The blood-testis barrier in the rat and the physiological compartmentation of the seminiferous epithelium. *Biology of Reproduction*, 3:308–326.

Dym M, Raj HGM, Lin YC et al. 1979. Is FSH required for maintenance of spermatogenesis in adult rats? *Journal of Reproduction and Fertility*, 26:175–181.

Fawcett DW. 1973. Observations on the organization of the interstitial tissue of the testis and the occluding cell junctions in the seminiferous epithelium. *Advances in Biosciences*, 10:83–99.

Fink G. 1988. Gonadotropin secretion and its control. In: *The Physiology of Reproduction*, E Knobil and JD Neill, Eds., Raven Press, New York, 1379–1392.

Fritz IB, Tung PS. 1986. Role of interactions between peritubular cells and Sertoli cells in mammalian testicular functions. In: *Gametogenesis and the Early Embryo*, JG Gall, Ed., Alan R. Liss Inc., New York.

Ganguly A, Misro MM, Das RP. 1994. Roles of FSH and testosterone in the initiation of spermatogenesis in prepubertal rats medically hypophysectomized by a GnRH antagonist. *Archives of Andrology*, 32:111–210.

George FW, Catt KJ, Neaves WB et al. 1978. Studies on the regulation of testosterone synthesis in the fetal rabbit testis. *Endocrinology*, 102(3):665–673.

Graf KM, Dias JA, Griswold MA. 1997. Decreased spermatogenesis as the result of an induced autoimmune reaction directed against the gonadotropin receptors in male rats. *Journal of Andrology*, 18:174–185.

Greep RO, Fevold HL, Hisaw FL. 1936. Effects of two hupophyseal gonadotropic hormones on the reproductive system of male rat. *Anatomical Record*, 65:261–270.

Grino PB, Griffin JE, Wilson JD. 1990. Testosterone at high concentrations interacts with the human androgen receptor similarity to dihydrotestosterone. *Endocrinology*, 126:1165–1172.

Gromoll J, Simoni M, Nieschlag E. 1996. An activating mutation of the follicle- stimulating hormone receptor autonomously sustains spermatogenesis in a hypophysectomized man. *Journal of Clinical Endocrinology Metabolism*, 81(4):1367–1370.

Goriely A, McVean GA, van Pelt AM et al. 2005. Gain-of-function amino acid substitutions drive positive selection of FGFR2 mutations in human spermatogonia. *Proceedings of the National Academy of Sciences of the United States of America*, 102:6051–6056.

Hall PF. 1988. Testicular steroid synthesis: Organisation and regulation. In: *The Physiology of Reproduction*, E Knobil, JD Neill, Eds., Raven Press, New York, 975–998.

Hayashida T. 1963. Inhibition of spermiogenesis, prostate and seminal vesicle development in normal animals with antigonadotrophic hormone serum. *Journal of Endocrinology*, 26:75–83.

Heckert L, Griswold MD. 1991. Expression of follicle-stimulating hormone receptor mRNA in rat testes and Sertoli cells. *Molecular Endorinology*, 5:670–677.

Hodgson Y, Robertson DM, de Kretser DM. 1983. The regulation of testicular function. In: *Reproductive Physiology IV, International Review of Physiology*, Vol. 27, R Greep, Ed., University Park Press, Baltimore, 275–327.

Hofmann MC. 2008. Gdnf signaling pathways within the mammalian spermatogonial stem cell niche. *Molecular and Cellular Endocrinology*, 288:95–103.

Jahnukainen K, Chrysis D, Hou M et al. 2004. Increased apoptosis occurring during the first wave of spermatogenesis is stage-specific and primarily affects midpachytene spermatocytes in the rat testis. *Biology of Reproduction*, 70:290–296.

Jeyakumar M, Suresh R, Krishnamurthy HN et al. 1995. Changes in testicular function following specific deprivation of LH in the adult male rabbit. *Journal of Endocrinology*, 147(1):111–120.

Kangasniemi M, Kaipia A, Toppari J et al. 1990. Cellular regulation of follicle-stimulating hormone (FSH) binding in rat seminiferous tubules. *Journal of Andrology*, 11:336–343.

Keeney DS, Mendis-Handagama SM, Zirkin BR et al. 1988. Effect of long term deprivation of luteinizing hormone on Leydig cell volume, Leydig cell number, and steroidogenic capacity of the rat testis. *Endocrinology*, 123(6):2906–2915.

Keeney DS, Ewing LL. 1990. Effects of hypophysectomy and alterations in spermatogenic function on Leydig cell volume, number, and proliferation in adult rats. *Journal of Andrology*, 11(4):367–378.

Kerr JB, Millar M, Mddocks S et al. 1993. Stage-dependent changes in spermatogenesis and Sertoli cells in relation to the onset of spermatogenic failure following withdrawal of testosterone. *Anatomical Record*, 235:547–559.

Khan H, Yarney TA, Sairam MR. 1993. Cloning of alternately spliced mRNA transcripts coding for variants of ovine testicular follitropin receptor lacking the G protein coupling domains. *Biochemistry Biophysics Research Communications*, 190(3):888–894.

Klinefelter GR, Hall PF, Ewing LL. 1987. Effect of luteinizing hormone deprivation *in situ* on steroidogenesis of rat Leydig cells purified by a multistep procedure. *Biology of Reproduction*, 36(3):769–783.

Kolho KI, Huhtaniemi I. 1989. Neonatal treatment of male rats with a gonadotropin-releasing hormone antagonist impairs ejaculation and fertility. *Physiology and Behavior*, 46:373–377.

Krishnamurthy H, Babu PS, Morales CM et al. 2001a. Delay in sexual maturity of the follicle-stimulating hormone receptor knockout mouse (FORKO). *Biology of Reproduction*, 65(2):522–531.

Krishnamurthy H, Danilovich N, Morales C et al. 2000a. Qualitative and quantitative decline in spermatogenesis of the follicle-stimulating hormone receptor knockout (FORKO) mouse. *Biology of Reproduction*, 62:1146–1159.

Krishnamurthy H, Kats R, Danilovich N et al. 2001b. Intercellular communication between Sertoli and Leydig cells in the absence of FSH receptor. *Biology of Reproduction*, 65(4):1201–1207.

Krishnamurthy H, Prasanna Kumar KM, Joshi V et al. 2000b. Alterations in sperm characteristics of (FSH)-immunized men are similar to those of FSH deprived infertile male bonnet monkeys. *Journal of Andrology*, 21(2), 316–327.

Kumar TR. 2009. FSHß knock out mouse model: A decade ago and into the future. *Endocrine*, 36(1):1–5.

Kumar TR, Low MJ, Matzuk MM. 1998. Genetic rescue of follicle-stimulating hormone ß-deficient mice. *Endocrinology*, 139(7):3289–3295.

Kumar TR, Wang Y, Lu N et al. 1997. Follicle stimulating hormone is required for ovarian follicle maturation but not male fertility. *Nature Genetics*, 15(2):201–204.

Laborde P, Barkey RJ, Belair L et al. 1996. Ontogenesis of LH and FSH receptors in postnatal rabbit testes: Age-dependent differential expression of long and short RNA transcripts. *Journal of Reproduction and Fertility*, 108:25–30.

Lei PP, Cheng CY, Mruk DD. 2009. Coordinating cellular events during spermatogenesis: A biochemical model. *Trends in Biochemical Sciences*, 34(7): 366–373.

Lei ZM, Mishra S, Zou W et al. 2001. Targeted disruption of luteinizing hormone/human chorionic gonadotropin receptor gene. *Molecular Endocrinology*, 15(1):184–200.

Lei ZM, Rao ChV. 2000. Endocrinology of trophoblast tissue. In: *Principals and Practice of Endocrinology and Metabolism*, 3rd edn., K Becker and R Rebar, Eds., Lippincott, Williams and Wilkins, Philadelphia.

Lerchl A, Sotiriadu S, Behre HM et al. 1993. Restoration of spermatogenesis by follicle-stimulating hormone despite low intratesticular testosterone in photoinhibited hypogonadotropic Djungarian hamsters (*Phodopus sungorus*). *Biology of Reproduction*, 49:1108–1116.

Libman J, Gabriel MS, Sairam MR et al. 2010. Catalase can protect spermatozoa of FSH receptor knock-out mice against oxidant-induced DNA damage in vitro. *International Journal of Andrology*, 33:818–822.

Lunn SE, Cowen GM, Fraser HM. 1997. Blockade of the neonatal increase in testosterone by a GnRH antagonist: The free androgen index, reproductive capacity and postmortem findings in the male marmoset monkey. *Journal of Endocrinology*, 154:125–131.

Ma X, Dong Y, Matzuk MM et al. 2004. Targeted disruption of luteinizing hormone beta-subunit leads to hypogonadism, defects in gonadal steroidogenesis, and infertility. *Proceedings of National Academy of Sciences USA*, 101(49):17294–17299.

Magre S, Jost A. 1991. Sertoli cells and testicular differentiation in the rat fetus. *Journal of Electron Microscopy Techniques*, 19:172–188.

Maguire SM, Tribley WA, Griswold MD. 1997. Follicle-stimulating hormone (FSH) regulates the expression of FSH receptor messenger ribonucleic acid in cultured Sertoli cells and in hypohysectomized rat testis. *Biology of Reproduction*, 56:1106–1111.

Marshall GR, Ramaswamy S, Plant TM. 2005. Gonadotropin-independent proliferation of the pale type A spermatogonia in the adult rhesus monkey (*Macaca mulatta*). *Biology of Reproduction*, 73(2):222–229.

Marshall GR, Wickings EJ, Nieshlag E. 1984. Testosterone can initiate spermatogenesis in an immature nonhuman primate, Macaca fascicularis. *Endocrinology*, 114:2228–2233.

Matikainen T, Toppari J, Vihko KK et al. 1994. Effects of recombinant human FSH in immature hypophysectomized male rats: Evidence for Leydig cell-mediated action on spermatogenesis. *Journal of Endocrinology*, 141:449–457.

Matsumoto AM. 1994. Hormonal therapy of male hypogonadism. *Endocrinology Metabolism Clinical North America*, 23:857–875.

Matsumoto AM, Karpas AE, Bremner WJ. 1986. Chronic human chorionic gonadotropin administration in normal men: Evidence that follicle stimulating hormone is necessary for the maintenance of quantitative normal spermatogenesis in men. *Journal of Clinical Endocrinology and Metabolism*, 62:1184–1190.

McCabe MJ, Allan CM, Foo CF et al. 2012. Androgen initiates Sertoli cell tight junction formation in the hypogonadal (hpg) mouse. *Biology of Reproduction*, 87(2):38.

McLachlan RI, Wreford NG, de Kretser DM et al. 1995. The effects of recombinant follicle-stimulating hormone on the restoration of spermatogenesis in the gonadotropin-releasing hormone-immunized adult rat. *Endocrinology*, 136:4035–4043.

McLachlan RI, Wreford NG, Meachem SJ et al. 1994. Effects of testosterone on spermatogenic cell population in the adult rat. *Biology of Reproduction*, 51:945–955.

Mendis-Handagama SM. 1997. Luteinizing hormone on Leydig cell structure and function. *Histology and Histopathology*, 12(3):869–882.

Migrenne S, Moreau E, Pakarinen P et al. 2012. Mouse testis development and function are differently regulated by follicle-stimulating hormone receptors signalling during fetal and prepubertal life. *PLoS One*, 7(12):e53257.

Milette JJ, Schwartz NB, Turek FW. 1988. The importance of follicle stimulating hormone in the initiation of testicular growth in photostimulated Djungarian hamsters. *Endocrinology*, 122:1060–1066.

Moudgal NR, Jagannadha Rao A, Maneckjee R et al. 1974. Gonadotropins and their antibodies. *Recent Progress in Hormone Research*, 30:47–77.

Moudgal NR, Murthy GS, Prasanna Kumar KM et al. 1997a. Responsiveness of human male volunteers to immunization with ovine follicle stimulating hormone vaccine: Results of a pilot study. *Human Reproduction*, 12:457–463.

Moudgal NR, Ravindranath N, Murthy GS et al. 1992. Long term contraceptive efficacy of vaccine of ovine follicle stimulating hormone in male bonnet monkeys (*Macaca radiata*). *Journal of Reproduction and Fertility*, 96:91–102.

Moudgal NR, Sairam MR. 1998. Is there a true requirement for follicle stimulating hormone in promoting spermatogenesis and fertility in primates? *Human Reproduction*, 13:916–919.

Moudgal NR, Sairam MR, Krishnamurthy HN et al. 1997b. Immunization of male bonnet monkeys (*M. radiata*) with a recombinant FSH receptor preparation affects testicular function and fertility. *Endocrinology*, 138:3065–3068.

Mullaney BP, Skinner MK. 1992. Basic fibroblast growth factor (bFGF) gene expression and protein production during pubertal development of the seminiferous tubule: Follicle-stimulating hormone-induced Sertoli cell bFGF expression. *Endocrinology*, 131:2928–2934.

Murty GS, Rani CS, Moudgal NR et al. 1979. Effect of passive immunization with specific antiserum to FSH on the spermatogenic process and fertility of adult male bonnet monkeys (*Macaca radiata*). *Journal of Reproduction and Fertility*, Suppl. 26:147–163.

Niklowitz P, Khan S, Bergmann M et al. 1997. Differential effects of follicle-stimulating hormone and luteinizing hormone on Leydig cell function and restoration of spermatogenesis in hypophysectomized and photoinhibited Djungarian hamsters (*Phodopus sungorus*). *Biology of Reproduction*, 41:871–880.

O'Donnell L, McLachlan RI, Wreford NG et al. 1996. Testosterone withdrawal promotes stage-specific detachment of round spermatids from the rat seminiferous epithelium. *Biology of Reproduction*, 55:895–901.

Orth JM. 1982. Proliferation of Sertoli cells in fetal and postnatal rats: A quantitative autoradiographic study. *Anatomical Record*, 203:485–492.

O'Shaughnessy PJ. 2014. Hormonal control of germ cell development and spermatogenesis. *Seminars in Developmental Biology*, 29:55–65.

O'Shaughnessy PJ, Monteiro A, Abel M. 2012. Testicular development in mice lacking receptors for follicle stimulating hormone and androgen. *PLoS One*, 7:e35136.

Phillip M, Arbelle JE, Segev Y et al. 1998. Male hypogonadism due to a mutation in the gene for the beta-subunit of follicle-stimulating hormone. *New England Journal of Medicine*, 338:1729–1732.

Pierce JG, Parsons TF. 1981. Glycoprotein hormones: Structure and function. *Annual Reviews of Biochemistry*, 50:465–495.

Ploen L, Setchell BP. 1992. Blood-testis barriers revisited. A homage to Lannart Nicander. *International Journal of Andrology*, 15:1–4.

Policastro PF, Daniels-McQueen S, Carle G et al. 1986. A map of the hCG beta-LH beta gene cluster. *Journal of Biological Chemistry*, 261(13):5907–5916.

Rannikko A, Penttila TI, Zhang FP et al. 1996. Stage-specific expression of the FSH receptor gene in the prepubertal and adult rat seminiferous epithelium. *Journal of Reproductive Endocrinology*, 151:29–35.

Rao ChV. 1982. Receptors for gonadotropins in human ovaries. In: *Recent Advances in Fertility Research: Developments in Reproductive Endocrinology*, Part A. TG Muldoon, VB Mahesh, B Perez-Ballester, Eds., Alan R. Liss, New York, 123–135.

Rao ChV. 1996. The beginning of a new era in reproductive biology and medicine: Expression of functional luteinizing hormone/human chorionic gonadotropin receptors in nongonadal tissue. *Journal of Physiology and Pharmacology*, 47:41–53.

Rao ChV. 1998. Novel concepts in neuroendocrine regulation of reproductive tract functions. In: *The Endocrinology of Pregnancy*, FW Bazer, Ed., Humana Press, Totowa, NJ, Chapter 5, 125–144.

Rao ChV. 1999. A paradigm shift in the targets of luteinizing hormone/human chorionic gonadotropin hormone actions in the body. *Journal of Bellevue Obstetrics and Gynecology Society*, 15:26–32.

Rebourcet D, O'Shaughnessy PJ, Monteiro A et al. 2014. Sertoli cells maintain Leydig cell number and peritubular myoid cell activity in the adult mouse testis. *PLoS One*, 9:e105687.

Remy JJ, Coutour L, Rabesona H et al. 1996. Immunization against exon 1 decapeptides from lutropin or choriogonadotropin receptor or the follitropin receptor as potential male contraceptive. *Journal Reproductive Immunology*, 32:37–54.

Rodriguez I, Ody C, Araki K et al. 1997. An early and massive wave of germinal cell apoptosis is required for the development of functional spermatogenesis. *EMBO Journal*, 16:2262–2270.

Rodway MR, Rao CV. 1995. A novel perspective on the role of human chorionic gonadotropin during pregnancy and in gestational trophoblastic disease. *Early Pregnancy*, 1:176–187.

Russell LD, Corbin TJ, Borg KE et al. 1993. Recombinant human follicle-stimulating hormone of exerting a biological effect in the adult hypophysectomized rat by reducing the numbers of degenerating germ cells. *Endocrinology*, 133:2062–2070.

Saez JM, Perrard-Sapori MH, Chatelain PG et al. 1987. Paracrine regulation of testicular function. *Journal of Steroid Biochemistry*, 27:317–329.

Sairam MR, Babu PS. 2007. The tale of follitropin receptor diversity: A recipe for fine tuning gonadal responses? *Molecular and Cellular Endocrinology*, 260–262:163–171.

Sairam MR, Dierich A, Monaco L et al. 1998. Targeted disruption of the FSH receptor leads to aberrant gametogenesis, hormonal imbalances causing infertility/reduced fertility. In: *80th Annual Meeting of the Endocrine Society*, New Orleans, Louisiana. Abstract #OR46–3.

Schlatt S, Arslan M, Weinbauer GF et al. 1995. Endocrine control of testicular somatic and premeiotic germ cell development in the immature testis of the primate *Macaca mulatta*. *European Journal of Endocrinology*, 137:107–117.

Schlatt S, Ehmcke J. 2014. Regulation of spermatogenesis: An evolutionary biologist's perspective. *Seminars in Cell and Developmental Biology*, 29:2–16.

Setchell BP, Voglmayr JK, Waites GM. 1969. A blood-testis barrier restricting passage from blood into rete testis fluid but not into 1ymph. *Journal of Physiology*, 200:73–85.

Sharpe RM. 1983. Local control of testicular function. *Quarterly Journal of Experimental Physiology*, 68:265–287.

Sharpe RM. 1994. Regulation of spermatogenesis. In: *The Physiology of Reproduction*, E Knobil and JD Neill, Eds., Raven Press, New York, 1363–1434.

Shetty G, Krishnamurthy H, Krishnamurthy HN et al. 1997. Use of norethesterone (NET) and estradiol (E) in mini doses as a contraceptive in the male: Efficacy studies in the adult male bonnet monkey (*M. radiata*). *Contraception*, 56:257–265.

Shetty J, Marathe GK, Dighe RR. 1996. Specific immunoneutralization of FSH leads to apoptotic cell death of the pachytene spermatocytes and spermatogonial cells in the rat. *Endocrinology*, 137:2179–2182.

Siegel ET, Kim HG, Nishimoto HK et al. 2013. The molecular basis of impaired follicle-stimulating hormone action: Evidence from human mutations and mouse models. *Reproductive Sciences*, 20:211–233.

Simoni M, Gromoll J, Nieschlag E. 1997. The follicle-stimulating hormone receptor: Biochemistry, molecular biology, physiology and pathophysiology. *Endocrine Review*, 18:739–773.

Simorangkir DR, Ramaswamy S, Marshall GR et al. 2009. A selective monotropic elevation of FSH, but not that of LH, amplifies the proliferation and differentiation of spermatogonia in the adult rhesus monkey (*Macaca mulatta*). *Human Reproduction*, 24(7):1584–1595.

Siiteri PK, Wilson JD. 1974. Testosterone formation and metabolism during sexual differentiation in the human embryo. *Journal of Clinical Endocrinology and Metabolism*, 38:113–125.

Singh J, Handelsman DJ. 1996. The effects of recombinant FSH on testosterone-induced spermatogenesis in gonadotrophin-deficient (hpg) mice. *Endocrinology*, 136:5311–5321.

Singh J, O'Neill C, Handelsman DJ. 1995. Induction of spermatogenesis by androgens in gonadotropin deficient (hpg) mice. *Endocrinology*, 136:5311–5321.

Skinner MK, Tung PS, Fritz IB. 1985. Cooperativity between Sertoli cells and testicular peritubular cells in the production and deposition of extracellular matrix components. *Journal of Cell Biology*, 100:1941–1947.

Smith LB, Walker WH. 2014. The regulation of spermatogenesis by androgens. *Seminars in Cell and Developmental Biology*, 30:2–13.

Sriraman V, Rao VS, Sairam MR et al. 2000. Effect of deprival of LH on Leydig cell proliferation: Involvement of PCNA, cyclin D3 and IGF-1. *Molecular and Cellular Endocrinology*, 162:113–120.

Suresh R, Medhamurthy R, Moudgal NR. 1995. Comparative studies on the effects of specific immunoneutralization of endogenous FSH or LH on testicular germ cell transformations in the adult bonnet monkey (*Macaca radiata*). *American Journal of Reproductive Immunology*, 34(1):35–43.

Tadokoro Y, Yomogida K, Ohta H et al. 2002. Homeostatic regulation of germinal stem cell proliferation by the GDNF/FSH pathway. *Mechanisms of Development*, 113:29–39.

Talaat M, Laurence KA. 1971. Impairment of spermatogenesis and libido through antibodies to luteinizing hormone. *Fertility Sterility*, 22:113–118.

Tapanainen JS, Aittomäki K, Min J et al. 1997. Men homozygous for an inactivating mutation of the follicle-stimulating hormone (FSH) receptor gene present variable suppression of spermatogenesis and fertility. *Nature Genetics*, 15:205–206.

Van Alphen MMA, Van de Kant HJG, De Rooij DG. 1988. Follicle-stimulating hormone stimulates spermatogenesis in the adult monkey. *Endocrinology*, 123:1449–1453.

Vihko KK, LaPolt PS, Nishimori K et al. 1991. Stimulatory effects of recombinant follicle-stimulating hormone on Leydig cell function and spermatogenesis in immature hypophysectomized rats. *Endocrinology*, 129:1926–1932.

Wang BY, Liang W, Cui YX et al. 2008. Follicle-stimulating hormone autoantibody is involved in idiopathic spermatogenic dysfunction. *Asian Journal of Andrology*, 10:915–921.

Weinbauer GF, Behre HM, Fingscheidt U et al. 1991. Human follicle-stimulating hormone exerts a stimulatory effect on spermatogenesis, testicular size, and serum inhibin levels in the gonadotropin-releasing hormone antagonist treated nonhuman primate (*Macaca fascicularis*). *Endocrinology*, 129:1831–1839.

Weinbauer GF, Nieschlag E. 1978. The effect of active immunization with testosterone on pituitary–gonadal feedback in the male rhesus monkey (*Macaca mulatta*). *Biology of Reproduction*, 18:602–607.

Weinbauer GF, Nieschlag E. 1993. Hormonal control of spermatogenesis. In: *Molecular Biology of the Male Reproductive System*, DM de Kretser, Ed., Academic Press, New York, 99–142.

Weinbauer GF, Nieschlag E. 1996. The Leydig cell as a target for male contraception. In: *The Leydig Cell*, AH Pyne, MP Hardy, LD Russell, Eds., Cache River Press, Clearwater, FL, 629–662.

Weinbauer GF, Nieschlag E. 1998. The role of testosterone in. In: *Testosterone, Action Deficiency Substitution,* E Nieschlag and HM Behre, Eds., Springer Verlag, Berlin, 143–168.

Weinbauer GF, Schlatt S, Walter V et al. 2001. Testosterone-induced inhibition of spermatogenesis is more closely related to suppression of FSH than to testicular androgen levels in the cynomolgus monkey model (*Macaca fascicularis*). *Journal of Endocrinology*, 168:25–38.

Willems A, De Gendt K, Allemeersch J et al. 2010. Early effects of Sertoli cell-selective androgen receptor ablation on testicular gene expression. *International Journal of Andrology*, 33:507–517.

Willems A, De Gendt K, Deboel L et al. 2011. The development of an inducible androgen receptor knockout model in mouse to study the postmeiotic effects of androgens on germ cell development. *Spermatogenesis*, 1(4):341–353.

Wilson JD, Lasnitzki I. 1971. Dihydrotestosterone formation in fetal tissues of the rabbit and rat. *Endocrinology*, 89:659–668.

Yeh S, Meng-Yin Tsai, Xu Q et al. 2002. Generation and characterization of androgen receptor knockout (ARKO) mice: An in vivo model for the study of androgen functions in selective tissues. *Proceedings of National Academy of Sciences USA*, 99:13498–13503.

Zhang FP, Poutanen M, Wilbertz J et al. 2001. Normal prenatal but arrested postnatal sexual development of luteinizing hormone receptor knockout (LuRKO) mice. *Molecular Endocrinology*, 15:172–183.

Ziecik AJ, Derecka-Reszka K, Rzucidlo SJ. 1992. Extragonadal gonadotropin receptors, their distribution and function. *Journal of Physiology Pharmacology*, 43(4 Suppl 1):33–49.

Chapter 5

Role of Apoptosis in Spermatogenesis

Amiya P. Sinha Hikim

Contents

5.1 Introduction....97
5.2 Germ Cell Apoptosis in Spermatogenesis....98
5.3 Major Pathways of Apoptosis....99
5.4 Key Signal Transduction Pathways in Male Germ Cell Apoptosis....101
5.4.1 Involvement of the Mitochondria-Dependent Pathway in Apoptotic Signalling of Testicular Germ Cells....101
5.4.2 Involvement of Fas Signalling in Male Germ Cell Apoptosis....103
5.4.3 Functional Role of Downstream Caspases in Male Germ Cell Apoptosis....104
5.4.4 Upstream Signalling Pathways for Testicular Germ Cell Apoptosis....104
5.4.5 Caspase 2 Is an Upstream Activator of p38 MAPK-Mediated Intrinsic Pathway Signalling....108
5.5 Conclusions....110
Acknowledgements....111
References....111

5.1 Introduction

Programmed cell death (apoptosis) is an evolutionarily conserved cell death process that plays a major role during normal development and homeostasis of multicellular organisms (Jacobson et al., 1997; Green, 2000; Hengartner, 2000; Reed, 2000;

Danial and Korsmeyer, 2004; Taylor et al., 2008; Youle and Strasser, 2008). The term *apoptosis* coined by Kerr and colleagues (Kerr et al., 1972) in the early 1970s is characterised by a series of dramatic perturbations to the cellular architecture that contribute not only to cellular demise, but also prepare cells for removal by phagocytes. The morphological features include cell shrinkage, plasma membrane blebbing, chromatin condensation and margination, nuclear fragmentation and formation of apoptotic bodies. The biochemical features associated with apoptosis include high molecular weight DNA fragmentation and formation of an oligonucleosomal ladder, externalisation of phosphatidylserine that is normally confined to the inner surface of the plasma membrane, and a disruption in the mitochondrial transmembrane potential. It is now widely recognised that apoptosis serves as a prominent force in sculpting body parts, deleting unneeded structures, maintaining tissue homeostasis, and as a defence mechanism to remove unwanted and potentially dangerous cells, such as self-reactive lymphocytes, virus infected cells and tumour cells (Thompson, 1995; Jacobson et al., 1997). A large number of studies have been published, as reflected in several recent reviews (Sinha Hikim et al., 2003a, 2011; Baum et al., 2005; Shaha, 2007; Sofikitis et al., 2008), on male germ cell apoptosis and its control. This review highlights the signal transduction pathways in inducing testicular germ cell apoptosis.

5.2 Germ Cell Apoptosis in Spermatogenesis

Spermatogenesis is an elaborate process of cell differentiation in which stem spermatogonia, through a series of events become mature spermatozoa and occurs continuously during the reproductive lifetime of the individual (Russell et al., 1990; Sharpe, 1994; Sinha Hikim et al., 2005). Stem spermatogonia undergo mitosis to produce two types of cells: additional stem cells and differentiating spermatogonia, which undergo rapid and successive mitotic divisions to form primary spermatocytes. The spermatocytes then enter a lengthy meiotic phase as preleptotene spermatocytes and proceed through two cell divisions (meiosis I and II) to give rise to haploid spermatids. These in turn undergo a complex process of morphological and functional differentiation resulting in the production of mature spermatozoa. The formation of spermatozoa takes place within the seminiferous epithelium, consisting of germ cells at various phases of development and supporting Sertoli cells. Of interest, the different generations of germ cells form associations with fixed composition or stages, which constitute the cycle of seminiferous epithelium (12 in the mouse and 14 in the rat). Not all germ cells, however, achieve maturity, and such spontaneous death of certain classes of germ cells by apoptosis appears to be a constant feature of normal spermatogenesis.

Programmed germ cell death plays an indispensable role during fetal, neonatal, postnatal and adult spermatogenesis (Mori et al., 1997; Rodriguez et al., 1997; Wang et al., 1998). For example increased numbers of germ cells, predominantly spermatogonia and spermatocytes, undergo apoptosis in mice between 2 and 4 weeks after birth. This early wave of spermatogonial apoptosis is thought to regulate the ratio of germ cells to Sertoli cells, thereby ensuring that adult Sertoli cell function is not compromised by excessive germ cells. Deregulation of this early wave of spermatogonial apoptosis leads to accumulation of premeiotic germ cells and causes derailment of spermatogenesis in the adult (Furuchi et al., 1996; Rodriguez et al., 1997; Russell et al., 2002; Wright et al., 2007).

In adult mammals, germ cell apoptosis occurs spontaneously during normal spermatogenesis or can be triggered by various external stresses, including deprivation of gonadotropins and intratesticular testosterone (T) by gonadotropin-releasing hormone antagonist (GnRH-A), exposure to local testicular heating, Sertoli cell toxicant, and chemotherapeutic agents (Sinha Hikim and Swerdloff, 1999; Sinha Hikim et al., 2003a, 2011; Baum et al., 2005; Shaha, 2007; Sofikitis et al., 2008). Earlier studies in humans have further demonstrated that both spontaneous (Sinha Hikim et al., 1998) and increased germ cell death in conditions of abnormal spermatogenesis (Dunkel et al., 1997; Pentikaainen et al., 2003; Wang et al., 2007) involve apoptosis.

5.3 Major Pathways of Apoptosis

As depicted in Figure 5.1, the signalling events leading to apoptosis can be divided into two major pathways, involving either mitochondria or death receptors (Green, 2000; Hengartner, 2000; Reed, 2000; Danial and Korsmeyer, 2004). The mitochondria or the intrinsic pathway for apoptosis involves the release of cytochrome c into the cytosol where it binds to apoptotic protease activating factor-1 (Apaf-1), resulting in the activation of the initiator caspase 9 and the subsequent proteolytic activation of the executioner caspases 3, 6 and 7. Members of the BCL-2 family of proteins play a major role in governing this mitochondria-dependent apoptotic pathway, with proteins such as BAX functioning as an inducer and proteins such as BCL-2 as suppressor of cell death. Additionally, SMAC (second mitochondria-derived activator of caspases), also known as DIABLO, is released from mitochondria into the cytosol following apoptotic stimuli and promotes apoptosis by antagonising inhibitor of apoptosis proteins (IAPs). The death receptor or the extrinsic pathway for apoptosis involves ligation of the death receptor (such as FAS) to its ligand, FASL. Binding of FASL to FAS induces trimerisation of FAS receptors, which recruit FADD (FAS-associated death domain) through shared death domains (DDs). FADDs also contain a 'death effector domain' or DED in its N-terminal region. The FAS/FADD complex then binds to the initiator caspase 8 or 10, through interactions between DED of the FADD and these caspase molecules. Crosstalk between these pathways does occur at some levels. In certain cells, caspase 8 through cleavage of BID, a proapoptotic BCL-2 family member, can induce cytochrome c release from mitochondria in FAS-mediated death signalling. Both these pathways converge on caspase 3 and other executioner caspases and nucleases that drive the terminal events of programmed cell death.

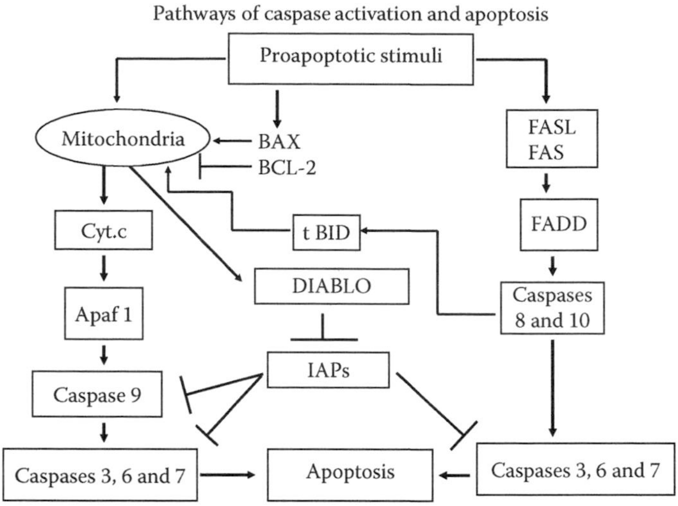

Figure 5.1 Schematic representation of apoptotic pathways. There are at least two major pathways that induce apoptosis in various cell systems. The first or the intrinsic pathway for apoptosis involves the release of cytochrome C (Cyt.c) from mitochondria into the cytosol where it binds to apoptotic protease activating factor-1 or Apaf-1. The complex of Apaf-1, cytochrome c and dATP results in the activation of the initiator caspase 9 and the subsequent activation of the executioner caspases 3, 6 and 7. The BCL-2 family of proteins usually governs this mitochondria-dependent pathway with protein such as Bax functioning as an inducer. The extrinsic pathway for apoptosis involves ligation of Fas to FasL, resulting in the activation of a different set of initiator caspases – caspases 8 and 10 – through interactions between death domains and death effector domains of an adopter molecule such as FADD and these caspases. Available evidence also suggests a cross-talk between these pathways. The protein BID (a proapoptotic BCL-2 family member) mediates cytochrome c release from mitochondria in response to Fas-mediated death signalling. BID exists in the cytosolic fraction of living cells that becomes activated upon cleavage by caspase 8. The truncated cleavage product (t BID) then translocates to mitochondria and induces cytochrome c release. Interestingly, the functions of the initiator caspase 9 and the effector caspases 3 and 7 are inhibited by another set of proteins, the IAPs (inhibitor of apoptosis proteins). IAPs are themselves regulated by mitochondrial proteins, such as DIABLO (also known as Smac which stands for second mitochondria-derived activator of caspases), which binds to IAPs and blocks caspase inhibition. Like cytochrome c, DIABLO is located in mitochondria and released into the cytosol when cells undergo apoptosis.

5.4 Key Signal Transduction Pathways in Male Germ Cell Apoptosis

5.4.1 Involvement of the Mitochondria-Dependent Pathway in Apoptotic Signalling of Testicular Germ Cells

Earlier studies, using murine models of testicular hyperthermia and hormone deprivation, have characterised the key molecular components of the effector pathways leading to caspase activation and increased germ cell death in the testis (Sinha Hikim et al., 2003a,b; Vera et al., 2004, 2006; Jia et al., 2009). Preleptotene and pachytene spermatocytes and round spermatids at mid (VII and VIII) stages are the most susceptible germ cells (by undergoing apoptosis) to a lack of hormonal stimulation (Sinha Hikim et al., 1995, 1997). In striking contrast to the hormone-deprivation model, short-term exposure (43°C for 15 min) of the rat testis to mild heat results, within 6 h, in stage- and cell-specific activation of germ cell apoptosis. Pachytene spermatocytes and early spermatids (steps 1–4) at stages I–IV and pachytene, diplotene and dividing spermatocytes at stages XII–XIV are most susceptible to heat (Lue et al., 1999, 2000; Yamamoto et al., 2000). Thus, the vulnerability of germ cells to apoptosis in these two paradigms is different.

These different but complementary models for induction of testicular germ cell apoptosis are routinely used to elucidate the key signal transduction pathways for male germ cells apoptosis. For example initiation of apoptosis was preceded by a redistribution of BAX from a cytoplasmic to perinuclear localisation in heat-susceptible germ cells and elevated levels of BCL-2 in the mitochondria. The relocation of BAX is accompanied by cytosolic translocation of cytochrome c and is associated with activation of the initiator caspase 9 and the executioner caspases 3, 6 and 7, and poly (ADP-ribose) polymerase (PARP) cleavage (Sinha Hikim et al., 2003b). Collectively, these data suggest the involvement of the mitochondria-dependent pathway for heat-induced male germ cell apoptosis.

To characterise the involvement of the intrinsic pathway signalling for induction of apoptosis in the hormone-deprivation model, in an earlier study, groups of adult male rats were given a daily injection of vehicle for 14 days or GnRH-A, acyline at a dose of 1.6 mg/kg BW for 2, 5 and 14 days. Within 2 days of GnRH-A treatment testicular concentrations of T declined markedly to 17.1% of control values and plasma T levels fell below detectable limits. Germ cell apoptosis, involving exclusively stages VII–VIII, was achieved by day 5 (Figure 5.2). Within the study paradigm, the highest number of dying cells occurred by day 14, at which time a modest but significant increase in the incidence of apoptosis was also noted at stages other than VII–VIII. As shown in Figure 5.3a, unlike hyperthermia model, initiation of germ cell apoptosis after hormone withdrawal was associated an increase in BAX and a decrease in BCL-2 expression in the mitochondrial fraction of testicular lysates. Such alteration in the BAX and BCL-2 ratio was accompanied by cytosolic

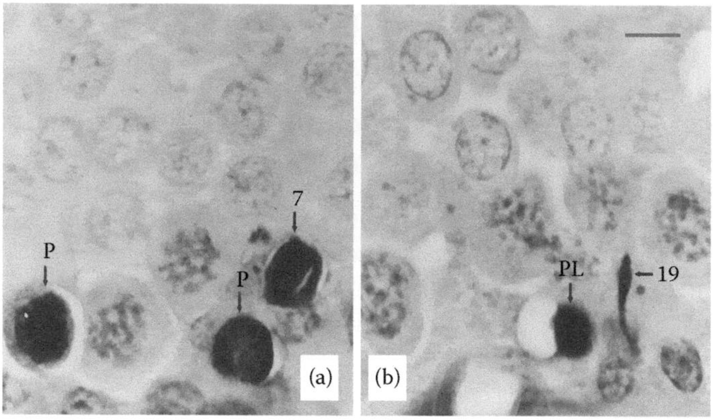

Figure 5.2 (See colour insert.) *In situ* 3′-end labelling of DNA strand breaks in apoptotic germ cells in glutaraldehyde-fixed, paraffin-embedded testicular sections. Methyl green was used as a counterstain. (a and b) Portions of stage VII tubules from a rat treated with GnRH-A for 5 days exhibiting apoptotic preleptotene (PL) and pachytene (P) spermatocytes, and step 7 (7) and 19 (19) spermatids. Scale bar, 10 μm.

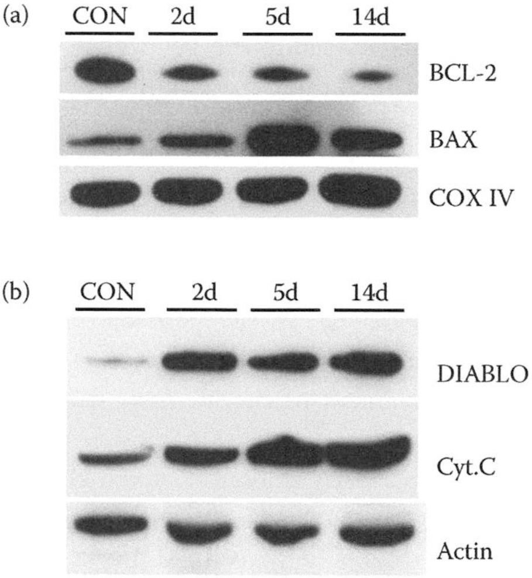

Figure 5.3 Hormonal deprivation results in activation of the intrinsic pathway signalling. (a) Western blot analysis shows an increase in BAX expression and a decrease in BCL-2 expression in the mitochondrial fraction of testicular lysates after GnRH-A treatment. COX IV in the immunoblot is shown as a loading control. (b) Representative Western blots of cytosolic fractions of testicular lysates from control and rats 2, 5 and 14 days after GnRH-A treatment show a marked accumulation of DIABLO and cytochrome c in the cytosolic fractions after hormone deprivation. Actin in the immunoblot is shown as a loading control. (From Vera Y et al. 2006, *Mol Endocrinol* 20: 1597–1609. With permission.)

translocation of mitochondrial cytochrome c and DIABLO (Figure 5.3b), activation of caspase 9 and caspase 3, and PARP cleavage. These results indicate that withdrawal of gonadotropins and consequently intratesticular T induces germ cell apoptosis in the testis also by stimulating the intrinsic pathway signalling (Vera et al., 2006).

Importantly, the same signalling pathway also constitutes a critical component of apoptotic signalling in testicular germ cells in monkeys (Jia et al., 2007) as well as in men (Vera et al., 2006). Together, these studies indicate that the mitochondria-dependent pathway appears to be the key apoptotic pathway for germ cell death in the testis across species.

5.4.2 Involvement of Fas Signalling in Male Germ Cell Apoptosis

To evaluate the involvement of the Fas signalling system in male germ cell apoptosis, earlier studies were carried out to examine whether *gld* and *lpr*cg mice, which harbour loss-of-function mutations in Fas ligand (Fas L) and Fas, respectively (Nagata and Golstein, 1995), would confer resistance to heat-induced germ cell apoptosis. Similar to the rat model, scrota of *gld* and *lpr*cg mice and their wild types (C57BL6J and MRL/Mpj, respectively) were exposed once to 22°C (control) or 43°C (heat treated) for 15 min and the animals were killed at 0.5, 2 or 6 h after heating. The incidence of germ cell apoptosis before and after heat treatment was similar in both wild type and mutant mice, suggesting that germ cells from wild type and mutant mice with loss-of-function mutations in Fas ligand and Fas, respectively, are equally sensitive to heat-induced apoptosis (Sinha Hikim et al., 2003a,b). Of note, the initiation of apoptosis was preceded by a redistribution of BAX from a cytoplasmic to perinuclear localisation in heat-susceptible germ cells (Vera et al., 2004). The relocation of BAX is further accompanied by sequestration of ultra-condensed mitochondria into perinuclear areas of apoptotic germ cells, cytosolic translocation of mitochondrial cytochrome c and DIABLO, and is associated with activation of the initiator caspase 9 and the executioner caspase 3 (Vera et al., 2004). Furthermore, there was absence of truncated BID in either cytosolic or mitochondrial fractions of heat-treated testicular lysates of both wild type and mice lacking functional FAS, suggesting that the caspase 8-mediated cleavage of BID is not responsible for the observed release of cytochrome c from the mitochondria (Vera et al., 2004).

Nair and Shaha (2003) showed the involvement of the mitochondria-dependent pathway, characterised by loss of mitochondrial membrane potential, BAX translocation to mitochondria, cytochrome c release from mitochondria and subsequent activation of the caspase 9 and caspase 3, and PARP cleavage, in diethylstilbestrol-induced testicular germ cell apoptosis in the rat. One other important finding that comes out from this study is the involvement of the Fas-FasL system, characterised by upregulation of FasL and Fas and activation of caspase 8 in germ cells. Theas and colleagues (2006) have further demonstrated the involvement of both death receptor and mitochondrial pathways in germ cell apoptosis in an experimental model of autoimmune orchitis. It would be interesting to know whether the link between these two pathways is the caspase 8-mediated cleavage of BID. Evidence exists that germ cells, in particular spermatocytes, are able to undergo tumour necrosis factor (TNF)-related apoptosis-inducing ligand (TRAIL)-induced apoptosis and that pretreatment with anti-DR5 antibody can increase their sensitivity to TRAIL-mediated apoptosis (McKee et al., 2006). Taken together, these results indicate that regulation of testicular germ cell apoptosis varies depending upon the nature of apoptotic stimulus and can be triggered by more than one pathway.

5.4.3 Functional Role of Downstream Caspases in Male Germ Cell Apoptosis

Given the importance of caspases in the apoptotic process in general, earlier studies have further investigated the functional role of these proteases in testicular germ cell apoptosis. Using a mouse model of testicular hyperthermia, it has been shown that pretreatment with Quinoline-Val-asp (Ome)-CH_2-O-Ph (Q-VD-OPH), a broad-spectrum pan caspase inhibitor markedly inhibits caspase 9 and caspase 3 activation, PARP cleavage, and attenuates heat-induced germ cell apoptosis in mice (Vera et al., 2005). The protection offered by Q-VD-OPH was independent of mitochondrial cytochrome c release and occurred downstream of mitochondria by inhibiting caspase activation (Vera et al., 2005). Collectively, this study further emphasises the role of caspase 9 and caspase 3 in testicular germ cell apoptosis.

5.4.4 Upstream Signalling Pathways for Testicular Germ Cell Apoptosis

Although there has been a spectacular progress in deciphering the downstream signalling pathways of germ cell apoptosis in the testis, elucidation of upstream signalling machinery that triggers these downstream apoptotic pathways has only just begun. Mitogen-activated protein kinases (MAPKs) comprise a family of serine/threonine kinases that function as critical mediators of a variety of extracellular signals. These kinases include the extracellular signal-regulated kinase (ERK), c-jun NH_2-terminal kinase (JNK), also known as stress-activated protein kinase (SAPK), and the p38 MAPK. In order to provide some insight into the upstream signalling pathways, Vera and colleagues (2006) examined the role of p38 MAPK and inducible nitric oxide synthase (iNOS) in apoptotic signalling of male germ cells in rats after hormone deprivation by a potent GnRH-A treatment. Activation of p38 MAPK, as evidenced by an increase in phospho-activating transcription factor-2 (ATF-2), was detected as early as 2 days after GnRH-A treatment and remained active thereafter throughout the treatment period (Figure 5.4a). Activation of p38 MAPK was also substantiated by immunohistochemistry and confocal microscopy. Compared to control, where no staining is detected, strong phospho-p38 MAPK immunoreactivity was noted in the condensed nuclei of apoptotic germ cells after hormone withdrawal (Figure 5.4b, panels I–III). Co-staining for TUNEL and for phospho-p38 MAPK further confirmed activation of p38 MAPK only in those germ cells undergoing apoptosis (Figure 5.4b, panels IV–VI). A similar profile in the induction of iNOS was noted after GnRH-A treatment. Most important, p38 MAPK activation and iNOS induction within 2 days after GnRH-A treatment indicate that these events are indeed upstream of activation of apoptosis, which was first detected 5 days after GnRH-A treatment. p38 MAPK activation and iNOS induction were further accompanied by a marked perturbation of the BAX/ BCL-2 rheostat, cytochrome c and DIABLO release from mitochondria, caspase

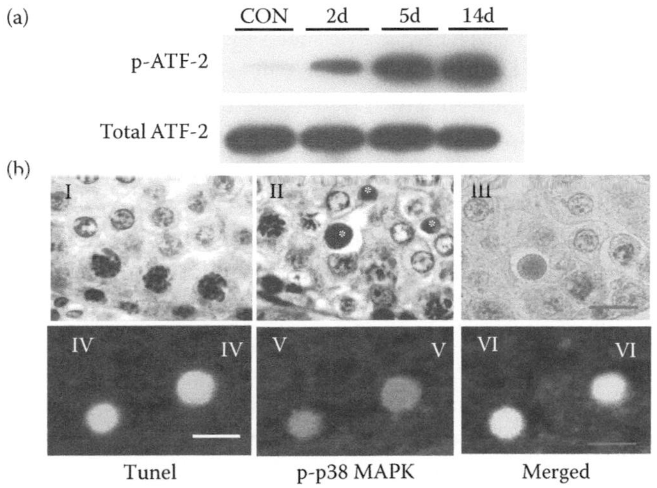

Figure 5.4 (See colour insert.) Activation of p38 MAPK in rat testes after GnRH-A treatment. (a) Analysis of p38 MAPK activation by Western blotting using phospho-ATF-2 (Thr 71) antibody in testicular lysates after GnRH-A treatment. Total ATF-2 in the immunoblot is shown as a loading control. (b) p38 MAPK activation visualised by immunocytochemistry and confocal microscopy. Portions of stage VII tubules from control (panel I) and rats treated with GnRH-A for 5 days (panel II) show a strong phospho-p38 MAPK immunoreactivity in the condensed nuclei of apoptotic germ cells (asterisk) after hormone withdrawal. A testicular section from a rat treated with GnRH-A for 5 days incubated with rabbit IgG (negative control) shows no such immunostaining in a stage VII tubule (panel III). Panels IV–VI, confocal images show TUNEL (green), active p38 MAPK (red) and co-localisation of TUNEL and active p38 MAPK (yellow) in apoptotic germ cells triggered by hormone deprivation. Scale bar, 15 μm (panels I–III) and 10 μm (panels IV–VI). (From Vera Y et al. 2006. *Mol Endocrinol* 20: 1597–1609. With permission.)

activation and PARP cleavage (Vera et al., 2006). Concomitant administration of aminoguanidine (AG), a selective iNOS inhibitor, significantly prevented hormone deprivation-induced germ cell apoptosis (Vera et al., 2006). Relevant to this is the demonstration that such hormone deprivation-induced male germ cell apoptosis can be effectively prevented by minocycline (Castanares et al., 2005), which suppresses p38 MAPK activation, iNOS induction and the cytochrome c-mediated death pathway in other systems (Zhu et al., 2002; Teng et al., 2004; Wei et al., 2005). Induction of germ cell apoptosis after hormone withdrawal is independent of JNK or ERK (Jia et al., 2009).

To characterise the upstream signalling pathways by which heat stress triggers male germ cell apoptosis, the contributions of the ERK, the JNK and the p38 MAPK to stage-specific activation of germ cell apoptosis triggered by testicular hyperthermia were examined (Jia et al., 2009). Like hormone deprivation model, testicular hyperthermia had no effect of JNK activation. Testicular hyperthermia, however, resulted in stage- and cell-specific activation of both p38 MAPK and ERK. Activation of p38 MAPK, as evidenced by a significant ($P < 0.05$) increase (by 5.3-fold) in phospho-p38 MAPK levels in testis lysates, was detected within half hour of heating and remained active thereafter throughout the treatment period. It is pertinent to note here, upon exposure of the testis to heat, the amount of BCL-2 increased significantly in the mitochondrial fraction, while BAX levels remained unchanged (Sinha Hikim et al., 2003b). Immunohistochemistry further revealed increase in BCL-2 expression on heat-susceptible germ cells (Yamamoto et al., 2000). The obvious question raised by these observations is why these germ cells are dying in spite of the enhanced expression of BCL-2 and with no change in BAX. Does this mean possible loss of its antiapoptotic function? A growing body of evidence indicates that serine phosphorylation of BCL-2 leads to its inactivation and its ability to form dimmers with BAX and, therefore, results in the loss of its antiapoptotic function (Halder et al., 1998; Fan et al., 2000; Rajah et al., 2002; Bu et al., 2006). Because phosphorylation of BCL-2 can be induced by p38 MAPK (Shimada et al., 2003; Bu et al., 2006), Jia and colleagues (2009) next examined whether the increased germ cell apoptosis after heat stress is associated with BCL-2 phosphorylation. Compared with control, where no staining was detected, a marked increase in the serine-phosphorylated form of inactive BCL-2 was noted only in heat-susceptible germ cells (Figure 5.5a, panels I and II). Co-staining for TUNEL and phospho-BCL-2 further confirmed phosphorylation of BCL-2 only in those germ cells undergoing apoptosis (Figure 5.5b, panels I–III). Most important, it has been shown that SB203580, a p38 MAPK inhibitor, effectively suppressed BCL-2 phosphorylation and cytochrome c release and significantly ($P < 0.05$) prevented heat-induced germ cell apoptosis (Jia et al., 2009). It is thus conceivable that the p38 MAPK signalling promotes heat-induced germ cell apoptosis by provoking BCL-2 phosphorylation, leading to its inactivation, thereby resulting in the perturbation of the BAX/BCL-2 rheostat in the mitochondria and the subsequent activation of the mitochondria-dependent death pathway.

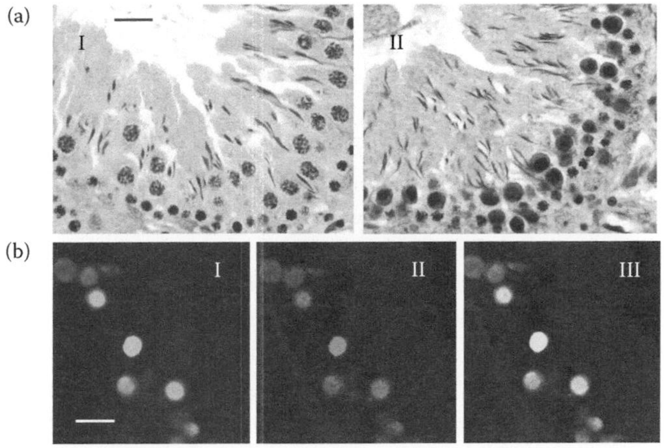

Figure 5.5 **(See colour insert.) Testicular hyperthermia results in serine phosphorylation of BCL-2 in germ cells. (a) Portions of stage XII tubules from control (panel I) and a rat that had been exposed to short-term local testicular heating (panel II) show serine phosphorylation of BCL-2 only in heat-susceptible late pachytene spermatocytes 6 h after heating. Scale bar, 25 μm. (b) (Panels I–III), confocal images of late pachytene spermatocytes at stage XII from a heat-treated rat show TUNEL (green), phospho-BCL-2 (red) and colocalisation of TUNEL and phospho-BCL-2 (yellow) in apoptotic germ cells 6 h after heat treatment. Scale bar, 15 μm.**

ERK was also found to be activated within half hour of heating in the Sertoli cells at heat-susceptible stages (Figure 5.6). Thus, the activation of ERK in the Sertoli cells is indeed upstream of activation of germ cell apoptosis, which was first detected 6 h after heating (Yamamoto et al., 2000; Sinha Hikim et al., 2003b). Inhibition of ERK by U0126, however, had no effect on the incidence of heat-induced germ cell apoptosis (Jia et al., 2009). The possible significance of these

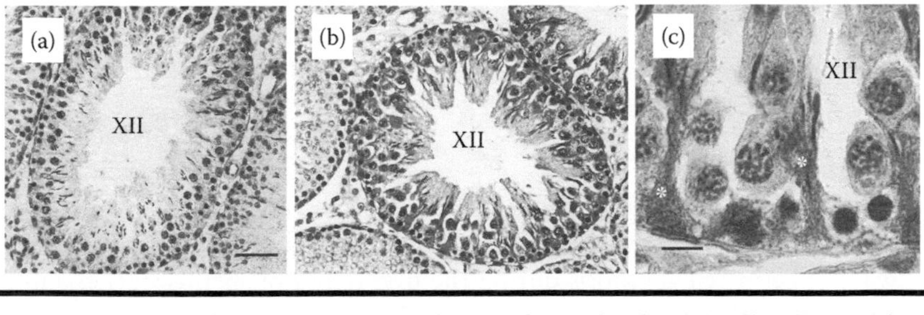

Figure 5.6 **(See colour insert.) Activation of ERK in the Sertoli cells. Testicular sections from (a) control and (b, c) rats that had been exposed once to short-term testicular heating show activation of ERK in Sertoli cells at stage XII (a heat sensitive stage) within half hour of heating. Scale bar (a, b) 50 μm and (c) 15 μm.**

findings is not known. These observations, however, do suggest that not only germ cells but also Sertoli cells may be affected by heat treatment. There have been studies indicating that heat treatment through activation of ERK induces dedifferentiation of adult Sertoli cells into immature states in monkeys (Zhang et al., 2006). Thus, it is possible that the affected Sertoli cells could have compromised functions, which in turn, sensitise these germ cells to apoptosis after heat stress. In this context, it is important to note that ERK plays an important role in regulating Sertoli-germ cell adherens junction (AJ) such as ectoplasmic specialisation (ES) in the testis (Wong and Cheng, 2005; Xia and Cheng, 2005). Furthermore, it is known that disruption of ES structure and function can induce male germ cell apoptosis (Xia and Cheng, 2005; Rao et al., 2006). Thus, the possibility that activation of ERK in the Sertoli cells after heat stress may perturb Sertoli-germ cell AJ dynamics at the heat-sensitive stages and promote, via-yet-to-be identified mechanisms, germ cell apoptosis cannot be excluded. The observation that inhibition of ERK has no effect on the incidence of heat-induced germ cell apoptosis, on the other hand, suggests that ERK signalling may be dispensable for heat-induced germ cell apoptosis in the testis. However, it remains possible that activation of ERK in the Sertoli cells, which precedes the initiation of apoptosis, could sensitise these germ cells to apoptosis through perturbation of Sertoli cell function and/or AJ dynamics.

In summary (Figure 5.7), in both testicular hyperthermia and hormone deprivation models the induction of germ cell apoptosis is triggered by a p38 MAPK-mediated death pathway (Vera et al., 2006; Jia et al., 2009). This pathway, through changes in the ratio of BAX and BCL-2 in the mitochondria, activates the intrinsic pathway signalling and promotes germ cell apoptosis in response to a lack of hormonal stimulation. The same signalling pathway is also the key pathway for heat-induced testicular germ cell apoptosis. However, unlike hormone deprivation model, this signalling pathway promotes germ cell apoptosis by provoking BCL-2 phosphorylation, leading to its inactivation, thereby resulting in the perturbation of the BAX/BCL-2 rheostat, and the subsequent activation of the mitochondria-dependent death pathway.

5.4.5 Caspase 2 Is an Upstream Activator of p38 MAPK-Mediated Intrinsic Pathway Signalling

Of all caspases discovered to date, caspase 2 is the most evolutionarily conserved and plays an important role in inducing apoptosis in various cell systems. Caspase 2-mediated intrinsic pathway signalling has recently been implicated in the initial wave of germ cell apoptosis during the first round of spermatogenesis in mice (Zheng et al., 2006). Lysiak and colleagues (2007) have demonstrated the involvement of caspase 2-mediated intrinsic pathway signalling in germ cell apoptosis in mice triggered by ischemia-reperfusion. To further explore the role of caspase 2, in a recent study, Johnson and colleagues sought to determine whether a specific

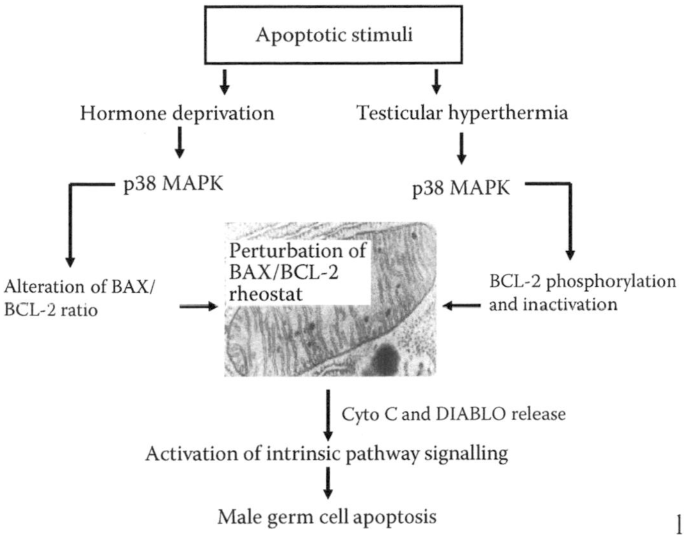

Figure 5.7 Upstream signalling pathway involved in germ cell apoptosis in rats triggered by mild testicular hyperthermia or by hormonal deprivation. In both the models the induction of apoptosis is triggered by a p38 MAPK-mediated pathway. Activation of p38 MAPK, through changes in the ratio of BAX and BCL-2 in the mitochondria, triggers cytochrome c and DIABLO release, which, in turn, activates the intrinsic pathway signalling and promotes germ cell apoptosis in response to a lack of hormonal stimulation. The same signalling pathway is also the key pathway for heat-induced testicular germ cell apoptosis. However, unlike the hormone deprivation model, this signalling pathway promotes germ cell apoptosis by provoking BCL2 phosphorylation, leading to its inactivation, thereby resulting in the perturbation of the BAX/BCL-2 rheostat, and the subsequent activation of the mitochondria-dependent death pathway.

inhibitor of caspase 2 (Z-VDAVDK-fmk) could prevent or attenuate heat-induced male germ cell apoptosis (Johnson et al., 2008). Rats were pretreated with intratesticular injections of DMSO (vehicle) or 50 or 100 μg of caspase 2 inhibitor (Z-VDA-VDK) and killed 6 h later, the earliest time associated with stage-specific increase in germ cell apoptosis (Yamamoto et al., 2000; Sinha Hikim et al., 2003b). Quantitation of the TUNEL-positive germ cells revealed that Z-VDAVDK significantly ($P < 0.05$) prevented heat-induced germ cells apoptosis by 68.8% and 70.1%, respectively. Most notably, protection offered by the caspase 2 inhibitor occurred upstream of mitochondria, involving suppression of p38 MAPK activation and, in turn, suppression of cytochrome c-mediated death pathway (Johnson et al., 2008). Interestingly, almost an identical level of protection (by 67.0%) of testicular germ cells from heat-induced apoptosis was noted in mice pretreated with Q-VD-OPH, a broad-spectrum pan caspase inhibitor (Vera et al., 2005). However, compared to Z-VDAVDK, the protection offered by Q-VD-OPH was independent of mitochondrial cytochrome c release and occurred by inhibiting caspase activation (Vera et al., 2005). Together, these studies indicate that caspase 2 activation is needed to fuel cytochrome c or DIABLO release from mitochondria.

5.5 Conclusions

Our understanding of the regulation of male germ cell apoptosis has greatly expanded in recent years. It is clear that male germ cell apoptosis is an important prerequisite for functional spermatogenesis. Much progress has been made towards unravelling the key signal transduction pathways in apoptotic signalling of male germ cells. However, significant gaps remain in our knowledge base. There is increasing evidence that null mutations of a number of genes in mice results in severe spermatogenic disruption and infertility through accelerated germ cell apoptosis. Most notably, it appears that null mutation of some genes, expressed in many tissues, including the testis, can have specific effects on germ cell apoptosis and spermatogenesis (reviewed in Sinha Hikim et al., 2011). Detailed characterisation of the underlying mechanism of those defects in these mutant mice will provide insight into the basic control mechanism of male germ apoptosis. However, it is possible that total bodywide knockout of a given gene can result in embryonic or neonatal lethality, which prevents an analysis of the role of that particular gene in spermatogenesis. A tissue-specific *in vivo* RNA interference (RNAi) approach has been used to elucidate the cell type-specific function of Wilms' tumour 1 (WT1) in regulating spermatogenesis (Rao et al., 2006). Mice depleted of WT1 in Sertoli cells exhibit increased germ cell apoptosis, loss of adherence junctions and impaired fertility. By substituting different lineage-specific or regulated promoters and stem-loops corresponding to different gene targets, this system has the potential to knock down the expression of virtually any gene in a cell-type specific and temporally regulated manner. This novel *in vivo* RNAi approach may avoid the frequently observed problems of early lethality or developmental redundancy.

Emerging evidence now suggests a more direct role of cellular metabolism in governing cell death through either activation of a specific death pathway or loss of a critical survival pathway. During spermatogenesis Sertoli-germ cell metabolic cooperation is essential for germ cell survival (Boussouar and Benahmed, 2004). Systemic glucose is taken by Sertoli cells, processed glycolytically into lactate, and is transported across the plasma membrane to the germ cells by specific monocarboxylate transporter. The challenge is now to characterise the metabolic networks, using stable isotope-based metabolic flux phenotyping in conjunction with gas chromatography and mass spectrometry (Lee, 2006), and the novel aspects of those networks that actually are necessary for male germ cell death. Metabolic profiling and its integration with signal transduction pathways inducing germ cell death will provide insight into how perturbation of the metabolic cooperation between Sertoli and germ cells affects germ cell survival.

Future efforts towards improved fertility control and clinical management of infertility associated with reduced sperm production in men are hampered by incomplete understanding of the processes responsible for normal germ cell homeostasis. Elucidation of the metabolic and molecular mechanisms by which various environmental stresses and male contraceptive approaches regulates germ cell death will fill a major gap in our knowledge of this fundamental biological process.

Acknowledgements

The work was supported by the grants from National Institutes of Health (RO1 HD 39293 and R25 GM 560902).

References

Baum JS, St. George JP, McCall GK. 2005. Programmed cell death in the germline. *Semin Cell Dev Biol* 16: 245–259.

Boussouar F, Benahmed M. 2004. Lactate and energy metabolism in male germ cells. *Trends in Endo Metab* 15: 345–350.

Bu SZ, Huang Q, Jiang YM, Min HB, Hou Y, Guo ZY, Wei JF, Wang GW, Ni X, Zheng SS. 2006. p38 Mitogen-activated protein kinase is required for counteraction of 2-methoxyestradiol to estradiol-stimulated cell proliferation and induction of apoptosis in ovarian carcinoma cells via phosphorylation of Bcl-2. *Apoptosis* 11: 413–425.

Castanares M, Vera Y, Erkkila K, Kyttanen S, Lue Y, Dunkel L, Wang C, Swerdloff RS, Sinha Hikim AP. 2005. Minocycline up-regulates BCL-2 levels in mitochondria and attenuates male germ cell apoptosis. *Biochem Biophys Res Commun* 337: 663–669.

Danial NN, Korsmeyer SJ. 2004. Cell death: Critical control points. *Cell* 116: 205–219.

Dunkel L, Taskinen S, Hovatta O, Tilly JL, Wikstrom S. 1997. Germ cell apoptosis after treatment of cryptorchidism with human chorionic gonadotropin is associated with impaired reproductive function in the adult. *J Clin Invest* 100: 2341–2346.

Fan M, Goodwin M, Vu T, Brantley-Finley C, Gaarde WA, Chambers TC. 2000. Vinblastine-induced phosphorylation of Bcl-2 and Bcl-XL is mediated by JNK and occurs in parallel with inactivation of the Raf-1/MEK/ERK cascade. *J Biol Chem* 29: 29980–29985.

Furuchi T, Masuko K, Nishimune Y, Obinata M, Matsui Y. 1996. Inhibition of testicular germ cell apoptosis and differentiation in mice misexpressing Bcl-2 in spermatogonia. *Development* 122: 1703–1709.

Green DR. 2000. Apoptotic pathways: Paper wraps stone blunts scissors. *Cell* 102: 1–4.

Halder S, Basu A, Croce CM. 1998. Serine-70 is one of the critical sites for drug-induced Bcl-2 phosphorylation in cancer cells. *Cancer Res* 58: 1609–1615.

Hengartner MO. 2000. The biochemistry of apoptosis. *Nature* 407: 770–776.

Jacobson MD, Weil M, Raff MC. 1997. Programmed cell death in animal development. *Cell* 88: 347–354.

Jia Y, Sinha Hikim AP, Swerdloff RS, Lue YH, Vera Y, Zhang X-S, Hu Z-Y, Li Y-C, Liu Y-X, Wang C. 2007. Signaling pathways for germ cell death in adult Cynomolgus monkeys

(*Macaca fascicularis*) induced by mild testicular hyperthermia and exogenous testosterone treatment. *Biol Reprod* 77: 83–92.

Jia Y, Castellanos J, Wang C, Sinha-Hikim I, Lue Y, Swerdloff RS, Sinha Hikim AP. 2009. Mitogen-activated protein kinase signaling in male germ cell apoptosis. *Biol Reprod* 80: 771–780.

Johnson CJ, Jia Y, Wang C, Lue Y, Swerdloff RS, Zhang X-S, Hu Z-Y, Li Y-C, Liu Y-X, Sinha Hikim AP. 2008. Role of caspase 2 in apoptotic signaling of primate and murine germ cells. *Biol Reprod* 79: 806–814.

Kerr JFR, Wyllie AH, Currie AR. 1972. Apoptosis: A basic biological phenomenon with wide-ranging implication in tissue kinetics. *Br J Cancer* 26: 239–257.

Lee WN. 2006. Characterizing phenotype with tracer based metabolomics. *Metabolomics* 2: 31–39.

Lue YH, Sinha Hikim AP, Swerdloff RS, Im P, Taing KS, Bui T, Leung A, Wang C. 1999. Single exposure to heat induces stage-specific germ cell apoptosis in rats: Role of intratesticular testosterone (T) on stage specificity. *Endocrinology* 140: 1709–1717.

Lue YH, Sinha Hikim AP, Wang C, Im M, Leung A, Swerdloff RS. 2000. Testicular heat exposure enhances the suppression of spermatogenesis by testosterone in rats: The 'two-hit' approach to male contraceptive development. *Endocrinology* 141: 1414–1424.

Lysiak JL, Zheng S, Woodson R, Turner TT. 2007. Caspase-9-dependent pathway to murine germ cell apoptosis: Mediation by oxidative stress, BAX, and caspase 2. *Cell Tissue Res* 328: 411–419.

McKee CM, Ye Y, Richburg JH. 2006. Testicular germ cell sensitivity to TRAIL-induced apoptosis is dependent upon p53 expression and is synergistically enhanced by DR5 agonistic antibody treatment. *Apoptosis* 11: 2237–2250.

Mori C, Nakamura N, Dix DJ, Fujioka M, Nakagawa S, Shiota K, Eddy EM. 1997. Morphological analysis of germ cell apoptosis during postnatal testis development in normal and Hsp 70-2 knockout mice. *Dev Dyn* 208: 125–136.

Nagata S, Golstein P. 1995. The Fas death factor. *Science* 267: 1449–1455.

Nair R, Shaha C. 2003. Diethylstilbestrol induces rat spermatogenic cell apoptosis in vitro through increased expression of spermatogenic cell Fas/FasL system. *J Biol Chem* 278: 6470–6481.

Pentikaainen V, Dunkel L, Erkkila K. 2003. Male germ cell apoptosis. *Endocr Dev* 5: 56–80.

Rajah R, Lee K-W, Cohen P. 2002. Insulin-like growth factor binding protein 3 mediates tumor necrosis factor-α-induced apoptosis: Role of Bcl-2 phosphorylation. *Cell Growth Differ* 13: 163–171.

Rao MK, Pham J, Imam S, MacLean J, Murali D, Furuta Y, Sinha Hikim AP, Wilkinson MF. 2006. Tissue-specific RNAi reveals that WT1 expression in nurse cells controls germ-cell survival and spermatogenesis. *Genes Dev* 20: 147–152.

Reed JC. 2000. Mechanisms of apoptosis. *Am J Pathol* 157: 1415–1430.

Rodriguez I, Ody C, Araki K, Garcia I, Vassalli P. 1997. An early and massive wave of germinal cell apoptosis is required for the development of functional spermatogenesis. *EMBO J* 16: 2262–2270.

Russell LD, Chiarini-Garcia H, Korsmeyer SJ, Knudson CM. 2002. Bax-dependent spermatogonia apoptosis is required for testicular development and spermatogenesis. *Biol Reprod* 66: 950–958.

Russell LD, Ettlin RA, Sinha Hikim AP, Clegg ED. 1990. *Histological and Histopathological Evaluation of the Testis*. Cache River Press, Clearwater, FL.

Shaha C. 2007. Modulators of spermatogenic cell survival. *Soc Reprod Fertil Suppl* 63: 173–186.

Sharpe RM. 1994. Regulation of spermatogenesis. In: Knobil E, Neill JD, editors, *The Physiology of Reproduction.* Raven Press, New York, 1363–1434.

Shimada K, Nakamura M, Ishida E, Kishi M, Konishi N. 2003. Roles of p38- and c-jun NH_2-terminal kinase-mediated pathways in 2-methoxyestradiol-induced p53 induction and apoptosis. *Carcinogenesis* 24: 1067–1075.

Sinha Hikim AP, Jia Y, Lue Y, Wang C, Swerdloff RS. 2011. Apoptotic signaling in male germ cells. In: Reed JC, Green D, editors, *Apoptosis: Physiology and Pathology of Cell Death.* Cambridge University Press, New York, 283–294.

Sinha Hikim AP, Lue Y, Diaz Romero M, Yen PH, Wang C, Swerdloff RS. 2003a. Deciphering the pathways of germ cell apoptosis in the testis. *J Steroid Mol Biol* 85: 175–182.

Sinha Hikim AP, Lue Y, Yamamoto CM, Vera Y, Rodriguez S, Yen PH, Soeng K, Wang C, Swerdloff RS. 2003b. Key apoptotic pathways for heat-induced programmed germ cell death in the testis. *Endocrinology* 144: 3159–3166.

Sinha Hikim AP, Rajavashisth TB, Sinha Hikim I, Lue YH Bonavera JJ, Leung A, Wang C, Swerdloff RS. 1997. Quantitative contribution of apoptosis to the temporal and stage-specific loss of germ cells in the adult rat after gonadotropin deprivation. *Biol Reprod* 57:1193–1201.

Sinha Hikim AP, Swerdloff RS. 1999. Hormonal and genetic control of germ cell apoptosis in the testis. *Rev Reprod* 4: 38–47.

Sinha Hikim AP, Swerdloff RS, Wang C. 2005. The Testis. In: Melmed S, Conn M, editors, *Endocrinology: Basic and Clinical Principles.* Humana Press, Totowa, NJ, 405–418.

Sinha Hikim AP, Wang C, Leung A, Swerdloff RS. 1995. Involvement of apoptosis in the induction of germ cell degeneration in adult rats after gonadotropin-releasing hormone antagonist treatment. *Endocrinology* 136: 2770–2775.

Sinha Hikim AP, Wang C, Lue Y, Johnson L, Wang X-H, Swerdloff RS. 1998. Spontaneous germ cell apoptosis in humans: Evidence for ethnic differences in the susceptibility of germ cells to programmed germ cell death. *J Clin Endocrinol Metab* 83: 152–156.

Sofikitis N, Giotitsas N, Tsounapi P, Baltogiannis D, Giannakis D, Pardalidis N. 2008. Hormonal regulation of spermatogenesis and spermiogenesis. *J Steroid Mol Biol* 109: 323–330.

Taylor RC, Cullen SP, Martin SJ. 2008. Apoptosis: Controlled demolition at the cellular level. *Nature Rev Mol Cell Biol* 9: 231–241.

Teng YD, Choi H, Onario RC, Zhu S, Desilets FC, Lan S, Woodard EJ, Snyder EY, Eichler ME, Friedlander RM. 2004. Minocycline inhibits contusion-triggered mitochondrial cytochrome c release and mitigates functional deficits after spinal cord injury. *Proc Natl Acad Sci USA* 101: 3071–3076.

Theas MS, Rival C, Jarazo Dietrich S, Guazzone VA, Lustig L. 2006. Death receptor and mitochondrial pathways are involved in germ cell apoptosis in an experimental model of autoimmune orchitis. *Hum Reprod* 21: 1734–1742.

Thompson CB. 1995. Apoptosis in the pathogenesis and treatment of disease. *Science* 267: 1456–1462.

Vera Y, Diaz-Romero M, Rodriguez S, Lue Y, Wang C, Swerdloff RS, Sinha Hikim AP. 2004. Mitochondria-dependent apoptotic pathway is involved in heat-induced male germ cell death: Lessons from mutant mice. *Biol Reprod* 70: 1534–1540.

Vera Y, Erkkila K, Wang C, Nunez C, Kyttanen S, Lue Y, Dunkel L, Swerdloff RS, Sinha Hikim AP. 2006. Involvement of p38 mitogen-activated protein kinase and inducible nitric oxide synthase in apoptotic signaling of murine and human male germ cells after hormone deprivation. *Mol Endocrinol* 20: 1597–1609.

Vera Y, Rodriguez A, Castanares M, Lue Y, Atienza V, Wang C, Swerdloff RS, Sinha Hikim AP. 2005. Functional role of caspases in heat-induced testicular germ cell apoptosis. *Biol Reprod* 72: 516–522.

Wang C, Cui YG, Wang XH, Jia Y, Sinha Hikim A, Lue YH, Tong JS et al. 2007. Transient scrotal hyperthermia and levonorgestrel enhance testosterone induced spermatogenesis suppression in men through increased germ cell apoptosis. *J Clin Endocrinol Metab* 92: 3292–3304.

Wang R-A, Nakane PK, Koji T. 1998. Autonomous cell death of mouse germ cells during fetal and postnatal period. *Biol Reprod* 58: 1250–1256.

Wei X, Zhao L, Liu J, Dodel RC, Farlow MR, Du Y. 2005. Minocycline prevents gentamicin-induced cytotoxicity by inhibiting p38 MAP kinase phosphorylation and caspase 3 activation. *Neuroscience* 131:13–521.

Wong C-H, Cheng CY. 2005. Mitogen-activated protein kinases, adherens junction dynamics, and spermatogenesis: A review of recent data. *Dev Biol* 286: 1–15.

Wright A, Reiley WW, Chang M, Jin W, Lee AJ, Zhang M. 2007. Regulation of early wave of germ cell apoptosis and spermatogenesis by deubiquitinating enzyme CYLD. *Dev Cell* 13: 705–716.

Xia W, Cheng CY. 2005. TGF-β3 regulates anchoring junction dynamics in the seminiferous epithelium of the rat testis via the Ras/ERK signaling pathway: An in vivo study. *Dev Biol* 280: 321–343.

Yamamoto CM, Sinha Hikim AP, Huynh PN, Shapiro B, Lue YH, Wang C, Swerdloff RS. 2000. Redistribution of Bax is an early step in an apoptotic pathway leading to germ cell death triggered by mild testicular hyperthermia. *Biol Reprod* 63: 1683–1690.

Youle RJ, Strasser A. 2008. The BCL-2 protein family: Opposing activities that mediate cell death. *Nat Rev Mol Cell Biol* 9: 47–59.

Zhang X-S, Zhang Z-H, Jin X, Wei P, Hu X-Q, Chen M, Lu C-L et al. 2006. Dedifferentiation of adult monkey Sertoli cells through activation of ERK1/2 kinase induced by heat treatment. *Endocrinology* 147: 1237–1245.

Zheng S, Turner TT, Lysiak JL. 2006. Caspase 2 activity contributes to the initial wave of germ cell apoptosis during the first round of spermatogenesis. *Biol Reprod* 74: 1026–1033.

Zhu S, Stavrovskaya IG, Drozda M, Kim BYS, Ona V, Li M, Sarang S et al. 2002. Minocycline inhibits cytochrome c release and delays progression of amyotrophic lateral sclerosis in mice. *Nature* 417: 74–78.

Chapter 6

The Epididymis: Structure and Function

Mohammad A. Akbarsha, Kunnathodi Faisal and Arumugam Radha

Contents

6.1 Introduction 116
6.2 Structure of Epididymis 117
 6.2.1 Gross Structure 117
 6.2.2 Epididymal Segments and Their Roles 117
 6.2.3 Cell Types in the Epididymal Epithelium 120
  6.2.3.1 Principal Cell 120
  6.2.3.2 Apical Cell 123
  6.2.3.3 Narrow Cell 124
  6.2.3.4 Clear Cell 125
  6.2.3.5 Basal Cell 126
  6.2.3.6 Halo Cells or Intra-Epithelial Lymphocytes/Intra-Epithelial Macrophages 127
  6.2.3.7 Pale Vacuolated Epithelial Cell 128
 6.2.4 Blood–Epididymis Barrier 130
 6.2.5 Epididymal Lumen 130
6.3 Functions of Epididymis 131
 6.3.1 Protein Secretion and the Roles of the Proteins 131
  6.3.1.1 Proteins which Associate with the Sperm 131
  6.3.1.2 Proteins Added to the Sperm Surface during Transit through the Epididymis 132

6.3.1.3 Sperm Surface Proteins Modified during Epididymal Transit 132
6.3.1.4 Sperm Surface Proteins which Migrate to Different Locations during Epididymal Transit 133
6.3.1.5 Semen Coagulum Proteins 134
6.3.2 Lipid Changes in Sperm during Epididymal Transit 135
6.3.3 Apocrine Secretion at Epididymis 135
6.3.4 Role in Constantly Changing Luminal Microenvironment, Segment-Specific Gene Expression and Fluid Resorption 137
6.3.5 Role in Protection of Luminal Sperm from the Adverse Reactive Oxygen Species 139
6.3.6 Role in Protection of Sperm: Antimicrobial Proteins 139
6.3.7 Role in Sperm Quality Control 140
6.3.8 Role in Passage of Sperm to Distal Parts 141
6.3.9 Role in Sperm Storage 142
6.4 Regulation of Epididymis Structure and Function 142
6.5 Organ and Cell Culture of the Epididymis 143
6.5.1 Organ Culture 143
6.5.2 Cell Culture 144
6.6 Epididymis and Male Fertility Problems 145
6.7 Epididymis as a Target to Toxicants 146
6.8 Epididymis as a Target to Male Contraceptive Attack 147
6.9 Conclusions 150
Acknowledgements 150
References 151

6.1 Introduction

The epididymis, also known as De Graaf's thread (named after Regnier De Graaf who was the first to uncoil the human epididymal duct), lies adjacent and adherent to the testis. Though apparently simple, the mammalian epididymis is a highly complex organ. This complexity is contributed by the heterogeneity of the cell types lining its lumen, the distribution, structure and function of which vary along the length of the duct. As a result, different segments are recognised, each playing different or overlapping roles. On the other hand, this complexity contributes to varied functions in relation to physiological maturation of the spermatozoa, which is the major role of the epididymis. This epididymal maturation process is a prerequisite for the sperm to acquire the ability to be motile and to fertilise the oocyte under physiological condition. According to Cooper (2007), every sperm function required for fertilisation seemed to be developed in the epididymis: motility, zona binding and membrane fusion. The epididymis has gained importance in view of the basic science pertaining to the intricacies in the post-testicular sperm maturation process, and the applied science of (1) it as a target to toxicants affecting male fertility, (2) its contribution to the increasing trend of male infertility and (3) it as a target for male contraceptive development.

6.2 Structure of Epididymis

6.2.1 Gross Structure

The epididymis is a single convoluted duct, the ductus epididymidis, connecting the vasa efferentia, which originate from the rete testis, to the vas deferens. It is lodged in the scrotum in scrotal mammals and lies subjacent to the testis on each side. Medially, the epididymis is attached to the testis by the epididymo-testicular connective tissue, and distally by both the caudal connective tissue and the epididymal fat pad. The ductus epididymidis measures from 3 to 80 m in different mammals. It is a highly coiled, contorted and convoluted duct and gets surrounded by a connective tissue capsule to form the organ, the epididymis, with septulae that divide it into a number of segments that are continuous along the length. The epididymis is derived from the Wolffian duct. At birth it consists mainly of mesenchymal tissue. During postnatal development, the epididymis undergoes considerable remodelling including duct elongation and convolution. By puberty the epididymis has acquired its fully differentiated state consisting of a highly tortuous duct lined by epithelial cells (Rodríguez et al. 2002).

6.2.2 Epididymal Segments and Their Roles

Investigations on the epididymal histology started in the mid-nineteenth century, and since then scientists have attempted to subdivide the epididymis into different segments basing their judgment on anatomy, histology and cytology (Figures 6.1 and 6.2). The epididymis was first divided into three anatomical segments: the head lying on the top of the testis, the body lying along the side, and the tail lying at its posterior aspect. They are respectively called caput, corpus and cauda of the epididymis (Hermo 1995). An initial segment (IS) as the earliest segment, lying ahead of caput, was discovered later (Robaire and Hermo 1988). Another segment, known as intermediate zone (IZ), between the IS and caput, was discovered in the rat (Hermo et al. 1991a). The initial segment and intermediate zone have not been reported in the human epididymis (Jelinsky et al. 2007). Several investigators have found more major or subtle differences in the histological organisation along the length of the ductus epididymidis and have identified various zones/segments in a variety of mammalian species and presented them in a numerical order (Nicander 1957, 1958; Reid and Cleland 1957; Holstein 1969; Nicander and Glover 1973; Jones et al. 1979; Orsi et al. 1980; Johnston et al. 2005). The various segments differ in terms of diameter of the duct, height of the single layer of pseudo-stratified epithelium and the cells that constitute it. In general, the diameter of the duct and

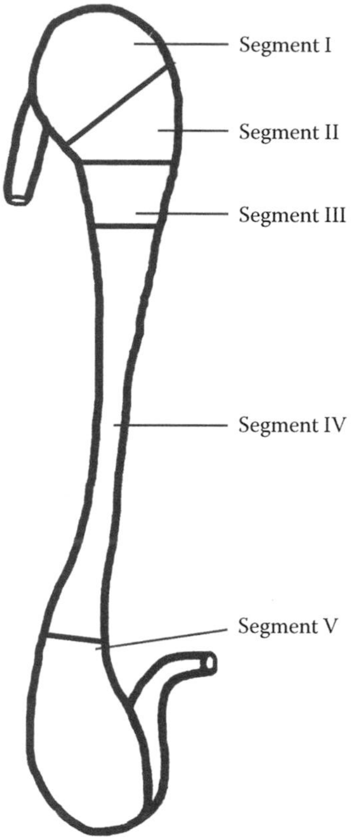

Figure 6.1 Diagram of mouse epididymis showing segments I (initial segment), II (intermediate zone), II (caput), IV (corpus) and V (cauda). Not drawn to scale.

size of the lumen increase from IS to cauda where as the height of the epithelium decreases (Figure 6.2). The distribution of the cell types in the epithelium is also dependent on the segment.

The anatomical/histological segmentation of the epididymis has a bearing on its role in the processing of the spermatozoa. The epididymis by and large performs such roles as sustenance and protection of spermatozoa, contribution to maturation and storage of spermatozoa and transport of spermatozoa to the distal parts. The IS is the most critical part of the epididymis contributing mainly to reabsorption of fluid arriving from the testis through the ductuli efferentes, leading to concentration of the spermatozoa such that the number of spermatozoa per milliliter of epididymal fluid is increased several times and the fluid content is correspondingly decreased (Da Silva et al. 2006; Belleannée et al. 2011; Dacheux et al. 2012). There is also considerable change in the ionic composition of the luminal fluid (Wong et al. 1981; Clulow et al. 1998). The IS secretes several proteins, glycoproteins and small molecular weight compounds (Hinton and Hernandez 1987; Robaire and

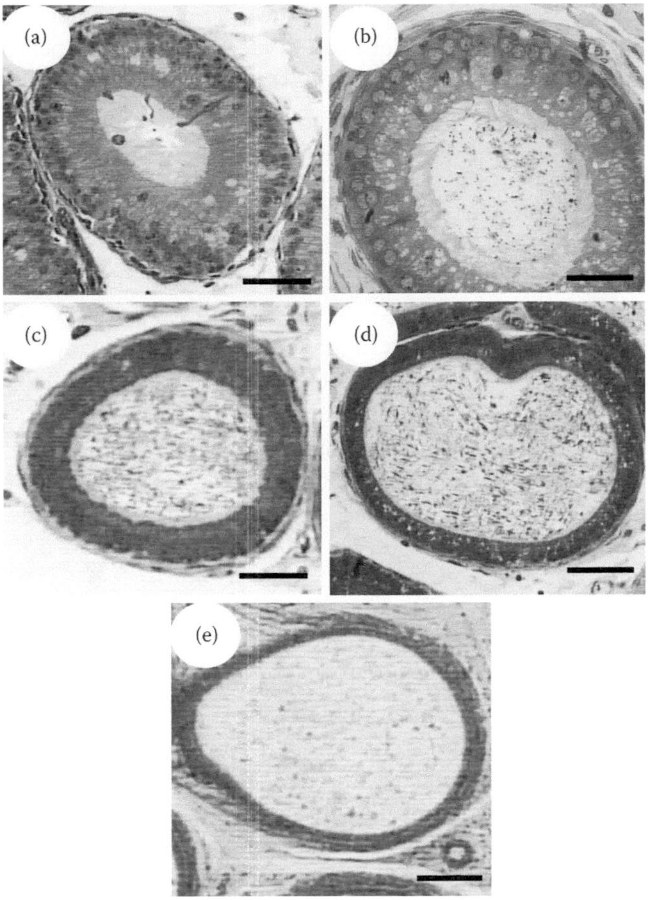

Figure 6.2 (See colour insert.) T.S. of different segments of rat epididymis. (a) Initial segment, (b) intermediate zone, (c) caput, (d) corpus, (e) cauda. Scale bar (a) 20 µm, (b, c, d) 18 µm, (e) 15 µm.

Hermo 1988; Cooper 1998a; Dacheux et al. 1998). There is also endocytic uptake of particulate as well as dissolved material from the lumen (Robaire and Hermo 1988; Hermo et al. 1991b, 1991c, 1992). The spermatozoa, thus concentrated and lying in a medium modified in ionic composition and containing the secretory products of the IS, arrive at the IZ, which is concerned with secretion as well as endocytosis (Hermo 1995). At the caput the epithelium further secretes proteins, glycoproteins and small molecular weight compounds many of which may be different from those secreted by the earlier segments. The spermatozoa interact with these proteins and small molecular weight compounds in various ways, which include addition of new proteins to the sperm surface, deletion of some of the existing proteins, translocation of the existing proteins to different sites and modification of the existing proteins (Cooper 1998a). In the process, the spermatozoa undergo considerable changes in the surface domains (Robaire and Hermo 1988). It is believed that these changes to the spermatozoa are essential for them to acquire motility and fertilising ability. The spermatozoa thus altered arrive at the corpus where the role of epididymal epithelium is not adequately known, but the interaction between spermatozoa and the luminal content towards the physiological maturation of the sperm continues (Brooks 1987; Yeung et al. 1997). The cauda is essentially considered as an organ of storage of sperm until ejaculation though it may also be concerned with cleansing the dead, defective and aged spermatozoa and the cytoplasmic droplets shed by the spermatozoa.

Thus, the epididymal duct is a highly ordered and segmented organ. Each of these segments represents a unique physiological compartment. Each compartment possesses distinctive gene expression profiles within the epithelium that result in segment-specific secretion of proteins into the luminal fluid directly or indirectly affecting sperm maturation (Cornwall 2009). In the human epididymis there is an initial segment-like tall epithelium but no specific initial segment region is present (Yeung et al. 1991).

6.2.3 Cell Types in the Epididymal Epithelium

The various epithelial cell types lining the duct are responsible for the histological differentiation and the changing luminal microenvironment of the ductus epididymidis. The pseudo-stratified epithelium of the ductus epididymidis is formed of several cell types such as principal cell (PC), narrow cell (NC), apical cell (AC), clear cell (CC), basal cell (BC) and intraepithelial leukocyte cell populations, named by the various authors as halo cells, intraepithelial lymphocytes (IELs) or intraepithelial macrophages (IEMs). Yet another cell type emphasised recently is pale vacuolated epithelial cell (Agnes and Akbarsha 2001).

6.2.3.1 Principal Cell

Principal cells (PCs) are the most abundant cell type in the epididymal lining, making up about 80% of the total cells in IS, IZ and caput, and declining to 69% in corpus, and 65% in cauda (Robaire and Hermo 1988). They are tall columnar cells, spanning the entire height of the epithelium (Figures 6.3 and 6.4), but decrease in height from initial segment to cauda. They are densely stereociliated towards the apical border, with the stereocilia decreasing in abundance and height from initial segment to cauda. From the basal portion of the stereocilia, coated and uncoated pits of the apical plasma membrane are produced. Laterally, the neighbouring cells establish intercellular association consisting of zonula occludens subjoining the apical border, which constitutes the blood epididymis barrier, followed by a series of zonula adherentes forming the tight junctions though there could still be intercellular spaces between the PCs. Basally the cells rest on a basement membrane.

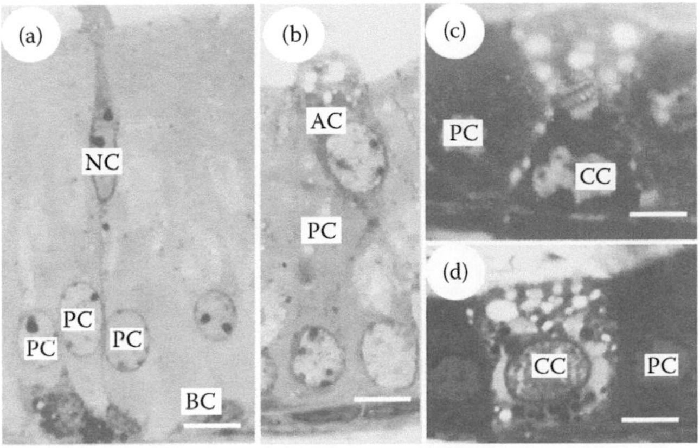

Figure 6.3 (See colour insert.) Epithelial organisation of mouse epididymis. LM photomicrographs. (a) Initial segment, (b) intermediate zone, (c) proximal cauda, (d) distal cauda. AC, apical cell; BC, basal cell; CC, clear cell; NC, narrow cell; PC, principal cell. Scale bar 4 μm.

The ultrastructural organisation of the PC and its variation in relation to the segmentation of the epididymis has been studied extensively by several investigators, and reviewed by Robaire and Hermo (1988) and others (Cyr et al. 1995; Hermo et al. 1998; Hermo and Robaire 2002; Sever 2010). The prominent spherical nucleus is located invariably in the basal one-third of the cell (Figures 6.4 and 6.5). Immediately apical to the nucleus is the prolifically developed Golgi apparatus, which in the IS, IZ and caput is arranged as several vertical stacks of sacs and discs (Figure 6.5b). There are smooth surfaced vesicles in the trans-face of the Golgi stacks. These are the secretory vesicles meant for intra-/extra-cellular transport of the secretory material. The perinuclear cytoplasm including the Golgi area has abundant endoplasmic reticulum, which are granular in the perinuclear zone, sparsely granular in the Golgi area and agranular (smooth ER) in the apical cytoplasm. Mitochondria are abundant. There are membrane-bound electron-dense granules in the basal cytoplasm. There are also secretory granules. These granules occur either individually or in clusters formed by the coalescence of their bounding membranes. Some of the electron-dense granules are attached to the plasma membrane by tubular extensions of their surrounding membrane. The contents of the electron-dense granules are believed to be discharged into the intercellular space. Lysosomes are variable in abundance, and are very high in the corpus.

The other diagnostic ultrastructural features of PCs relate to the membrane-bound vesicles – coated as well as uncoated – containing the material endocytosed from the lumen (the pinocytic vesicles) or from within, and the multivesicular bodies (Robaire and Hermo 1988; Goyal and Williams 1991). The apical region of the principal cell possesses an endocytic apparatus with endosomes, multivesicular bodies, lysosomes, coated and uncoated pits and vesicles. Probably, the coated pits

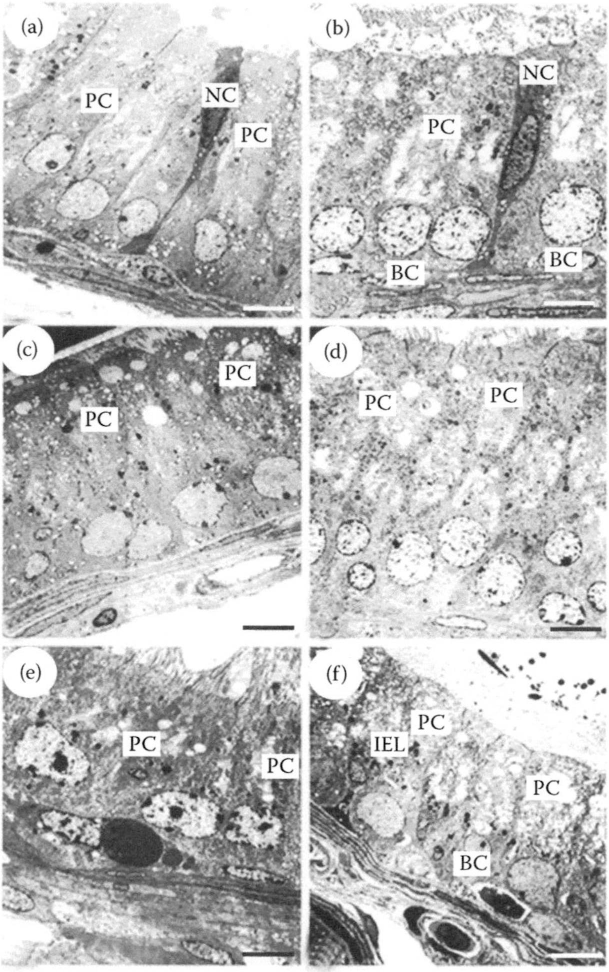

Figure 6.4 TEM of mouse epididymal epithelim. (a) Initial segment, (b) intermediate zone, (c) caput, (d) corpus, (e) proximal cauda, (f) distal cauda. BC, basal cell; IEL, intra-epithelial leukocyte; NC, narrow cell; PC, principal cell. Scale bar (a, c) 6 μm, (b, d) 4 μm, (e) 3 μm, (f) 5 μm.

became coated vesicles in the cytosol. The coated pits are seen in the luminal border. A coated pit is braced by a coat of protein molecules and bears surface receptors that bind specific extracellular ligands. In most cases the coat protein is clathrin. Resorption is an important aspect of epididymal control of the intraluminal environment, and many proteins are internalised by cells through receptor-mediated endocytosis (Ramos-Ibeas et al. 2013). Here, the protein binds to a specific receptor on the cell surface and is, subsequently, internalised through clathrin-coated pits along the plasma membrane (Schimming and Vicentini 2008).

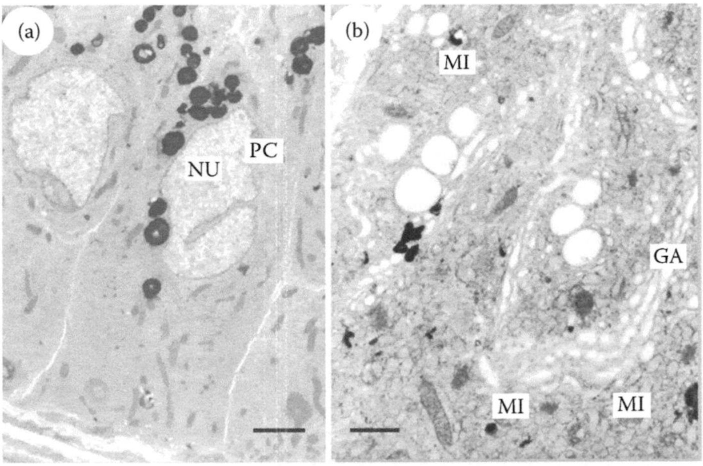

Figure 6.5 Ultrastructural organisation of rat epididymal epithelial principal cell. (a) Three principal cells in a row; (b) a portion of a principal cell showing the prolific nature of Golgi apparatus. GA, Golgi apparatus; MI, mitochondria; NU, nucleus; PC, principal cell. Scale bar (a) 2 μm, (b) 0.5 μm.

The PC is attributed with the following important roles in respect of processing of spermatozoa and modulation of luminal content, in a segment-specific manner:

- Secretion of proteins, simple as well as conjugated, most of which are associated with physiological maturation of spermatozoa
- Besides the classical exocrine secretory process, contributes to the apocrine secretion, which contains small membranous vesicles, the epididymosomes, proteins from which are added onto spermatozoa and some are concerned with ubiquitination and/or removal of dead/defective spermatozoa and the proteins released from them
- Fluid-phase as well as receptor-mediated endocytic uptake of materials from the lumen
- Absorption of fluid and secretion from/into the lumen thereby modifying the composition of luminal fluid and concentration of spermatozoa
- Contribution to the constantly varying luminal microenvironment
- Phagocytosis of dead and/or defective spermatozoa, etc.

6.2.3.2 Apical Cell

Apical cells (ACs) (Robaire and Hermo 1988) or apical mitochondria-rich cells (Palacios et al. 1991) are confined to the initial segment (Clermont and Flannery 1970; Sun and Flickinger 1979) or present even in the intermediate zone (Adamali and Hermo 1996; Adamali et al. 1999a). The goblet-shaped ACs (Figure 6.6a) comprise about 10.7% of the total epithelial population in the proximal IS, but only about

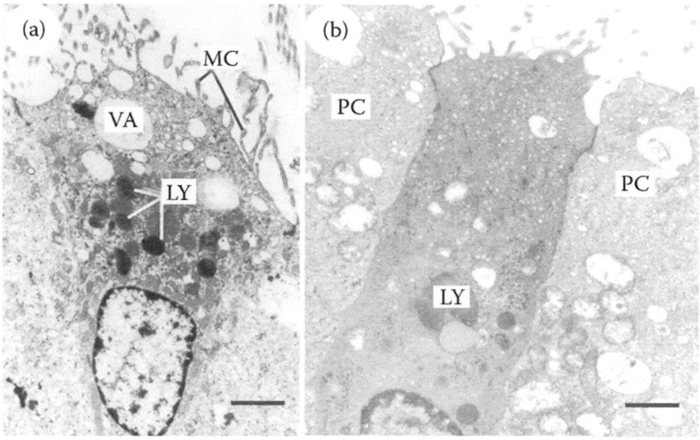

Figure 6.6 TEM of apical (a) and clear cells (b) of rat epididymis. MC, cytoplasmic flaps; VA, vacuoles; NU, nucleus; LY, lysosomes. Scale bar 2 μm.

1.3% in the IZ of rat epididymis (Adamali and Hermo 1996). They are similar to PCs but can be distinguished from the location of their round or oblong nuclei, which are in the upper half of the cells, and the dome-shaped luminal border with a few or without stereocilia but with luminal flaps and folds. The cytoplasm is pale, with numerous mitochondria (hence the name 'mitochondria-rich cell') and vacuoles of varying sizes in the apical cytoplasm. In the rat, these cells possess a few C-shaped vesicles (Robaire and Hermo 1988; Akbarsha et al. 2000). Some of these cells protrude deep into the lumen and such protrusions are often pinched off into the lumen (Akbarsha et al. 2000). Pinocytic/coated vesicles are absent (Goyal and Williams 1991).

ACs are intensely reactive for Yf-subunit of glutathione-S transferase, suggesting a role for ACs in protection of luminal sperm from the attack of electrophiles (Adamali and Hermo 1996). In possessing carbonic anhydrase activity and vacuolar proton-pumping adenosine tri-phosphatase activities, the ACs may be concerned with luminal acidification (Brown et al. 1992; Cooper 1998b). The β-hexosaminidase and cathepsin D activities indicate a role in lysosomal degradation of endocytosed proteins (Adamali and Hermo 1996). In mouse epididymis these cells are involved in the synthesis of a glycoprotein, which binds to sperm tail and prevent tail-to-tail sperm agglutination (Feuchter et al. 1988). Immunocytochemical studies indicate ACs to possess positive sites for estradiol receptors and cytokeratin. Mitochondrial aggregation in the apical cytoplasm suggests these cells play a role in the generation of ATP that is required for the transport of H^+ and Cl^- ions across the cell membrane, which is an aspect of the epididymal sperm maturation process (Goyal and Williams 1991).

6.2.3.3 Narrow Cell

Narrow cells (NCs) are found only in the initial segment (Serre and Robaire 1998) or in the intermediate zone also (Adamali and Hermo 1996) and in the latter case they increase in counts from 2.8% in the IS to 6.3% in the IZ. These cells (Figure 6.4a,b) are identified by their deep-staining cytoplasm, dense elongated nucleus located in the upper half of the cell and an ill-defined narrow base contacting the basement membrane through a peduncle (Sun and Flickinger 1979; Adamali and Hermo 1996). Clermont and Flannery (1970) considered NCs as non-stereociliated cells. They are characterised by electron-dense cytoplasmic matrix containing numerous cup-shaped vesicles in the most apical cytoplasm and abundant mitochondria and multivesicular bodies in the supranuclear cytoplasm (Sun and Flickinger 1979). They may be concerned with degradation of proteins within the lysosomes, protecting the spermatozoa from the changing environment of harmful electrophiles, modification of pH in the lumen of the ductus epididymidis and maintenance of a quiescent state in the spermatozoa (Adamali and Hermo 1996). NCs secrete protons via the vacuolar H+-ATPase (V-ATPase) and contribute to the acidification of the lumen (Shum et al. 2009). NCs are known to be endocytic, but little is known of what they endocytose; the intense localisation of the lysosomal enzyme hexosaminidase, which is a marker of NC, is a clear indication of a role for this cell in endocytosis (Adamali and Hermo 1996; Adamali et al. 1999a,b).

6.2.3.4 Clear Cell

Clear cells (CCs) are found in the corpus and cauda epididymides, and have a very pale-staining cytoplasm. They are highly vacuolated in the apical region and possess dense granules above the nucleus (Figure 6.7). The basal region contains pale or moderately dense bodies. Nucleus is variable in position, round, pale staining and with a prominent nucleolus. According to Abe et al. (1984) and Akbarsha and

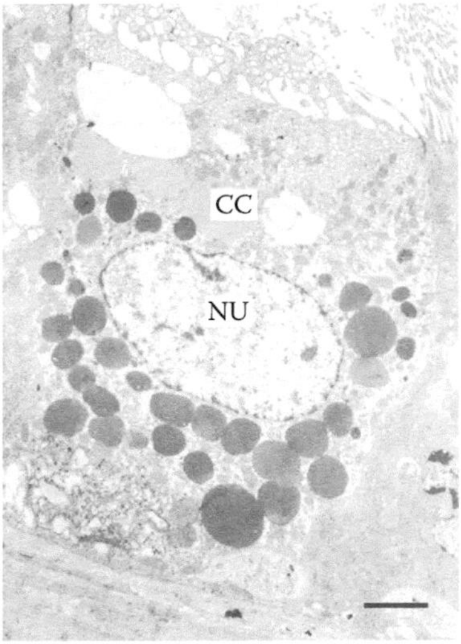

Figure 6.7 TEM of clear cell (CC) of rat epididymis. NU, nucleus. Scale bar 1 μm.

Averal (1999a), the CCs are stereociliated whereas according to Cooper (1999), they are not stereociliated. Păunescu et al. (2014), however in a very recent study adopting high-resolution helium ion microscopy, have found the CCs under control condition to be microvillated and also to possess relatively short microplicae or small membrane ruffles. CCs play a role in the removal of the disintegration products of the cytoplasmic droplets released from the sperm, which are further acted upon by the lysosomal enzymes for their ultimate removal (Hermo et al. 1998; Akbarsha and Averal 1999b). They may also be concerned with fluid-phase endocytosis of proteins and their further processing through the lysosomes (Cooper 1998b). The carbonic anhydrase of the CCs may be concerned with acidification of the lumen (Cooper 1998b; Pastor-Soler et al. 2008) in which H(+)-ATPase, with its different but specific isoforms, would play a critical role (Shum et al. 2009; Zuo et al. 2010).

6.2.3.5 Basal Cell

Basal cells (BCs) are the second largest cell population of the epididymal epithelium along the entire length (Figure 6.8a,b). They are flat, elongated or triangular in shape and form a network beneath the principal cells (Yeung et al. 1994). In the adult rat they have been seen to extend slender processes towards the lumen but during postnatal development, especially PD 7, these processes contact the lumen (Shum et al. 2008, 2009, 2013, 2014). In the epididymis in which luminal acidification is crucial for sperm maturation and storage, these projections contain the angiotensin II type 2 receptor (AGTR2). Activation of AGTR2 by luminal angiotensin II increases proton secretion by adjacent clear cells which are devoid of AGTR2 (Shum et al. 2008), probably with a role to vacuolar H^+-ATPase or V-ATPase (Shum et al. 2013). BCs are reactive for glutathione S-transferase (GST) and Cu-Zn superoxide dismutase (SOD). Therefore, BCs may play a role in detoxification mechanisms (Nonogaki et al. 1992; Veri et al. 1993). Lysosomes filled with lipofuscin pigment in these cells suggest that these cells have a role in scavenging

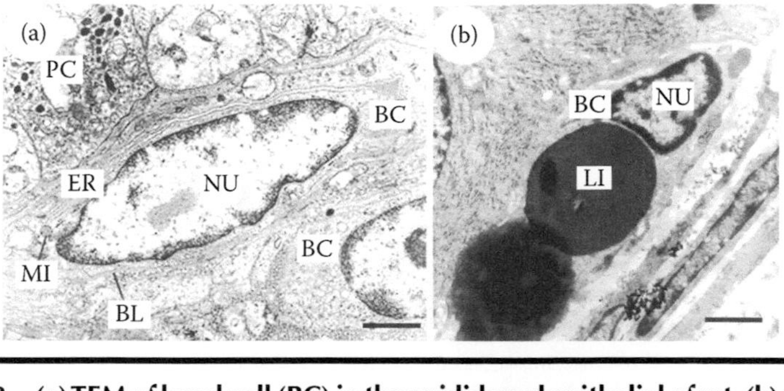

Figure 6.8 (a) TEM of basal cell (BC) in the epididymal epithelial of rat. (b) A portion of a basal cell showing lipofuscin inclusions. BL, basal lamina; ER, endoplasmic epithelium; LI, lipfuscin; MI, mitochondria; NU, nucleus. Scale bar 0.8 μm.

reactive oxygen species (Bajpai et al. 1985; Yeung et al. 1994). In the murine and human epididymis, BCs express macrophage antigens (Yeung et al. 1994; Seiler et al. 1998). Through local formation of prostaglandins, BCs may regulate electrolyte and water transport by the PCs. This regulatory process involves two proteins which are exclusively expressed by the BCs. They are the transient receptor potential (Trp) proteins, which serve as transmembrane pathways for Ca^{2+} influx, and cyclooxygenase 1 (COX-1), a key enzyme in the formation of prostaglandins. These two proteins would play a key role in the integrated functions of BCs as humoral regulators of PCs (Leung et al. 2004).

In recent years, several reports have ascribed a role to epididymal epithelial BCs in surveillance of sperm antigens when there is obstruction to passage of sperm (Yeung et al. 1994; Seiler et al. 1998, 1999, 2000; Holschbach and Cooper 2002; Aruldhas et al. 2006). Obstruction of sperm passage leads to increase in the number of BCs and a dense accumulation of lipofuscin (LF) material in their cytoplasm. These are indications of arrival of sperm antigens into the epithelium through transcytosis across the PCs and also through leaky junctions between the tall columnar cells. These changes result in increased abundance of BCs and also in increased content of LF material in them (Figure 6.8b). Apparently, the sperm antigens are acquired by the PCs and processed into LF material, which is then acquired by the BCs and accumulated as dense bodies (Seiler et al. 2000; Aruldhas et al. 2006).

6.2.3.6 *Halo Cells or Intra-Epithelial Lymphocytes/Intra-Epithelial Macrophages*

Intra-epithelial lymphocytes (IELs) and intra-epithelial macrophages (IEMs), together known as halo cells, are found at all levels of the epididymal epithelium (Figure 6.9a,b). The nature of these cells has been a subject of controversy since

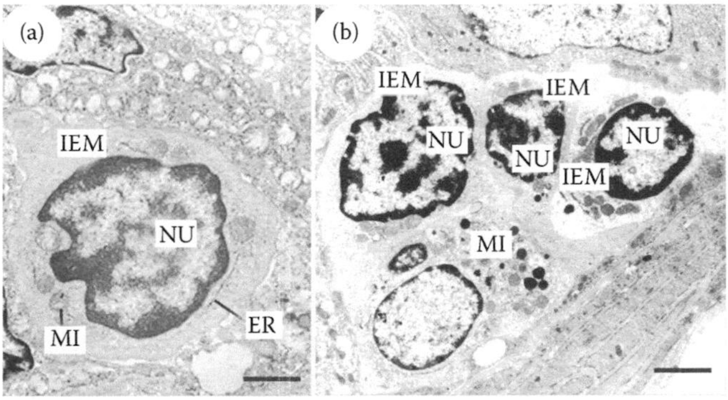

Figure 6.9 TEM of intra-epithelial macrophages (IEM) in the epididymal epithelial of rat. (a) A single IEM in a control animal; (b) a group of IEMs in a toxicant-treated animal. BL, basal lamina; ER, endoplasmic epithelium; MI, mitochondria; NU, nucleus. Scale bar (a) 0.5 μm, (b) 0.6 μm.

their discovery by Reid and Cleland (1957). They can be distinguished by the dense nucleus with patches of peripheral condensed chromatin and pale-staining cytoplasm (Robaire and Hermo 1988). They are migratory cells and found at different heights along the epithelium in clear spaces. They are believed to form an immunological barrier (Wang and Holstein 1983; Robaire and Hermo 1988). The IELs and IEMs differ in their location in the epithelium. According to Flickinger et al. (1997), IELs in the epididymal epithelium are $CD4^+$ and $CD8^+$ leukocytes and, thus, lymphocyte derivatives, and those in the interstitium are macrophages expressing Mac 1 and Mac F4/80 antigens and, thus, monocyte derivatives (Seiler et al. 1998). Recently, antibodies labelling the three main types of immunocompetent cells have been used to resolve the distribution of T lymphocytes, B lymphocytes, monocytes and macrophages in the reproductive tract. The ED1 antibody distinguishes an intracytoplasmic antigen in monocytes, tissue macrophages and free macrophages (Damoiseaux et al. 1994). In the epididymal epithelium of young Brown Norway rats, the number of cells that stain for antibody against monocyte-macrophages ($ED1^+$), helper T lymphocytes ($CD4^+$) and cytotoxic T lymphocytes ($CD8^+$) is equivalent to the number of halo cells (Serre and Robaire 1999).

6.2.3.7 Pale Vacuolated Epithelial Cell

Pale vacuolated epithelial cells (PVECs) have been reported in the epididymal epithelium of normal mice aged 8–16 weeks (Toshimori et al. 1990) and arise as a result of ligation of the epididymal duct at the junction between segments III and IV of 20–30-day-old mice (Abe et al. 1982). Studies of the effect on the epididymis of androgen deprivation in goats (Goyal et al. 1994) and effect on the epididymis of mice receiving high-dose testosterone implants (Itoh et al. 1999) also showed development of such cells. Treatment of mice with aflatoxin B1 (AFB1) resulted in the development of small or large vacuoles in the epithelial lining of all segments of the epididymis (Figure 6.10a,b,c). Such vacuoles are enclosed in large pale epithelial cells, which are quite different in organisation from the other epididymal epithelial cell types. The lumen of the vacuole contains spermatozoa and debris or an amorphous to dense PAS-positive material, or all three materials. Short microvilli extend from the pale epithelial cell into the vacuolar space. The vacuole appears to arise as a result of the degeneration of a principal cell that leads to fistula formation, during which the contents of the ductal lumen and the principal cell fistula merge and spermatozoa from the ductal lumen enter into the fistula. The neighbouring intact principal cells bend over the degenerating principal cell, cutting off its continuity with the ductal lumen. A pale epithelial cell that apparently develops from a basal cell presumably encloses the disintegrating principal cell including the spermatozoa (Agnes and Akbarsha 2001).

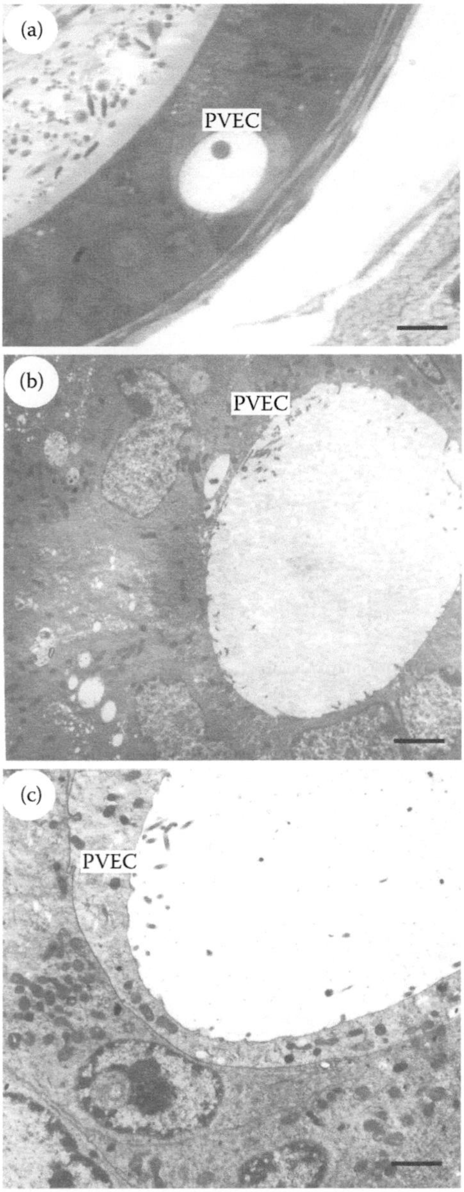

Figure 6.10 (See colour insert.) (a) LM and (b, c) TEM of pale vacuolated epithelial cell (PVEC) in rat epididymal epithelium.

6.2.4 Blood–Epididymis Barrier

Since the intraluminal environment of the epididymis is the site of spermatozoa maturation, it requires strict regulation and control which are taken care of by the tight junctions between tall columnar epithelial cells by the formation of the blood–epididymis barrier. The junctional complexes between adjacent epididymal tall columnar cells are composed of an apically located gap, adherens and tight junctions. The tight junctions form the blood–epididymis barrier, which plays a critical role in secluding sperm antigens from the blood under physiological conditions so as to prevent autoimmune response to the sperm antigens. However, epididymis has also developed other defence mechanisms to selectively permeate the compounds to enter into the lumen for the protection of sperm from the immune system, xenobiotics and reactive oxygen species. The mechanism by which the epididymis renders protection to the sperm may include synthesis and secretion of specific proteins, antioxidants and conjugating enzymes. Gap junctions make the luminal compartment and the intercellular spaces as two distinct physiological compartments by sealing the spaces between the epithelial cells at their apical aspect, thereby aiding in the communication between continuous zonule around the cell. In the initial segment, the junctional complex between the principal cells is mainly comprised of a tight junction that spans a considerable length of the adjacent plasma membranes, but contains relatively few desmosomes, while in the other regions of the epididymis, the extent of the tight junctions is reduced and there are more desmosomes (Robaire and Hermo 1988; Robaire et al. 2006; Mital et al. 2011).

6.2.5 Epididymal Lumen

The lumen of the epididymis contains spermatozoa and a fluid arriving from the testis, adequately modified during transit through the efferent ductules (Hess 2002), and the secretory products of the epididymal epithelium. The composition of the luminal fluid with respect to ions, small organic molecules, proteins and other large macromolecules has clearly demonstrated that the environment to which spermatozoa are exposed during their epididymal transit is continuously undergoing major changes (Hinton and Palladino 1995; Dacheux et al. 1998; Wong et al. 2002; Cooper 2007; Dacheux and Dacheux 2014). The region-specific gene expression in the epididymal epithelium may contribute to such continually changing microenvironment (Kirchhoff 1999; Jervis and Robaire 2001; Yamazaki et al. 2006) which plays a major role in sperm maturation (Hinton and Palladino 1995; Leung et al. 1998; Fouchécourt et al. 1999; Jensen et al. 1999). The epididymal tubule and its luminal compartment are like no other extracellular environment in the body in that sperm maturation must occur despite a surrounding environment that favours protein misfolding and aggregation. The unusual composition and extraordinary levels of some components in the epididymal fluid might represent an extracellular environment that has gone to extreme measures to control problems/toxicities that are known to exist with aggregate structures.

6.3 Functions of Epididymis

6.3.1 Protein Secretion and the Roles of the Proteins

The importance of understanding epididymal function and sperm maturation is emphasised by the fact that up to 40% of infertile men exhibit idiopathic infertility that may reflect sperm maturational disorders. The significance of the lack of understanding regarding the role of epididymis in sperm maturation is also underscored by the lack of contraceptives for men. In this connection, interest has been evinced to identify epididymal molecules that could serve as targets for non-steroidal-based male contraceptives with the idea that sperm production would occur normally, but the spermatozoa would be non-functional (Cornwall 2009). The principal mechanism by which the epididymal epithelium contributes to sperm maturation is through secretion of hundreds of proteins and other molecules which play different roles towards the physiological maturation of spermatozoa. The secretory substances arrive at the lumen in two routes: merocrine and apocrine. The merocrine route is classical and involves regulated gene expression as transcripts, their translation, post-translational modification in the Golgi apparatus via endoplasmic reticulum and generation of secretory vesicles that will release the proteins into the epididymal lumen by exocytosis. In the second and more recently discovered route the proteins are assembled in small membrane-bound vesicles called aposomes, which form apical blebs towards the lumen and are discharged in an apocrine mechanism. Once in the luminal fluid the aposomes disintegrate and release small membrane vesicles called epididymosomes which contain proteins and other molecules for delivery to defined destinations (Belleannée et al. 2013; Sullivan and Saez 2013).

6.3.1.1 Proteins which Associate with the Sperm

During epididymal maturation the spermatozoa undergo considerable changes in the surface domains (Hamilton et al. 1986; Eddy and O'Brien 1994; Cooper et al. 1998). Remodelling processes include uptake of secreted epididymal glycoproteins, removal or utilisation of specific phospholipids from the inner leaflet of the bilayer, processing of existing or acquired glycoproteins by endoproteolysis, and repositioning of both protein and lipid components to different membrane domains (Cooper 1998b; Jones 1998). It is believed that these changes to the sperm are essential for them to acquire motility and fertilising ability. Epididymal proteins that are known to interact with spermatozoa may be involved in their maturation as inferred from studies in experimental animals, and these proteins include CRISP family proteins, P26h, P34h, SPAG11, Eppin and others (Pujianto et al., 2013). Though hundreds of proteins have already been identified from different species, about 20 proteins constitute 80%–90% of the total epididymal luminal proteins and many of these are common between different species (Dacheux and Dacheux 2014). Several epididymal luminal proteins that have implication in sperm maturation have been identified in the lumen of the human epididymis including ALB (43.8%), CLU (7.6%), NPC2 (6%), LTF (5.9%), extra-cellular matrix protein (3.2%), α1-antitrypsin (2.7%), PTGDS (3.9%), actin (1.2%) (Dacheux et al. 2006), the CRISP (D/E) proteins (Krätzschmar et al. 1996), P34h short-chain dehydrogenase/reductase (Boué et al. 1994; Boué and Sullivan 1996), β-hexosaminidase (Miranda et al. 1995), clusterin (O'Bryan et al. 1994), ADAM 7 (Liu et al. 2000) and others (Lasserre et al. 2001). CLU is the most commonly secreted epididymal protein and it constitutes around 30% of the total epididymal secretions (Dacheux and Dacheux 2014). By the use of a tissue-specific cloning strategy, several novel genes were first identified in the human epididymis, which were then studied further in other mammalian species. These include HE1 (encoding a putative cholesterol binding protein), HE2 (a β-defensin), HE3 (with unknown function), HE4 (a member of the four disulfide core or whey acidic protein [WAP] family of proteinase inhibitors), HE5, CD52 (implicated in human immunological infertility) and HE6 (a new member of the LNB-7TM subfamily of GPCR) (Kirchhoff 2002; Cornwall 2009).

6.3.1.2 Proteins Added to the Sperm Surface during Transit through the Epididymis

Mouse sperm maturation antigen 4 (SMA-4) is a surface-component of sperm tail, and may be secreted by principal cells of the distal caput epididymidis and bound to spermatozoa as they pass through that region of the duct. E-3, an epididymis-specific secretory glycoprotein, was detected in rat epididymal fluid and in sperm. The E-3 mRNA was only predominantly expressed in the corpus and cauda of the epididymis, but not in caput. The 27 kDa (EP-1) protein identified in chimpanzee cauda epididymal fluid is maturation-related protein that may be involved in the initiation of motility or in the attainment of fertilising capacity of sperm. In humans, the HE 1 mRNA is abundant in the epididymis. HE1 has been suggested to be associated with epididymal spermatozoa. The localisation of HE2 at acrosomal and equatorial regions of ejaculated sperm head suggests the involvement of this protein in sperm maturation. The other proteins that are added to spermatozoa during the epididymal transit include CRISP1, cathepsin, ADAM7, EPPIN, MAN2A2, SPAM1, MFGW8, GPX5, CLU and MIF (Dacheux and Dacheux 2014).

6.3.1.3 Sperm Surface Proteins Modified during Epididymal Transit

There are proteins the sperm carry from the testis which are modified during the epididymal transit. For example, PH-20 is initially present on the whole membrane and subsequently becomes restricted to a particular domain by some mechanism not yet defined. The 2B1 on rat spermatozoa is a homologue of mouse/human PH-20. The 2B1 is expressed post-meiotically in the testis as a precursor glycoprotein of approximately 60 kDa that first appears on the plasma membrane of stage 6 to 8 round spermatids. As spermatozoa pass through the caput epididymidis, 2B1 is endoproteolytically cleaved at a specific arginine residue (Arg 312) to produce a heterodimeric glycoprotein (40 kDa and 19 kDa) containing intra-molecular disulphide bridges (Jones et al. 1996). Rat sperm plasma membrane mannosidase is synthesised in the testis in an enzymatically inactive or less active prescursor form of 135 kDa, which is proteolytically processed to an enzymatically active mature form of 115 kDa during development in the testis and maturation in the epididymis (Tulsiani et al. 1995). Similarly, mouse sperm antigen M42, rat sperm glycoprotein CE9 and antigen 2B1 undergo proteolysis during sperm maturation. ADAM family members, fertilin and cyritestin, CE9, α-mannosidase and many others are modified such as by deglycosylation or proteolytic processing during epididymal transit (Cornwall 2009). The major maturation-dependent sperm membrane glycoproteins identified in five segments of primate epididymis include O-linked 170, 150, 86 and 60/58 kDa glycoproteins; N-linked 68, 56, 48 and 38 kDa glycoproteins; and N and O-linked 116 kDa glycoprotein, and all of these exhibited marked differences in the degree of glycosylation between immature and mature sperm surfaces (Srivastav 2000).

In addition, modifications of existing sperm proteins by phosphorylation, glycosylation, etc. are some of the post-translational modifications that sperm proteins undergo during epididymal transit which are important for sperm maturation and acquisition of fertilising ability (Cornwall 2014).

In rat epididymis, two proteins (D and E, Mr 27 and 28 kDa, respectively) are associated with spermatozoa during epididymal transit. Proteins D and E belong to the CRISP family and may be associated with surface receptors responsible for gamete recognition. The D/E molecules interact with specific membrane proteins, and are subsequently covalently bound to the surface of spermatozoa via a glycosylphosphatidyl inositol linkage. Two populations of D/E, a major, loosely bound population that is released during capacitation, and a minor strongly bound population that remains after capacitation which migrates to the equatorial segment (ES) with the acrosome reaction, and thus corresponds to the one with a role in gamete fusion (Cohen et al. 2000; Ellerman et al. 2002), remain covalently bound to spermatozoa even after deposition in the female reproductive tract, an observation which focuses their physiological function in the fertilisation process (Tubbs et al. 2002).

6.3.1.4 Sperm Surface Proteins which Migrate to Different Locations during Epididymal Transit

There are proteins which migrate to different locations. For example, fertilin from seven regions of epididymis showed migration from anterior head to posterior head after proteolytic processing of full-length fertilin ß processor (85 kDa pro ß-form) to 75 kDa intermediate proß to 25–28 kDa mature form (Hunnicutt et al. 1997). Acrin 1 is initially distributed throughout dorsal matrix in immature sperm but, during maturation, it becomes more restricted in the spherical bodies (Yoshinaga et al. 2000). The actin regulatory proteins (thymosh d10, destrin and a testis-specific actin-capping protein), involved in controlling the balance between actin monomers (G-actin) and actin filaments (F-actin), are localised to the acrosomal domain of bull sperm but during epididymal maturation they become confined to the equatorial segment. CE 9 is a domain-specific integral plasma membrane protein of rat sperm. On testicular spermatozoa CE 9 is concentrated within the posterior tail domain of the plasma membrane, whereas on vas deferens spermatozoa, it is concentrated within the anterior tail domain. SPAM1 and β-galactosyltransferase exhibit new localisation patterns during epididymal transit which may in part be triggered by proteolytic processing (Phelps et al. 1990). Suryawanshi et al. (2011) noticed a domain shift of protein UQCRC2 (ubiquinol-cytochrome-c reductase complex core protein 2) from the midpiece in testicular sperm to the principal piece in caudal sperm.

6.3.1.5 Semen Coagulum Proteins

The semen coagulum proteins semenogelin I and semenogelin II are degraded to low molecular mass fragments by prostate-specific antigen (PSA) during semen liquefaction. Recently, it has been shown that PSA is rendered active owing to chelation of inhibiting Zn^{2+} ions by the semenogelin molecules, which thereby initiate their own destruction. The major source of semenogelins is the seminal vesicle, but the proteins are also made in the epididymis, where they might regulate PSA and other peptidases that could be of importance during the maturation of spermatozoa. Studies on the evolution of semen coagulum proteins have revealed that most of them carry an exon that displays a rapid and unusual evolution. As a consequence, homologous proteins in rodents and primates show almost no conservation in primary structure. The progenitor of the semen coagulum proteins was probably a protease inhibitor that might have displayed antimicrobial activity. The semenogelin locus on chromosome 20 contains at least 17 related genes encoding probable protease inhibitors with homology to semen coagulum proteins. All of these are highly expressed in the epididymis where they, like the semenogelins, could affect the maturation of spermatozoa or display antibacterial properties (Lundwall 2007). Three recently identified genes – SPINT3, SPINT4 and SPINT5 – by their similarity in organisation, chromosomal location and site of expression are homologous with the genes of WFDC-type protease inhibitors and semen coagulum proteins, despite the lack of similarity in primary structure of their protein products. Their restricted expression to the epididymis hints at their role in sperm maturation (Clauss et al. 2011).

6.3.2 Lipid Changes in Sperm during Epididymal Transit

Major sperm membrane lipids comprise phospholipid (approx. 75% w/w), neutral lipid (approx. 15% w/w) and glycolipid (approx. 10% w/w). During epididymal maturation there is a significant decline in the total lipids, phospholipid and glycolipid contents of sperm membrane. On the contrary, the mature cauda-sperm membrane shows greater neutral lipid content than that of the immature caput sperm. Phosphatidylcholine (PC), phosphatidylethanolamine (PE) and sphingomyelin are the phospholipids of the sperm membrane, the former two being the major lipids. Phosphatidylethanolamine decreases most strikingly during sperm maturation. The sperm maturity is associated with marked increase of sterol and steryl ester and decrease of the other membrane-bound neutral lipids. The fatty acid profile of the various membrane lipids undergoes marked alteration during the epididymal transit of the sperm. Cholesterol, phospholipid and saturated and unsaturated fatty acids ratio increase greatly in the maturing sperm membrane (Gupta, 2005). Murine cauda spermatozoa, in comparison to caput spermatozoa, show higher degree of unsaturation and enhanced plasmalogen and lysolipid contents (Pyttel et al. 2014).

6.3.3 Apocrine Secretion at Epididymis

The epididymal epithelium, notably caput region, contributes to the apocrine secretion apart from its classical merocrine secretion (Hermo and Jacks 2002). This secretion pathway involves formation of apical cytoplasmic blebs (aposomes) that detach into the intraluminal compartment (Figure 6.11a,b). The apical blebs disintegrate, liberating their content which contains small membranous vesicles, the epididymosomes (Figure 6.11b, insert). Epididymosomes have been described in a number of mammalian species (Frenette and Girrouard 2007). Epididymosomes are mainly concerned with the transfer of various proteins that lack signal sequences to spermatozoa where the proteins are GPI anchored (Gatti et al. 2005; Sullivan et al. 2007; Cornwall 2009). Towards transfer of the protein content the epididymosomes may even fuse with the sperm plasma membrane (Păunescu et al. 2014).

Analysis of proteins associated with epididymosomes reveals protein profiles quite different from that of proteins in the lumen. Proteins associated with epididymosomes include P26h (believed to be involved in zona pellucida binding), HE5, macrophage migration inhibitory factor, ubiquitin and glutathione peroxidase, all of which have been shown to be transferred to spermatozoa in the epididymis (Frenette et al. 2003; Saez et al. 2003; Cornwall 2009). A recent study suggests that cathepsin D is also transported to the spermatozoa via epididymosome (Asuvapongpatana et al. 2013). It is possible that epididymosomes evolved to ensure safe delivery of some proteins to the sperm cells and perhaps to particular sperm domains without possible damage by luminal proteases. Alternatively, given

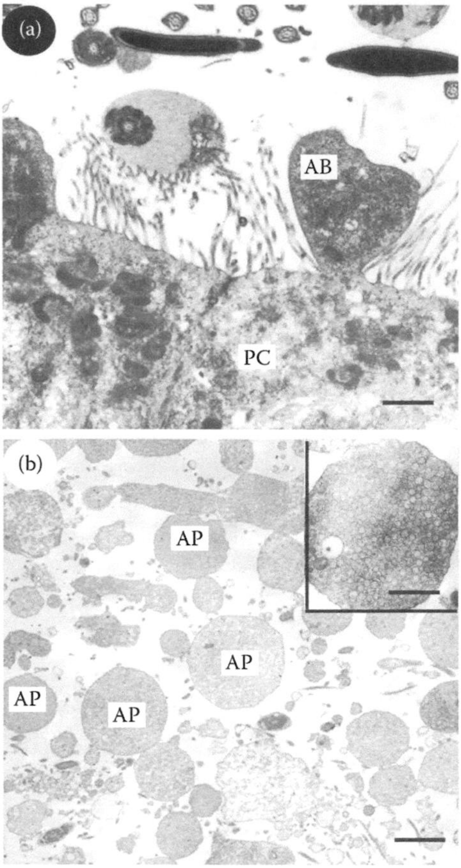

Figure 6.11 TEM of principal cell (PC) of rat epididymal epithelium showing (a) protrusion of apical bleb (AB) into the lumen and (b) the free aposomes (AP) in the lumen. Insert: An apsosome magnified showing epididymosomes. Scale bar (a, b) 0.4 μm; insert 0.2 μm.

the complexities of sperm maturation and the vast multitude of cellular and extracellular events that the epididymis must carry out for maturation to occur, it is also possible that the epididymis has developed new strategies to deliver cellular proteins to the sperm surface. It is also possible that, in addition to the delivery of proteins to the sperm surface, the epididymosomes act as signalling centres or scaffolds within the luminal compartment affecting protein function independently of that associated with spermatozoa (Cornwall 2009). The protein(s) acquired by the spermatozoa from the epididymosomes may mark them for elimination, as a mechanism to protect the spermatozoa from the deleterious effect of molecules released from the dead and damaged ones (D'Amours et al. 2012). Epididymosomes are heterogeneous with different protein compositions depending on downstream functions.

Not all proteins associated with epididymosomes are transferred to spermatozoa, suggesting that only some proteins have the ability to be transferred. Complete fusion and transfer of vesicles to spermatozoa do not occur (Sullivan et al. 2005).

6.3.4 Role in Constantly Changing Luminal Microenvironment, Segment-Specific Gene Expression and Fluid Resorption

Electrolyte and water transport are an important function of the epididymis as it has an immediate effect on sperm because the latter are bathed in a milieu created by the epithelium. The microenvironment of the epididymis affects the sperm concentration. It is a well-established fact that formation of the unique epididymal environment is the result of reabsorption and secretion throughout the length of the reproductive tract. The epithelial cells lining the epididymis contribute extensively to the fluid reabsorption most of which occurs at the proximal parts (Wong and Yeung 1978). This results in tremendous increase in the luminal sperm concentration (10^8 spermatozoa per mL in the testis to 10^9 in the ductus deferens) and protein concentration in the luminal fluid (2–4 mg/mL in rete testis to 50–60 mg/mL in distal caput) and then returns to 20–30 mg/mL in the cauda (Dacheux et al. 2012; Dacheux and Dacheux 2014). The epididymal epithelial cells express various iso-forms of ion exchangers. NHE2, which is present in the basolateral membrane of the epididymal cells (Leung et al. 2001), is engaged in Na^+ reabsorption. Apart from the electro-neutral Na^+/K^+ exchanger, electrogenic sodium reabsorption also takes place in the epididymis. Leung et al. (1996) have provided the functional evidence for the presence of the Na^+/glucose co-transporter in mouse and a Na^+/nucleoside co-transporter in rat epididymis, which contribute to the transport of Na^+ and fluid. The epithelial cells secrete chloride and bicarbonate upon stimulation by adrenergic agonists, purinergic agonists and peptide hormones. Therefore, it is quite evident that these absorptive and secretory processes influence the movement of water across the epithelium because of the expression of the aquaporins in the epithelial cells.

The mammalian aquaporins (AQPs) are a family of 13 trans-membrane channel proteins that are involved in the transport of water in numerous organs. Epididymal epithelial cells express several members of the AQP family with a region-specific pattern. AQP9 is abundantly expressed in the efferent ducts, the epididymis and the vas deferens, where it could represent an important apical pathway for transmembrane water and solute movement (Da Silva et al. 2006; Arrighi and Aralla 2014). While AQP9 is the main AQP of the epididymis, the mRNAs specific for Aqp2, 5, 7 and 11 are also expressed in epididymal epithelial cells. AQP5 protein co-localises with AQP9 in the apical membrane of a subpopulation of principal cells in the corpus and cauda regions. AQP2 protein was, however, detected only in the distal cauda of young rats (between the second and fourth postnatal week).

In addition, epididymal epithelial cells express significant amounts of the mRNAs coding for AQP7 and 11 (Da Silva et al. 2006). The differential expression of different subsets of AQP could be one of the principal bases of the changing luminal microenvironment along the duct (Klein et al. 2013).

The reabsorption in the epididymis is well balanced and held at a constant tonic rate due to the circulating level of testosterone and mineralocorticoids. On the other hand, secretion in the epididymis is influenced by the neurohormonal factors. Thus, the tonic reabsorptive machinery and secretory machinery synchronise with each other by the regulation of autocrine and paracrine factors as well as neurotransmitters to maintain the fluid microenvironment of the epididymis.

Cornwall (2009) suggested differential expression pattern of the specific genes along the tubule to be associated with different epididymal regions. Though the division between regions is anatomical, it is now established that discrete functions take place in the various segments. On a physiological level, this is evidenced by studies of sperm populations taken from discrete regions of the adult epididymis in which developing spermatozoa exhibit segment-related acquisition of motility and fertilising ability (Lacham and Trounson 1991). Segment-specific expression of genes encoding signalling molecules, regulatory proteins, transporters and receptors also contribute to the formation of special microenvironments by allowing the epithelium to respond uniquely to different stimuli such as hormones and other regulatory factors. Identifying and determining the function of segment-specific proteins is of paramount importance for understanding epididymal sperm maturation. For this reason a number of gene profiling studies have been carried out in attempts to identify genes exhibiting regionalised expression in the epididymis (Jervis and Robaire 2001; Penttinen et al. 2003; Hsia and Cornwall 2004; Jelinsky et al. 2007; Thimon et al. 2007) as well as proteomic studies to identify proteins (Chaurand et al. 2003; Dacheux et al. 2006). These studies have yielded large data sets including novel sequences as well as genes with known identities but previously not known to be expressed by the epididymis. Gene sequences include those as potential proteases and protease inhibitors, defensins, transporters, transcription factors, as well as genes associated with metabolism, cell signalling and part of the antioxidant system. Recently, it has also been reported that epididymosomes transport different microRNAs into the intraluminal fluid in a segment-specific manner (Belleannée et al. 2013) and the expression of microRNA regulates the pattern of gene expression in the changing microenvironment of the epididymis (Belleannée et al. 2012). Studies of the cystatin CRES in the epididymal luminal fluid demonstrated that it was present not only in monomeric forms but also in soluble SDS-sensitive and SDS-resistant forms as well as insoluble forms as defined by its precipitation following high-speed centrifugation (Von Horsten et al. 2007). Within the epididymis CRES is synthesised and secreted by the proximal caput epididymidal epithelium and accumulates in the lumen of the mid-caput, but abruptly disappears from the distal caput epididymidal lumen (Cornwall and Hann 1995).

6.3.5 Role in Protection of Luminal Sperm from the Adverse Reactive Oxygen Species

It is widely recognised that epididymal spermatozoa are highly susceptible to damage resulting from reactive oxygen species (ROS) (Aitken 1999). Spermatozoa are very poor in intracellular antioxidant enzymes and have high concentrations of polyunsaturated fatty acids which are vulnerable to peroxidative damage (Aitken and Vernet 1998). Spermatozoa themselves produce hydrogen peroxide and superoxide that are essential for capacitation and chromatin condensation. Thus, the epididymal epithelium must be able to protect spermatozoa and itself from oxidative damage. The epididymis is a rich source of antioxidant enzymes (Zini and Schlegel 1997; Schwaab et al. 1998). Several of the antioxidant-related genes are expressed along the tissue. Particularly, the expression of GSTpi, a class of GST that has been immunolocalised in the epididymis of the rat (Hermo et al. 1994), decreases greatly between the IS and the corpus. The GSTpi functions in the metabolism of xenobiotics. This high expression in the initial segment of the epididymis may indicate the presence of a preferred substrate in the proximal epididymis (Jervis and Robaire 2001). On the other hand, Jun et al. (2014) report about the necessity of ROS1 kinase activity and the ensuing MEK1/2-ERK1/2 signalling for the normal postnatal development of the IS epithelium.

Antioxidant strategies involving the glutathione system (glutathione peroxidase, glutathione-S-transferase, c-glutamyl transpeptidase, etc.) and the superoxide dismutase/catalase system have been studied in laboratory species (Hinton et al. 1995; Weir and Robaire 2007) and are known to exist in the human epididymis (Potts et al. 1999; Noblanc et al. 2012). The relative importance of the two antioxidant systems for the protection of sperm is not known, but the presence of at least two major systems in the epididymis suggests an importance in prolonged sperm survival.

6.3.6 Role in Protection of Sperm: Antimicrobial Proteins

A variety of antimicrobial proteins are expressed in the lumen of epididymis. These proteins are found to bind sperm thereby rendering epididymal immunity besides their role in sperm maturation. The antimicrobial proteins expressed in the male reproductive tract include human cationic antimicrobial protein (hCAP18, a cathelicidin) (Malm et al. 2000), defensins (Com et al. 2003; Palladino et al. 2003; Dorin and Barratt 2014), the epididymal β-defensin member Bin1b (Li et al. 2001), cystatins (Blankenvoorde et al. 1998; Hamil et al. 2002), lactoferrin (Jin et al. 1997, Pearl and Roser 2014), seminalplasmin (Reddy and Bhargava 1979), seminogelin-derived peptides (Bourgeon et al. 2004), lipocalins, CRES (Wang et al. 2012), epididymis protein 2 (EP2) (Avellar et al. 2004) and HE2 (Hamil et al. 2000). The mode of action of the individual antimicrobial proteins differs from one another. Defensins, the most predominant class of antibacterial proteins known to be expressed in epididymis, rapidly kill microorganisms by permeating the microbial membranes and impairing the microorganism's ability to carry out metabolic processes (White et al. 1995). The recent identification in the rat epididymis of a new peptide – Bin1b – that exhibits structural characteristics and antimicrobial activity similar to that of β-defensins (Malm et al. 2000) together with data on other antimicrobial proteins in rat and mouse epididymis indicate that this secretory organ is an important site of antimicrobial peptide production in the male reproductive system.

6.3.7 Role in Sperm Quality Control

Epididymis plays a critical role in the sperm quality control as it has the inherent capacity to recognise and remove dead or defective spermatozoa in normal animals (Axnér 2006) and also when there is an instance of sperm death signals (Jones 2004). Flickinger (1975) proposed the possible mechanisms by which the epididymis eliminates the defective spermatozoa after vasectomy, which include (a) phagocytosis of spermatozoa by epithelium of the epididymis and vas deferens, (b) migration of phagocytes through the epithelium and phagocytosis of spermatozoa in the lumen, possibly following alterations in the duct epithelium, (c) degeneration and dissolution of spermatozoa in the lumen of the male ducts with uptake of soluble material by the epithelium and (d) rupture of the duct system with formation of spermatic granulomas and phagocytosis of spermatozoa by macrophages outside the confines of the male reproductive system. These mechanisms would apply to conditions other than vasectomy as well (Barth and Oko 1989; Ramos-Ibeas et al. 2013). According to Priesel (1924), the non-ejaculated abnormal spermatozoa undergo regressive changes leading to their death and are liquefied within the epididymis and are reabsorbed. Yet another mechanism of sperm quality control is the uptake of spermatozoa or parts thereof by the epididymal epithelium (Alexander 1972; Flickinger 1972; Sutovsky et al. 2001; Aruldhas et al. 2006; Axnér 2006). Apart from these aforementioned proposed mechanisms, lymphatic system also has its role in the elimination of defective spermatozoa by the uptake and transport of the sperm components to the regional lymph nodes (Barratt and Cohen 1987). The extent of each of these mechanisms may vary with the species, although in some, such as the hamster, several of these mechanisms operate (Flickinger 1982).

The defective spermatozoa become surface-ubiquitinated during epididymal maturation and are in part removed during the descent down the corpus and cauda epididymides by phagocytosis (Sutovsky et al. 2001). Precisely, the percentage of ubiquitinated sperm increases during their transit from rete-testis to the corpus of epididymis, and decreases from the corpus to the cauda epididymidis. Although it is likely that defective sperm are ubiquitinated because of structural damage, the mechanism by which such sperm are recognised by the ubiquitination machinery and the means of disposal after phagocytosis are not yet clearly understood. One possible explanation could be that the process of ubiquitination of epididymis is at the intersection, or perhaps at the common end point of several apoptotic mechanisms operating in the testis (Sinha Hikim and Swerdloff 1999) and such mechanisms may recognise structural damage of sperm DNA (Sakkas et al. 1999) and/or damaged sperm accessory structures. Sutovsky et al. (2001) suggested that a whole sperm cell can become ubiquitinated on its surface and eventually phagocytosed. ELSPBP1 is a protein secreted by principal cells in association with epididymosomes, which specifically transfer ELSPBP1 to dead spermatozoa in the presence of zinc, as a tag for removal of the dead spermatozoa during epididymal passage (D'Amours et al. 2012; Sullivan et al. 2015). Similar to the proteolysis of endocytosed receptors (Strous and Govers 1999) the destruction of phagocytosed sperm seems to occur by means of lysosomal/vacuolar proteolysis. Thus, in the epididymis, pathways that normally function within the cell may also occur in the extracellular environment and as a result provide the mechanisms to appropriately maintain the luminal environment ultimately protecting the maturing spermatozoa (Cornwall 2009).

6.3.8 Role in Passage of Sperm to Distal Parts

The mammalian epididymis serves to propel spermatozoa from the testes to the epididymal cauda region. Passage of the spermatozoa through the epididymal duct is a prerequisite for epithelium-dependent sperm-maturation processes, ensuring acquisition of fertilising potential and progressive motility. Net transport rates are estimated to be most rapid in the efferent ducts and proximal epididymis, where fluid is non-viscous and water is being rapidly absorbed from the lumen; transport rates decrease in the more distal tubule where the lumen content becomes more viscous. Sperm are moved through the epididymis in part by hydrostatic pressures originating from fluids secreted in the seminiferous tubules and by peristalsis-like contractions of the tubules. It is brought about by spontaneous phasic contractions of periductal muscle layers, which are predominantly myogenic and function generally under the principle of autorhythmicity (Markkula-Viitanen et al. 1979; Cosentino and Cockett 1986; Mewe et al. 2006). Contractions of the tunica albuginea of testis also potentially play a role in the generation of positive fluid pressure in the head of the epididymis. Peristalsis-like contractions of the peritubular myoid cells surrounding the epididymal tubule plus the positive hydrostatic pressure from the caput, aid in moving the luminal content down to the more distal region of the duct.

Epididymal processing of the sperm is known to occur over a specific length of time. It is known that mouse and rat sperm spend 10–11 days in the ductus epididymidis when they undergo various changes towards the physiological maturation. They remain viable in the cauda in the absence of ejaculation for up to 5 days in the rat and 6 days in the rabbit (Robaire and Hermo 1988). Data are also available on the duration for which sperm remain in the different segments of the epididymis (Jones and Clulow 1987).

6.3.9 Role in Sperm Storage

The role of epididymis in sperm storage has not been given much attention. Approximately 55%–65% of total epididymal sperm in the human are stored in the cauda epididymidis (Amann 1981). As a proportion of testicular output this compares favourably with other species (Bedford 1994). However, it is only the equivalent of three average ejaculates (Johnson and Varner 1988). In some species, the cauda contains sperm that are sufficient for many ejaculates (Curtis and Amann 1981). Though sperm can pass through the human cauda within a couple of days, fertile sperm can be stored for several weeks in both man (Bedford 1994) and other mammals (Jones and Murdoch 1996). For how long effective storage may be there in the human is uncertain, but sperm motility in the ejaculate of young men can be preserved for up to 7–8 weeks after the last ejaculation (Bedford 1994; Turner 2008). Although the critical features of the storage environment remain unknown, specific proteins secreted into the lumen by the more proximal epithelium, intraluminal ionic concentrations controlled by the epithelium, the subsequent reduction of luminal pH and the high osmolality of the lumen fluid have all been shown to play a role in maintaining cauda sperm quiescent (Jones and Murdoch 1996; Verma 2001).

6.4 Regulation of Epididymis Structure and Function

The epididymis is highly dependent on androgens. Testosterone is the principal androgen that regulates epididymal structure and function. It accesses the epididymis in two routes, as the circulating hormone and as a luminal form from the testis through the rete testis and efferent ductules, bound to androgen-binding protein (ABP). Testosterone has two modes of action at the target cells: genomic and non-genomic. Towards the genomic action testosterone (T) enters the cell and either directly binds to the androgen receptor (AR) or is converted to dihydrotestosterone (DHT) by steroid 5α-reductases after which it binds to the AR. The androgen-bound AR dimerises, and then associated co-receptors are modified. The dimerised receptor translocates to the nucleus, where it acts as a transcription factor to modulate gene expression. The non-genomic action is immediate and rapid, and is not prevented by inhibiting transcription (or) translation. Rather it involves membrane-bound receptors (perhaps not AR) and is linked to signal transduction cascade involving SRC, MEK, ERK1/2 and KREB (Hamzeh and Robaire 2011).

The epididymal weight is reduced to 25% following castration. Restoration of circulating testosterone reverses the cellular changes in the caput, corpus and cauda epididymides, but not in the initial segment (Ezer and Robaire 2002; Hu et al. 2014). The maintenance of initial segment morphology requires components in the luminal fluid, that is lumicrine regulation including spermatozoa (Scheer and Robaire 1980; Hinton et al. 1998; Turner and Riley 1999). One or many testicular factors are required to maintain initial segment function. Administration of FGF2 but not EGF to efferent duct-ligated rats restored gamma-glutamyl transpeptidase mRNA IV (Ggt_pr4 mRNA) and gamma-glutamyl transpeptidase protein in the initial segment to control levels (Lan et al. 1998). There may be either several testicular factors each differentially regulating specific subsets of genes or one or a few testicular factors may mediate different downstream effects via the activation of multiple signalling pathways and subsequent effector molecules. One such molecule could be ABP itself (Hamzeh and Robaire 2011). Other molecules which appear critical for the maintenance of the epididymal epithelium and subsequently sperm maturation include oxysterols, derivatives of cholesterol. Thus, in addition to proteinaceous factors, lipids also play an important role in the regulation of epididymal function (Cornwall 2009). Other studies suggest that in addition to factors in the fluid, sperm cells themselves may regulate the epididymal epithelium (Garrett et al. 1990). Cell–cell communication between spermatozoa and the epithelium may direct epididymal function (Cornwall 2009).

6.5 Organ and Cell Culture of the Epididymis

6.5.1 Organ Culture

Earlier attempts to study the sperm maturation *in vitro* involved the incubation of immature epididymal spermatozoa along with the epididymal proteins. Orgebin-Crist and Jahad (1979) showed that when crude extracts from sperm-free rabbit epididymis were applied to spermatozoa collected from the distal corpus region it increased their *in vitro* fertilising ability. Studies of González et al. (1984) in hamster provided evidence for the increased fertilising capacity of sperm upon incubation of the sperm suspension of various segments of epididymis with an epididymal protein for 5 h. These discoveries could provide evidence for the influence of epididymal glycoprotein secretions in the sperm survival and fertilising capacity.

The experimental procedure for investigating the sperm maturation and storage in the epididymis employs combinatorial approaches. For instance, through direct micro-puncture and canulation analysis, it is possible to analyse the epididymal luminal milieu wherein the spermatozoa are suspended, but the difficult sampling procedure does not favour this approach. Also, spermatozoa can be directly extracted from various regions to study their physiological and biochemical status. The *in vivo* and *in vitro* fertilising methods in combination with surgical manipulative methods such as ligation, surgical bypass, etc., may be employed to elucidate the influence of various perturbations of the tract on the sperm maturation process *in situ* (Temple-Smith et al. 1998). But, these procedures have their own merits and demerits and also these methods could provide only correlative evidence for interactions of epididymal function and the processes of sperm maturation and storage. To overcome these limitations and also to investigate the processes of sperm maturation and storage in a more direct way, efforts have been under way to culture epididymal duct and cells. Much of what is known about the functioning of the human epididymis has been derived from a tubule culture system in which human epididymal tubules are maintained for up to 8 days (Tezón and Blaquier 1981). This method allows study of the whole tubule with metabolic contributions from the peritubular muscle and intraluminal spermatozoa as well as the epithelium.

There are several advantages to using organ culture for the study of hormone-dependent tissues such as epididymis. First, the endocrine environment of the tissue can be precisely defined and easily manipulated in organ culture. Thus, the direct effect of a single hormone, or the interaction of several compounds, can be studied under controlled and reproducible conditions. Second, tissues that are hormone-sensitive *in vivo* usually retain their hormone responsiveness in organ culture and anatomic relationships between various cells are persevered in organ culture.

Hoffman et al. (1976) attempted organ culture of the corpus epididymal tubules in the absence of androgens which led to the regression of the cells through induced changes in the stereocilia and swelling of Golgi cristae. But, the addition of testosterone and dihydrosterone to the culture delayed this degeneration to a great extent. This led to the requirement of androgen in the medium to retain the epithelial function *per se*, *in vivo* (Vázquez et al. 1986) and the epididymal cell culture techniques used included androgen in the medium. Nowadays, testosterone and dihydrosterone are added to the medium. Apart from these, α-estrogen receptors have also been identified in the efferent ducts and initial segment of the rat epididymis which might be involved in the regulation of fluid resorption (Hess et al. 1997); this suggests the necessity of androgens and estrogens in the medium for the maintenance of epididymal function.

Klinefelter and Hamilton (1984, 1985) established epididymal organ culture which involves the perfusion of rat caput epididymal tubule segments in custom-made apparatus. This method maintained structural and functional integrity of the interacting epithelial, luminal and spermatozoal compartments. During the first three days of culture, spermatozoa became progressively motile on dilution, implying that certain aspects of maturation had occurred.

6.5.2 Cell Culture

Considering the complexity and difficulty in the maintenance of epididymal duct under culture conditions, defined cell culture methods were investigated. The cell culture technique favours the study of the specific function of the epithelium including directional secretion, endocytosis, ion transport and vectorial secretion by epithelial cells from the rat efferent ducts (Byers et al. 1985), caput and corpus epididymides (Byers et al. 1986; Cooper et al. 1989) and cauda epididymidis (Cuthbert and Wong 1986; Cooper et al. 1989). Cell culture systems permit study of isolated cells in the absence of other cell types. This has been done for human fetal epididymal cells (Harris and Coleman 1989) and feline epididymal epithelial cells (Buff et al. 2005) on plastic. Buff et al. (2005) described a tissue culture system in which cells retain defined ultrastructural and functional characteristics to provide a basis for functional investigations of the epididymal duct in the cat. Primary culture of mixed cell preparation from rat epididymis was used as a tool for monitoring pro-inflammatory signals (Chapin et al. 2014). In regard to nutrient access and the polarity of the cells, however, growth on permeable supports is superior to that on plastic (Byers et al. 1986) and it also offers the possibility of determining directional secretion, absorption and transcellular transport. Being responsible for the preparation of spermatozoa for fertilisation, the study of epididymal function may serve in the treatment of infertility or might be useful in developing post-testicular contraceptive agents. Therefore, for studying the human epididymis, epithelial cultures can be employed to overcome the shortcomings of other procedures associated with the paucity of the tissue and the experimental restrictions due to ethical issues.

Although one loses organisational integrity by dissociating a tissue into isolated cells, it is possible to obtain more specific information with respect to the activity and functions of an individual cell type within the tissue. It is also possible to maintain the cells for longer periods of time and to study the hormonal regulation of the different cell types. Several attempts have been made to develop methods for the isolation of purified populations of viable epididymal epithelial cells. Disaggregation of the epididymis by enzymatic and/or mechanical methods has been successful in yielding viable, yet heterogeneous, suspensions of epididymal cells. Defined cell culture procedures that reduce the epididymal epithelium to a purified single cell suspension of principal cells were also developed (Kierszenbaum et al. 1980; Joshi 1985). Unfortunately, the disruption of the epithelial architecture and the lack of basal lamina and other peritubular elements rapidly compromise epithelial cells that de-differentiate and lose function in culture. To overcome these setbacks, various techniques have been adopted to maintain normal epithelial cell function including the use of semi-permeable support and extracellular matrix to maintain polarised monolayers. Thus, pioneering attempts were made by Cooper et al. (1990) to culture the epididymal epithelial cells as a monolayer culture on permeable support.

The first stable epithelial cell lines transformed with SV40 large T antigen (LTAg) from two regions of the adult human epididymis have recently been established. The cells possess ultrastructural characteristics of human epididymal principal cells *in vivo*. The cell lines retained some of the characteristics of *in vivo* principal cells. They exhibited similar ultrastructure, were diploid, and expressed a variety of epididymal markers, at least at the RNA level, which are specifically expressed or highly expressed by human epididymal principal cells. These cell lines are an excellent tool to study the regulation of junctional proteins in the epididymis (Dubé et al. 2010).

6.6 Epididymis and Male Fertility Problems

Obstruction of the epididymal duct is a major issue. Kroovand and Perlmutter (1981) described eight different types of developmental anomalies of the epididymis including complete absence of the excurrent ducts, absence of a connection between the testis and caput epididymidis, agenesis of different regions or segments of the epididymis, and extensive disconnection of the corpus and/or cauda of the epididymis from the testis allowing, respectively, a 'looping' epididymis or one that can angle sharply away from the testis. According to Turner (2008), in the case of an obstructed epididymis, the normal distribution of sperm in the epididymis changes depending on the level of obstruction. Typically, there will be no sperm in the lumen below the site of obstruction, but above the obstruction sperm will have accumulated in the lumen, potentially backing up all the way to the efferent ducts or even to the rete testis. Near the site of obstruction, the dense sperm pack in the tubule lumen will consist of senescent and degenerating sperm, and in cases of sperm granuloma these will be mixed with macrophages and neutrophils, which may be seen in the interstitium, epithelium or even in the lumen of the duct (Wang and Holstein 1983; Pöllänen and Cooper 1994). In almost 300 cases of obstructive azoospermia, Girgis et al. (1969) found that a dysgenesis or atresia of the tubule had occurred between the caput epididymidis and the testis in 21% of cases and in the cauda epididymidis or vas deferens in 39% of cases. The histologic appearance of the cryptorchid epididymis also reflects a failure of normal development of the epididymal epithelium as well as the failure of the peritubular musculature of the cauda to develop properly (DeMiguel et al. 2001).

Approximately 95% of men with clinical cystic fibrosis have congenital absence of the vas deferens (Wong 1998), which is commonly accompanied by absence of the cauda and corpus epididymides as well. Inflammation and oxidative stress are important clinical entities in the epididymis because epididymitis is the fifth most common urological diagnosis in men within their reproductive years. Although many cases of epididymitis are bacterial in origin, most studies show such cases to be in minority (Tracy et al. 2008; Turner 2008).

Oxidative stress occurs due to the imbalance between free radicals and antioxidants and has been implicated in many cases of male infertility. Cells under oxidative stress are more likely to have altered protein expression and in this regard several proteins that could possibly act as biomarkers of oxidative stress-induced male infertility are suggested, including DJ-1, PIP, lactotransferrin and peroxiredoxin (Agarwal et al. 2014). Oxidative stress is a major cause of DNA fragmentation in spermatozoa and that in turn is a determinant in the etiology of male infertility (Wright et al. 2014).

6.7 Epididymis as a Target to Toxicants

Epididymis as a target to toxicants has received lesser attention than other regions of the male reproductive system. One of the problems has been the difficulty in separating direct versus indirect effects of the toxicants on the excurrent ducts, because altered testicular function indirectly alters functions of the ductal epithelium downstream. There are many pathways by which toxicants can affect the epididymis and we have only just begun to understand these mechanisms for a limited number of chemicals. Over the years reproductive and toxicological studies have identified at least 20 toxicants having direct or indirect effects on the epididymal epithelium or the epididymal spermatozoa. Since the epididymis is a very versatile organ, it is prone to toxic inflictions which will negatively affect the epididymal function. For example, exposure of male rats to the α-adrenergic antagonist prazosin resulted in reduced numbers of ejaculated sperm, reduced numbers of sperm in the distal vas deferens and an accumulation of cauda epdidymidal sperm indicating disrupted transit of sperm from the cauda epididymidis to the vas deferens (Solomon et al. 1997). Indeed, there are now numerous chemicals that produce epididymis-specific reductions in cauda epididymal sperm numbers with no change in testicular sperm numbers, suggesting accelerated sperm transit. These chemicals include the pesticide methoxychlor (Gray et al. 1989), PCB 169 (Gray 1998), the synthetic antiandrogen hydroxyflutamide (Klinefelter and Suarez 1997), the antiandrogenic pesticide vinclozolin (Gray et al. 1994) and the structural analogs of ethylene glycol such as ethane dimethanesulphonate (EDS; Klinefelter et al. 1990) and chloroethylmethanesulphonate (CEMS; Klinefelter et al. 1994), vincristine (Stanley and Akbarsha 1992; Averal et al. 1996; Akbarsha and Averal 1998; Akbarsha and Averal 1999a, 1999b; Akbarsha et al. 2000), phosphamidon (Akbarsha and Sivasamy 1998), andrographolide (Akbarsha and Murugaian 2000), aflatoxin (Agnes and Akbarsha 2001, 2003; Faisal et al. 2008), chromium (Aruldhas et al. 2004, 2006; Akbarsha et al. 2006) and epichlorohydrin (Lee et al. 2013). Lead has been shown to inflict damage to seminiferous tubules, Leydig cells, epididymal luminal sperm and epididymal epithelial luminal stereocilia in rat, and these changes were reversed when vitamin E was co-administered with lead (Fahim et al. 2013).

6.8 Epididymis as a Target to Male Contraceptive Attack

During the past two decades, considerable attention has been focused on the epididymis as an extragonadal site for the control of male fertility. This emerged from the recent understanding about the role of epididymis in sperm maturation, which is achieved by a number of unique absorptive and secretory functions of the duct resulting in an optimal environment for maturation. Interference with the epididymal functions should lead to alteration of the processes whereby the spermatozoon attains its fertilising capacity. A review of various post-testicular approaches to contraception (Cooper 1992) talks about three possible major sites of action of a putative antifertility agent: (1) action on peritubular muscle-hastening sperm transport should lead to ejaculation of spermatozoa that have not matured as a consequence of their young age or reduced time for contact with necessary epididymal secretions; (2) action on the epithelium-altering the transport and synthesis of ions and organic solutes should modify the production and composition of epididymal fluid with consequences for the spermatozoa bathed in it; and (3) action on the spermatozoa (mature and immature) may be susceptible to chemical agents because of their highly differentiated state and characteristic enzyme systems. By removing sympathetic innervations from the distal epididymis, normal sperm transport through the epididymis is affected (Ricker and Chang 1996) and infertility stemming from azoospermia can result because spermatozoa accumulate in the epididymis. Although this is an interesting model and has produced evidence that neurotransmitters may influence epithelial function (Ricker et al. 1996), it is a form of chemical vasectomy that would suffer from problems like irreversibility, ductal rupture and antibody production. By reducing or blocking known epididymal secretions, infertility can be introduced. High concentrations of organic solutes and an array of hydrolytic enzymes characterise epididymal fluid from many species, but their exact roles in fertilisation are not clear.

A male contraceptive pill must be effective, safe and reasonably quick acting, and these criteria can be attained best by agents that attack maturation and storage of spermatozoa in the epididymis to render them incapable of fulfilling their role in fertilisation. Since cell division is complete by this stage, the possibility of mutagenic effect is reduced and the likelihood of interfering with normal hormonal status is much less than for agents acting at the testicular level. However, the most obvious advantage of the epididymal approach is its speed of action; an effect on stored spermatozoa could produce infertility within days while an effect on the testis would need several weeks to completely suppress spermatogenesis. One such drug that affects epididymal maturation of spermatozoa is chlorohydrin. At low doses it has no effect on the testis, libido or accessory glands. The action of this drug is species specific. For example, rapidly reversible antifertility properties of the drug were demonstrated in male rats, guinea pigs, monkeys and sheep, but not in mice and rabbits. When given at low doses, chlorohydrin interferes with the functions of spermatozoa, causing a reversible phase of sterility. The metabolism and motility of testicular and epididymal sperm were inhibited. Although much work has been done on chlorohydrin, interest in its further development as male contraceptive has declined following the discovery of toxic effects in rhesus monkey at its therapeutic dose. Early successes with chlorohydrin prompted search for other compounds that may have the same site of action. Based on the observation that the glycolytic effect of chlorohydrin in sperm is closely linked to antifertility activity, scientists in the United Kingdom have investigated the effects of a number of chlorodeoxysugars on male fertility and sperm glycolysis. Since epididymal functions and sperm maturation are highly dependent on circulating androgens, antagonising the effect of androgens at the epididymis (receptor) level would result in interference with epididymal function and, hence, sperm dysfunction. The most well-known antiandrogen is cyproterone acetate. It was shown that continuous release of low doses of cyproterone acetate from silastic capsules implanted in rats led to infertility without affecting libido or spermatogenesis. Epididymal sperm appeared to have lost normal function after long-term exposure to the drug. However, clinical trial of cyproterone acetate has revealed that it also reduced plasma testosterone levels in men by inhibiting gonadotropin release by the pituitary. Consequently, the use of cyproterone acetate as a male contraceptive has been discouraged.

Rat protein DE is an androgen-dependent cysteine-rich secretory protein (CRISP) synthesised by proximal epididymal regions. DE, also known as CRISP-1, is localised on the equatorial segment of acrosome-reacted spermatozoa and participates in gamete fusion through binding to egg complementary sites. Immunisation of rats with DE inhibits fertility and sperm fusion ability, suggesting that DE represents a good epididymal contraceptive target (Cohen et al. 2007). DE (CRISP-1) was first described in 1976 (Cameo and Blaquier 1976). This 32 kDa protein contains 10% carbohydrates (Garberi et al. 1982), and is synthesised in an androgen-dependent manner by the proximal segments of the epididymis; it is associated with the sperm surface during epididymal maturation. Originally localised in the dorsal region of the acrosome, DE migrates to the equatorial segment as the acrosome reaction occurs (Rochwerger and Cuasnicú 1992). The relocation of DE to the equatorial segment, the region through which sperm fuse with the egg (Bedford et al. 1979; Yanagimachi 1994), opened the possibility of a role for DE in sperm–egg fusion.

Sequential extraction of proteins from epididymal sperm revealed the existence of two populations of DE bound to sperm: (1) a major (70%) population loosely associated with sperm by ionic interactions, which is released from the cells during capacitation and, therefore, proposed to act as a decapacitation factor (Kohane et al. 1980; Cohen et al. 2000) and (2) a minor (30%) tightly bound population, which behaves as an integral protein, and remains on sperm after capacitation and corresponds to the protein that migrates to the equatorial segment and participates in gamete fusion. Immunisation of rats with DE produced specific antibodies against the protein in over 90% of the animals and a reversible inhibition of fertility in both sexes (Cuasnicú et al. 1990). Thus, the results so far obtained indicate that epididymal protein DE/CRISP-1 fulfils many of the requisites for an epididymal contraceptive target: (1) it is an epididymal specific protein; (2) it is localised on the sperm surface being accessible for its blockage in the male tract; (3) it is relevant for fertility, as demonstrated by the immunisation studies; (4) it plays a specific role in fertilisation (sperm egg fusion) and capacitation; (5) its active site has been identified and resides in a discrete region of the molecule (12 amino acids); and (6) it is a member of a highly evolutionarily conserved family (CRISP) with functional homologues in other species, such as mouse and human.

Another promising target is the epididymal protease inhibitor (EPPIN), a cysteine-rich protein that has antimicrobial properties and is thought to play a critical role in sperm motility. EPPIN mRNA and the protein are abundantly found both in testis and epididymis of rats. During ejaculation, semenogelin (SEMG1) from seminal vesicles binds to EPPIN, initiating a series of events including the modulation of prostate-specific antigen (PSA) activity, provision of antimicrobial protection, engendering inhibition of sperm motility. As PSA hydrolyses SEMG1, spermatozoa gain progressive motility. To find the suitability of EPPIN as prospective male contraceptive, attempts are being made to develop compounds that inhibit EPPIN function (O'Rand et al. 2011; Silva et al. 2012; Garside et al. 2013).

A recent study showed that the sclerosing agent sodium tetradecyl sulfate, a widely used sclerosant for the treatment of telangiectatic veins and has been approved for use in the United States by the Food and Drug Administration, when injected into the epididymal duct, induced sterilisation in male rats through ductal occlusion. However, this is a preliminary short-term study of rats. Hence, a further detailed study of large animals and a clinical study are required (Park et al. 2014).

6.9 Conclusions

The necessity of understanding the normal processes of epididymal sperm maturation is emphasised by the fact that up to 40% of infertile men exhibit idiopathic infertility, which in many cases can reflect maturational disorders. The importance of epididymal study is also underscored by the continued lack of development of a male contraceptive. Targeting specific molecules in the epididymis would be a more attractive approach than disrupting spermatogenesis because the contraceptive effects would be rapid and more readily reversible and side effects associated with alterations in hormones would be avoided.

Clearly, if we are to improve human health by developing new and better ways to improve as well as prevent fertility, further research into the epididymis is needed (Cornwall 2009). With these as the priorities, epididymal biologists have identified key areas of epididymal research that need further emphasis including studying spermatozoa and associated maturational changes during epididymal transit, the specifics of sperm–protein interactions as well as sperm interactions with non-protein molecules in the epididymal lumen, studies of the cytoplasmic droplet and its role in maturation, the development of better *in vitro* systems to facilitate study of sperm maturation, understanding the mechanisms of sperm storage and maintenance of sperm activity, and examination of interstitial–epithelial cell interactions as well as identifying functions for the vast number of epididymal secretory proteins that are part of the epididymal luminal microenvironment. The epididymis is a versatile organ capable of overcoming adverse situations through taking up newer roles (Aruldhas et al. 2004).

Acknowledgements

We gratefully thank the Department of Science and Technology (DST), Government of India, New Delhi, for the grant to MAA (No. SP/SC-C-31/98; No. SR/SO/AS-59/2004) for research on epididymis as a target to toxicants. The

Visiting Professorship to MAA from King Saud University, Riyadh, Kingdom of Saudi Arabia, is heartily acknowledged. The contributions, by way of the illustrations in this chapter, of Dr. V.F. Agnes, Dr. R. Girija and Dr. A. Faridha, are gratefully acknowledged.

References

Abe, K., H. Takano and T. Ito. 1982. Appearance of peculiar epithelial cells in the epididymal duct of the mouse ligated epididymis. *Biol Reprod* 26:501–09.

Abe, K., H. Takano and T. Ito. 1984. Interruption of the luminal flow in the epididymal duct of the corpus epididymid is in the mouse, with special reference to differentiation of the epididymal epithelium. *Arch Histol Jpn* 47:137–47.

Adamali, H. I. and L. Hermo. 1996. Apical and narrow cells are distinct cell types differing in their structure, distribution, and functions in the adult rat epididymis. *J Androl* 17:208–22.

Adamali, H. I., I. H. Somani, J. Q. Huang, D. Mahuran, R. A. Gravel, J. M. Trasler and L. Hermo. 1999a. I. Abnormalities in cells of the testis, efferent ducts, and epididymis in juvenile and adult mice with beta-hexosaminidase A and B deficiency. *J Androl* 20:779–802.

Adamali, H. I., I. H. Somani, J. Q. Huang, R. A. Gravel, J. M. Trasler and L. Hermo. 1999b. II. Characterization and development of the regional- and cellular-specific abnormalities in the epididymis of mice with beta-hexosaminidase A deficiency. *J Androl* 20:803–24.

Agarwal, A., D. Durairajanarayagam, J. Halabi, J. Peng and M. Vazquez-Levin. 2014. Proteomics, oxidative stress and male infertility. *Reprod Biomed Online* 29:32–58.

Agnes, V. F. and M. A. Akbarsha. 2001. Pale vacuolated epithelial cells in the epididymis of aflatoxin-treated mice. *Reproduction* (formerly *J Reprod Fertil*) 122:629–41.

Agnes, V. F. and M. A. Akbarsha. 2003. Spermatotoxic effect of aflatoxin B_1 in albino mouse. *Food Chem Toxicol* 41:119–30.

Aitken, R. J. 1999. The amoroso lecture: The human spermatozoon- a cell in crisis? *J Reprod Fertil* 115:1–7.

Aitken, R. J. and P. Vernet. 1998. Maturation of redox regulatory mechanisms in the epididymis. *J Reprod Fertil Supp* 53:109–18.

Akbarsha, M. A. and H. I. Averal. 1998. Male reproductive toxicity of vincristine: Ultrastructural changes in the epididymal principal cell. *Biomed Lett* 57:159–69.

Akbarsha, M. A. and H. I. Averal. 1999a. Epididymis as a target for the toxic manifestation of vincristine: Ultrastructural changes in the narrow cell. *Biomed Lett* 59:113–20.

Akbarsha, M. A. and H. I. Averal. 1999b. Epididymis as a target for the toxic manifestation of vincristine: Ultrastructural changes in the clear cell. *Biomed Lett* 59:149–59.

Akbarsha, M. A. and P. Murugaian. 2000. Aspects of the male reproductive toxicity/male antifertility property of andrographolide in albino rat: Effect on testis and cauda epididymidal spermatozoa. *Phytother Res* 14:1–4.

Akbarsha, M. A. and P. Sivasamy. 1998. Reproductive toxicity of phosphamidon: Histopathological changes in the epididymis. *Indian J Exp Biol* 36:34–38.

Akbarsha, M. A., H. I. Averal, R. Girija, S. Anandhi and B. A. Faritha. 2000. Male reproductive toxicity of vincristine: Ultrastructural changes in the epididymal epithelial apical cell. *Cytobios* 102:85–93.

Akbarsha, M. A., S. Subramanian and M. M. Aruldhas. 2006. Hazards of chromium pollution on male reproduction: Review of TEM and biochemical studies on testis, epididymis and sperm of monkey. *Embryo Talk* 1:55–61.

Alexander, N. J. 1972. Vasectomy: Long-term effects in the Rhesus monkey. *J Reprod Fertil* 31:399–406.

Amann, R. P. 1981. A critical review of methods for evaluation of spermatogenesis from seminal characteristics. *Andrologia* 2:37–58.

Arrighi, S. and M. Aralla. 2014. Immunolocalization of aquaporin water channels in the domestic cat male genital tract. *Reprod Domest Anim* 49:17–26.

Aruldhas, M. M., S. Subramanian, P. Sekhar, G. C. Hasan, P. Govindarajulu and M. A. Akbarsha. 2004. Microcanalization in the epididymis to overcome ductal obstruction caused by chronic exposure to chromium – A study in the mature bonnet monkey (*Macaca radiata* Geoffroy). *Reproduction* 128:127–37.

Aruldhas, M. M., S. Subramanian, P. Sekhar, G. Vengatesh, P. Govindarajulu and M. A. Akbarsha. 2006. In vivo spermatotoxic effect of chromium and the consequent responses in the epididymal epithelial basal cells and intraepithelial macrophages: Study in a non-human primate (*Macaca radiata* Geoffroy). *Fertil Steril* 86:1097–105.

Asuvapongpatana, S., A. Saewu, C. Chotwiwatthanakun, R. Vanichviriyakit and W. Weerachatyanukul. 2013. Localization of cathepsin D in mouse reproductive tissues and its acquisition onto sperm surface during epididymal sperm maturation. *Acta Histochem* 15:425–33.

Avellar, M. C., L. Honda, K. G. Hamil, S. Yenugu, G. Grossman, P. Petrusz, F. S. French and S. H. Hall. 2004. Differential expression and antibacterial activity of epididymis protein 2 isoforms in the male reproductive tract of human and rhesus monkey (Macaca mulatta). *Biol Reprod* 71:1453–560.

Averal, H. I., A. Stanley, P. Murugaian, M. Palanisamy and M. A. Akbarsha. 1996. Specific effect of vincristine on epididymis. *Indian J Exp Biol* 34:53–56.

Axnér, E. 2006. Sperm maturation in the domestic cat. *Theriogenology* 66:14–24.

Bajpai, V. K., A. C. Shipstone, K. B. V. Ratna, J. Qaisar and B. S. Setty. 1985. Ultra structure of the epididymal epithelium of rhesus monkey (*Macacamulatta*). *Acta Eur Fertil* 16:207–17.

Barratt, C. L. and J. Cohen. 1987. Quantitation of sperm disposal and phagocytic cells in the tract of short- and long-term vasectomized mice. *J Reprod Fertil* 81:377–84.

Barth, A. D. and R. J. Oko. 1989. Photomicrographic features of bovine sperm cell abnormalities. In *Abnormal Morphology of Bovine Spermatozoa,* 89–129. Iowa State University Press, Ames, IA.

Bedford, J. M. 1994. The status and the state of the human epididymis. *Hum Reprod* 9:2187–99.

Bedford, J. M., H. D. M. Moore and L. E. Franklin. 1979. Significance of the equatorial segment of the acrosome of the spermatozoon in eutherian mammals. *Exp Cell Res* 119:119–26.

Belleannée, C., V. Labas, A. P. Teixeira-Gomes, J. L. Gatti, J. L. Dacheux and F. Dacheux. 2011. Identification of luminal and secreted proteins in bull epididymis. *J Proteomics* 74:59–78.

Belleannée, C., Z. Calvo, J. Caballero and R. Sullivan. 2013. Epididymosomes convey different repertoires of micrornas throughout the bovine epididymis. *Biol Reprod* 89: 1–11.

Belleannée, C., E. Calvo, V. Thimon, D. G. Cyr, C. Légaré, L. Garneau and R. Sullivan. 2012. Role of microRNAs in controlling gene expression in different segments of the human epididymis. *PLoS One* 7: e34996.

Blankenvoorde, M. F., W. van't Hof, E. Walgreen-Weterings, T. J. van Steenbergen, H.S. Brand, E.C. Veerman and A. V. Nieuw Amerongen. 1998. Cystatin and cystatin-derived peptides have antibacterial activity against the pathogen *Porphyromonas gingivalis*. *Biol Chem* 379:1371–75.

Boué, F. and R. Sullivan. 1996. Cases of human infertility are associated with the absence of P34H, and epididymal sperm antigen. *Biol Reprod* 54:1018–24.

Boué, F., B. Bérubé, E. De Lamirande, C. Gagnon and R. Sullivan. 1994. Human sperm-zona pellucida interaction is inhibited by an antibody against a hamster sperm protein. *Biol Reprod* 51:577–87.

Bourgeon, F., B. Evrard, M. Brillard-Bourdet, D. Colleu, B. Jégou and C. Pineau. 2004. Involvement of semenogelin-derived peptides in the antibacterial activity of human seminal plasma. *Biol Reprod* 70:768–74.

Brooks, D. E. 1987. Androgen-regulated epididymal secretory proteins associated with post-testicular sperm development. *Ann N Y Acad Sci* 513:179–94.

Brown, D., B. Lui, S. Gluck and I. Sabolić. 1992. A plasma membrane proton ATPase in specialized cells of rat epididymis. *Am J Physiol* 263:C913–6.

Buff, S., V. Lambert, T. Marchal and P. Guérin. 2005. Isolation, culture and characteristics of epididymal epithelial cells from adult cats. *Theriogenology* 64:1603–18.

Byers, S. W., M. A. Hadley, D. Djakiew and M. Dym. 1986. Growth and characterization of polarized monolayers of epididymal epithelial cells and Sertoli cells in dual environment culture chambers. *J Androl* 7:59–68.

Byers, S. W., N. A. Musto and M. Dym. 1985. Culture of ciliate dand nonciliated cells from rat ductuli efferentes. *J Androl* 6:271–78.

Cameo, M. S. and J. A. Blaquier. 1976. Androgen-controlled specific proteins in rat epididymis. *J Endocrinol* 69:317–24.

Chapin, R. E., T. R. Winton, W. S. Nowland, S. W. Kumpf, S. Davenport, D. Karanian, R. D. Streck, T. M. Coskran, E. G. Barbacci-Tobin, C. Houle and S. N Campion. 2014. Primary cell cultures for understanding rat epididymal inflammation. *Birth Defects Res B Dev Reprod Toxicol* 101:325–32.

Chaurand, P., S. Fouchécourt, B. B. DaGue, B. J. Xu, M. L. Reyzer, M. C. Orgebin-Crist and R. M. Caprioli. 2003. Profiling and imaging proteins in the mouse epididymis by imaging mass spectrometry. *Proteomics* 3:2221–39.

Clauss, A., M. Persson, H. Lilja and A. Lundwall. 2011. Three genes expressing Kunitz domains in the epididymis are related to genes of WFDC-type protease inhibitors and semen coagulum proteins in spite of lacking similarity between their protein products. *BMC Biochem* 12:1–13.

Clermont, Y. and J. Flannery. 1970. Mitotic activity in the epithelium of the epididymis in young and old adult rats. *Biol Reprod* 3:283–92.

Clulow, J., R. C. Jones, L. A. Hansen and S. Y. Man. 1998. Fluid and electrolytere absorption in the ductuli efferentestestis. *J Reprod Fertil Suppl* 53:1–14.

Cohen, D. J., V. G. DaRos, D. Busso, D. A. Ellerman, J. A. Maldera, N. Goldweic and P. S. Cuasnicú. 2007. Participation of epididymal cysteine-rich secretory proteins in sperm-egg fusion and their potential use for male fertility regulation. *Asian J Androl* 9:528–32.

Cohen, D. J., D. A. Ellerman and P. S. Cuasnicu. 2000. Mammalian sperm–egg fusion: Evidence that epididymal protein DE plays a role in mouse gamete fusion, *Biol Reprod* 63:462–68.
Com, E., F. Bourgeon, B. Evrard, T. Ganz, D. Colleu, B. Jégou and C. Pineau. 2003. Expression of antimicrobial defensins in the male reproductive tract of rats, mice, and humans. *Biol Reprod* 68:95–104.
Cooper, N. J., M. K. Holland and W. G. Breed. 1998. Extra testicular sperm maturation in the brush-tail possum, *Trichosurus vulpecula*. *J Reprod Fertil Suppl* 53:221–26.
Cooper, T. G. 1992. The epididymis as a site of contraceptive attack. In *Spermatogenesis, Fertilization, Contraception, Molecular, Cellular and Endocrine Events in Male Reproduction*, eds. E. Nieschlag and U. F. Habenicht, 419–60. Springer-Verlag, Berlin.
Cooper, T. G. 1998a. Interactions between epididymal secretions and spermatozoa. *J Reprod Fertil Suppl* 53:119–36.
Cooper, T. G. 1998b. Epididymis. In *Encyclopedia of Reproduction*, eds. J. D. Neill and E. Knobil, 1–17. Academic Press, San Diego.
Cooper, T. G. 1999. Immunology of the epididymis. *Andrologia* 31:322.
Cooper, T. G. 2007. Sperm maturation in the epididymis: A new look at an old problem. *Asian J Androl* 9:533–39.
Cooper, T. G., C. H. Yeung and R. Meyer. 1989. Immature rat epididymal epithelial cells grown in static primary monolayer culture on permeable supports. I. Vectorial secretion. *Cell Tissue Res* 256:567–72.
Cooper, T. G., C. H. Yeung, R. Meyer and H. Schulze. 1990. Maintenance of human epididymal epithelial cell function in monolayer culture. *J Reprod Fertil* 90:81–91.
Cornwall, G. A. 2009. New insights into epididymal biology and function. *Hum Reprod Update* 15:213–27.
Cornwall, G. A. 2014. Role of posttranslational protein modifications in epididymal sperm maturation and extracellular quality control. *Adv Exp Med Biol* 759:159–80.
Cornwall, G. A. and S. R. Hann. 1995. Transient appearance of CRES protein during spermatogenesis and caput epididymal sperm maturation. *Mol Reprod Dev* 41: 37–46.
Cosentino, M. J. and A. T. Cockett. 1986. Structure and function of the epididymis. *Urol Res* 14:229–40.
Cuasnicu, P. S., D. Conesa and L. Rochwerger. 1990. Potential contraceptive use of an epididymal protein that participates in fertilization. In *Gamete Interaction: Prospects for Immunocontraception*, eds. N. J. Alexander, D. Griffin, J. M. Spieler and G. M. H. Waites, 143–215. Wiley-Liss, New York.
Curtis, S. K. and R. P. Amann. 1981. Testicular development and establishment of spermatogenesis in Holstein bulls. *J Anim Sci* 53:1645–57.
Cuthbert, A. W. and P. Y. Wong. 1986. Electrogenic anion secretion in cultured rat epididymal epithelium. *J Physiol* 378:335–45.
Cyr, D. G., B. Robaire and L. Hermo. 1995. Structure and turnover of junctional complexes between principal cells of the rat epididymis. *Microsc Res Tech* 30:54–66.
D'Amours, O., G. Frenette, L. J. Bordeleau, N. Allard, P. Leclerc. P. Blondin and R. Sullivan. 2012. Epididymosomes transfer epididymal sperm binding protein 1 (ELSPBP1) to dead spermatozoa during epididymal transit in bovine. *Biol Reprod* 87:1–11.
Da Silva, N., C. Silberstein, V. Beaulieu, C. Piétrement, A. N. Van Hoek, D. Brown and S. Breton. 2006. Postnatal expression of aquaporins in epithelial cells of the rat epididymis. *Biol Reprod* 74:427–38.

Dacheux, J. L. and F. Dacheux. 2014. New insights into epididymal function in relation to sperm maturation. *Reproduction* 147:27–42.

Dacheux, J. L., M. Belghazi, Y. Lanson and F. Dacheux. 2006. Human epididymal secretome and proteome. *Mol Cell Endocrinol* 250:36–42.

Dacheux, J. L., C. Belleannee, B. Guyonnet, V. Labas, A. P. Teixeira-Gomes, H. Ecroyd, X. Druart, J. L. Gatti and F. Dacheux. 2012. The contribution of proteomics to understanding epididymal maturation of mammalian spermatozoa. *Syst Biol Reprod Med* 58:197–210.

Dacheux, J. L., X. Druart, S. Fouchecourt, P. Syntin, J. L. Gatti, N. Okamura and F. Dacheux. 1998. Role of epididymal secretory proteins in sperm maturation with particular reference to the boar. *J Reprod Fertil Suppl* 53:99–107.

Damoiseaux, J. G., E. A. Döpp, W. Calame, D. Chao, G. G. MacPherson and C. D. Dijkstra. 1994. Rat macrophage lysosomal membrane antigen recognized by monoclonal antibody ED1. *Immunology* 83:140–47.

DeMiguel, M. P., J. M. Mariño, P. Gonzalez-Peramato, M. Nistal and J. Regadera. 2001. Epididymal growth and differentiation are altered in human cryptorchidism. *J Androl* 22:212–25.

Dorin, J. R. and C. L. R. Barratt. 2014. Importance of b-defensins in sperm function. *Mol Hum Reprod* 20:821–26.

Dubé, É., J. Dufresne, P. T. K. Chan, L. Hermo and D. G. Cyr. 2010. Assessing the role of Claudins in maintaining the integrity of epididymal tight junctions using novel human epididymal cell lines. *Biol Reprod* 82:1119–28.

Eddy, E. M. and D. A. O'Brien. 1994. The spermatozoon. In *The Physiology of Reproduction*, eds. E. Knobil and J. D. Neill, 2nd edn., 29–77. Raven Press, New York.

Ellerman, D. A., V. G. Da Ros, D. J. Cohen, D. Busso, M. M. Morgenfeld and P. S. Cuasnicú. 2002. Expression and structure-function analysis of DE, a sperm cysteine-rich secretory protein that mediates gamete fusion. *Biol Reprod* 67:1225–31.

Ezer, N. and B. Robaire. 2002. Androgenic regulation of the structure and functions of the epididymis. In *The Epididymis: From Molecules to Clinical Practice*, eds. B. Robaire and B. T. Hinton, 297–316. Kluwer Academic/Plenum Publishers, New York.

Fahim, M. A., S. Tariq and E. Adeghate. 2013. Vitamin E modifies the ultra structure of testis and epididymis in mice exposed to lead in toxication. *Ann Anat* 195:272–27.

Faisal, K., V. S. Periasamy, S. Sahabudeen, A. Radha, R. Anandhi and M. A. Akbarsha. 2008. Spermetotoxic effect of aflatoxin B1 in rat: Extrusion of outer dense fibers and axonemal microtubule doublets of sperm flagellum. *Reproduction* 135:303–10.

Feuchter, F. A., A. J. Tabet and M. F. Green. 1988. Maturation antigen of the mouse sperm flagellum I: Analysis of its secretion, association with sperm and function. *Am J Anat* 181:67–76.

Flickinger, C. J. 1972. Alterations in the fine structure of the rat epididymis after vasectomy. *Anat Rec* 173:277–300.

Flickinger, C. J. 1975. Fine structure of the rabbit testis after vasectomy. *Biol Reprod* 13:61–67.

Flickinger, C. J. 1982. The fate of sperm after vasectomy in the hamster. *Anat Rec* 202:231–39.

Flickinger, C. J., L. A. Bush, S. S. Howards and J. C. Herr. 1997. Distribution of leukocytes in the epithelium and interstitium of four regions of the Lewis rat epididymis. *Anat Rec* 248:380–90.

Fouchécourt, S., F. Dacheux and J. L. Dacheux. 1999. Glutathione-independent prostaglandin D2 synthase in ram and stallion epididymal fluids: Origin and regulation. *Biol Reprod* 60:558–66.

Frenette, G. and J. Girouard. 2007. Epididymosomes are involved in the acquisition of new sperm proteins during epididymal transit. *Asian J Androl* 9:483–91.

Frenette, G., C. Lessard, E. Madore, M. A. Fortier and R. Sullivan. 2003. Aldose reductase and macrophage migration inhibitory factor are associated with epididymosomes and spermatozoa in the bovine epididymis. *Biol Reprod* 69:1586–92.

Garberi, J. C., J. D. Fontana and J. A. Blaquier. 1982. Carbohydrate composition of specific rat epididymal protein. *Int J Androl* 5:619–26.

Garrett, J. E., S. H. Garrett and J. Douglass. 1990. A spermatozoa-associated factor regulates proenkephalin gene expression in the rat epididymis. *Mol Endocrinol* 4:108–18.

Garside, D. A., A. M. Gebril, M. Alsaadi, N. Nimmo, A. Mullen and V. Ferro. 2013. An update on the potential for male contraception: Emerging options. *Open Access J Contracep* 4:1–11.

Gatti, J. L., S. Métayer, M. Belghazi, F. Dacheux and J. L. Dacheux. 2005. Identification, proteomic profiling, and origin of ramepididymal fluidexosome-likevesicles. *Biol Reprod* 72:1452–65.

Girgis, S. M., A. N. Etriby, A. A. Ibrahim and S. A. Kahil. 1969. Testicular biopsy in azoospermia. A review of the last ten years' experiences of over 800 cases. *Fertil Steril* 20:467–77.

González, E. F., P. S. Cuasnicú, A. Piazza, L. Piñeiro and J. A. Blaquier. 1984. Addition of an androgen-free epididymal protein extract increases the ability of immature hamster spermatozoa to fertilize in vivo and in vitro. *J Reprod Fertil* 71:433–37.

Goyal, H. O., V. Hutto and M. A. Maloney. 1994. Effects of androgen deprivation in the goat epididymis. *Acta Anat (Basel)* 150:127–35.

Goyal, H. O. and C. S. Williams. 1991. Regional differences in the morphology of the goat epididymis: A light microscopic and ultrastructural study. *Am J Anat* 190:349–69.

Gray, L. E. Jr. 1998. Xenoendocrine disrupters: Laboratory studies on male reproductive effects. *Toxicol Lett* 102–103:331–35.

Gray, L. E. Jr., J. Ostby, J. Ferrell, G. Rehnberg, R. Linder, R. Cooper, J. Goldman, V. Slott and J. Laskey. 1989. A dose-response analysis of methoxychlor-induced alterations of reproductive development and function in the rat. *Fundam Appl Toxicol* 12:92–108.

Gray, L. E., J. Ostby and W. R. Kelce. 1994. Developmental effects of an environmental antiandrogen: The fungicide vinclozolin alters sex differentiation of the male rat. *Toxicol Appl Pharmacol* 129:46–52.

Gupta, G. S. 2005. *Proteomics of Spermatogenesis*. Springer, New York.

Hamil, K. G., Q. Liu, P. Sivashanmugam, S. Yenugu, R. Soundararajan, G. Grossman, R. T. Richardson et al. 2002. Cystatin 11: A new member of the cystatin type 2 family. *Endocrinol* 143:2787–96.

Hamil, K. G., P. Sivashanmugam, R. T. Richardson, G. Grossman, S. M. Ruben, J. L. Mohler, P. Petrusz, M. G. O'Rand, F. S. French and S. H. Hall. 2000. HE2beta and HE2gamma, new members of an epididymis-specific family of androgen-regulated proteins in the human. *Endocrinol* 141:1245–53.

Hamilton, D. W., J. C. Wenstrom and A. Moore. 1986. Characterization of a sperm membrane glycoprotein. *Adv Exp Med Biol* 205:121–29.

Hamzeh, M. and B. Robaire. 2011. Androgens activate mitogen-activated protein kinase via epidermal growth factor receptor/insulin-like growth factor 1 receptor in the mouse PC-1 cell line. *J Endocrinol* 209:55–64.

Harris, A. and L. Coleman. 1989. Ductal epithelial cells cultured from human foetal epididymis and vasdeferens: Relevance tosterility in cystic fibrosis. *J Cell Sci* 92:687–90.

Hermo, L. 1995. Structural features and functions of principal cells of the intermediate zone of the epididymis of adult rats. *Anat Rec* 242:515–30.

Hermo, L., K. Barin and B. Robaire. 1992. Structural differentiation of the epithelial cells of the testicular excurrent duct system of rats during postnatal development. *Anat Rec* 233: 205–28.

Hermo, L., J. Dworkin and R. Oko. 1998. Role of epithelial clear cells of the rat epididymis in the disposal of the contents of cytoplasmic droplets detached from spermatozoa. *Am J Anat* 183:107–24.

Hermo, L., H. Green and Y. Clermont. 1991a. Golgiapparatus of epithelial principal cells of the epididymal initial segment of the rat: Structure, relationship with endoplasmic reticulum, and role in the formation of secretory vesicles. *Anat Rec* 229:159–76.

Hermo, L. and D. Jacks. 2002. Nature's ingenuity: Bypassing the classical secretory route via apocrine secretion. *Mol Reprod Dev* 63:394–410.

Hermo, L., R. Oko and C. R. Morales. 1991c. Endocytosis and secretion of proteins in the extratesticular duct system of the adult male rat. *Bull Assoc Anat (Nancy)* 75:147–51.

Hermo, L., S. Papp and B. Robaire. 1994. Developmental expression of the Yf subunit of glutathione S-transferase P in epithelial cells of the testis, efferent ducts, and epididymis of the rat. *Anat Rec* 239:421–40.

Hermo, L. and B. Robaire. 2002. Epididymal cell types and their functions. In *The Epididymis: From Molecules to Clinical Practice,* eds. B. Robaire and B. T. Hinton, 81–102. Plenum Press, New York.

Hermo, L., J. Wright, R. Oko and C. R. Morales. 1991b. Role of epithelial cells of the male excurrent duct system of the rat in the endocytosisor secretion of sulfated glycoprotein-2 (clusterin). *Biol Reprod* 44:1113–31.

Hess, R. A. 2002. The efferent ductules: Structure and functions. In *The epididymis: From molecules to clinical practice*, eds. B. Robaire and B. T. Hinton, 49–80. Plenum, New York.

Hess, R. A., D. H. Gist, D. Bunick, D. B. Lubahn, A. Farrell, J. Bahr, P.S. Cooke and G. L. Greene. 1997. Estrogen receptor (alpha and beta) expression in the excurrent ducts of the adult male rat reproductive tract. *J Androl* 18:602–11.

Hinton, B. T. and H. Hernandez. 1987. Neutral amino acid absorption by the rat epididymis. *Biol Reprod* 37:288–92.

Hinton, B. T. and M. A. Palladino. 1995. Epididymal epithelium: Its contribution to the formation of aluminal fluid micro environment. *Microsc Res Tech* 30:67–81.

Hinton, B. T., Z. J. Lan, D. B. Rudolph, J. C. Labus and R. J. Lye. 1998. Testicular regulation of epididymal gene expression. *J Reprod Fertil Suppl* 53:47–57.

Hinton, B. T., M. A. Palladino, D. Rudolph and J. C. Labus. 1995. The epididymis as protector of maturing spermatozoa. *Reprod Fertil Dev* 7:731–45.

Hoffman, L. H., N. Jahad and M. C. Orgebin-Crist. 1976. The effects of testosterone, 5a-dihydrotestosterone, 3a-androstanediol, and 313-androstanediol on epithelial fine structure of the rabbit epididymis in organ culture. *Cell Tissue Res* 167:493–514.

Holschbach, C. and T. G. Cooper. 2002. A possible extratubular origin of epididymal basal cells in mice. *Reprod* 123:517–25.

Holstein, A. F. 1969. Morphologische Studien am Nebenhoden des Menschen. In *Zwanglose Abhandlungen aus dem Gebiet der normalen und pathologischen Anatomie,* eds. W. Bargmann and W. Doerr, 1–91. Georg Thieme Verlag, Stuttgart.

Hsia, N. and G. A. Cornwall. 2004. DNA microarray analysis of region-specific gene expression in the mouse epididymis. *Biol Reprod* 70:448–57.

Hu, S. G., M. Zou, G.-X. Yao, W.-B. Yao, Q.-L. Zu, X.-Q. Li, Z.-J. Chen and Y. Sun. 2014. Androgenic regulation of beta-defensins in the mouse epididymis. *Reprod Biol Endocrinol* 12:1–9.

Hunnicutt, G. R., D. E. Koppel and D. G. Myles. 1997. Analysis of the process of localization of fertil to the sperm posterior head plasma membrane domain during sperm maturation in the epididymis. *Dev Biol* 191:146–59.

Itoh, M., K. Miyamoto, I. Satriotomo and Y. Takeuchi. 1999. Spermatic granulomata are experimentally induced in epididymides of mice receiving high-dose testosterone implants. I. A light-microscopical study. *J Androl* 20:551–58.

Jelinsky, S. A., T. T. Turner, H. J. Bang, J. N. Finger, M. K. Solarz, E. Wilson, E. L. Brown, G. S. Kopf and D. S. Johnston. 2007. The rat epididymal transcriptome: Comparison of segmental gene expression in the rat and mouse epididymides. *Biol Reprod* 76: 561–70.

Jensen, L. J., A. K. Stuart-Tilley, L. L. Peters, S. E. Lux, S. L. Alper and S. Breton. 1999. Immunolocalization of AE2 anion exchanger in rat and mouse epididymis. *Biol Reprod* 61:973–80.

Jervis, K. M. and B. Robaire. 2001. Dynamic changes in gene expression along the rat epididymis. *Biol Reprod* 65:696–703.

Jin, Y. Z., S. Bannai, F. Dacheux, J. L. Dacheux and N. Okamura. 1997. Direct evidence for the secretion of lactoferrin and its binding to sperm in the porcine epididymis. *Mol Reprod Dev* 47: 490–96.

Johnson, L. and D. D. Varner. 1988. Effect of daily spermatozoan production but not age on transit time of spermatozoa through the human epididymis. *Biol Reprod* 39: 812–17.

Johnston, D. S., S. A. Jelinsky, H. J. Bang, P. DiCandeloro, E. Wilson, G. S. Kopf and T. T. Turner. 2005. The mouse epididymal transcriptome: Transcriptional profiling of segmental gene expression in the epididymis. *Biol Reprod* 73: 404–13.

Jones, R. 1998. Plasma membrane structure and remodelling during sperm maturation in the epididymis. *J Reprod Fertil Suppl* 53:73–84.

Jones, R. 2004. Sperm survival versus degradation in the mammalian epididymis: A hypothesis. *Biol Reprod* 71:1405–11.

Jones, R. C. and J. Clulow. 1987. Regulation of the elemental composition of the epididymal fluids in the tammar, *Macropus eugenii*. *J Reprod Fertil* 81:583–90.

Jones, R., D. W. Hamilton and D. W. Fawcett. 1979. Morphology of the epithelium of the extratesticular rete testis, ductuli efferentes and ductus epididymidis of the adult male rabbit. *J Anat* 156:373–400.

Jones, R., A. Ma, S. T. Hou, R. Shalgi and L. Hall. 1996. Testicular biosynthesis and epididymal endoproteolytic processing of rat sperm surface antigen 2B1. *J Cell Sci* 109:2561–70.

Jones, R. C. and R. N. Murdoch. 1996. Regulation of the motility and metabolism of spermatozoa for storage in the epididymis of eutherian and marsupial mammals. *Reprod Fertil Dev* 8:553–68.

Joshi, M. S. 1985. Isolation and cell culture of the epithelial cells of cauda epididymidis of the bull. *Biol Reprod* 33:187–200.

Jun, H. J., J. Roy, T. B. Smith, L. B. Wood, K. Lane, S. Woolfenden, D. Punko, R. T. Bronson, K. M. Haigis, S. Breton and A. Charest. 2014. ROS1 signaling regulates epithelial differentiation in the epididymis. *Endocrinol* 155:3661–73.

Kierszenbaum, A. L., M. Feldman, O. Lea, W. A. Spruill, L. L. Tres, P. Petrusz and F. S. French. 1980. Localization of androgen-binding protein in proliferating Sertoli cells in culture. *Proc Natl Acad Sci USA* 77:5322–26.

Kirchhoff, C. 1999 Gene expression in the epididymis. *Int Rev Cytol* 188:133–202.

Kirchhoff, C. 2002. Specific gene expression in the human and non-human primate epididymis. In *The Epididymis: From Molecules to Clinical Practice*, eds. B. Robaire and B. T. Hinton, 201–18. Kluwer Academic/Plenum Publishers, New York.

Klein, C., M. H. Troedsson and J. Rutllant. 2013. Region-specific expression of aquaporin subtypes in equine testis, epididymis, and ductus deferens. *Anat Rec* 296:1115–26.

Klinefelter, G. R. and D. W. Hamilton. 1984. Organ culture of rat caput epididymal tubules in a perifusion chamber. *J Androl* 5:243–58.

Klinefelter, G. R. and D. W. Hamilton. 1985. Synthesis and secretion of proteins by perifused caput epididymal tubules, and association of secreted proteins with spermatozoa. *Biol Reprod* 33:1017–27.

Klinefelter, G. R. and J. D. Suarez. 1997. Toxicant-induced acceleration of epididymal spermtransit: Androgen-dependent proteins may be involved. *Reprod Toxicol* 11:511–19.

Klinefelter, G. R., J. W. Laskey, W. R. Kelce, J. Ferrell, N. L. Roberts, J. D. Suarez and V. Slott. 1994. Chloroethyl methane sulfonate-induced effects on the epididymis seem unrelated to altered Leydig cell function. *Biol Reprod* 51:82–91.

Klinefelter, G. R., J. W. Laskey, N. R. Roberts, V. Slott and J. D. Suarez. 1990. Multiple effects of ethane dimethanesulfonate on the epididymis of adult rats. *Toxicol Appl Pharmacol* 105:271–87.

Kohane, A. C., M. S. Cameo, L. Pineiro, J. C. Garberi and J. A. Blaquier. 1980. Distribution and site of production of specific proteins in the rat epididymis. *Biol Reprod* 23:181–87.

Krätzschmar, J., B. Haendler, U. Eberspaecher, D. Roosterman, P. Donner and W. D. Schleuning. 1996. The human cysteine-rich secretory protein (CRISP) family. Primary structure and tissue distribution of CRISP-1, CRISP-2 and CRISP-3. *Eur J Biochem* 236:827–36.

Kroovand, R. L. and A. D. Perlmutter. 1981. Congenital anomalies of the vas deferens and epididymis. In *Pediatric Andrology*, eds. S. J. Kogan and E. S. E. Hafez, 173–80. Martinus Nijhoff Publishers, Boston.

Lacham, O. and A. Trounson. 1991. Fertilizing capacity of epididymal and testicular spermatozoa microinjected under the zona pellucida of the mouse oocyte. *Mol Reprod Dev* 29:85–93.

Lan, Z. J., J. C. Labus and B. T. Hinton. 1998. Regulation of gamma-glutamyl transpeptidase catalytic activity and protein level in the initial segment of the rat epididymis by testicular factors: Role of basic fibroblast growth factor. *Biol Reprod* 58:197–206.

Lasserre, A., R. Barrozo, J. G. Tezón, P. V. Miranda and M. H. Vazquez-Levin. 2001. Human epididymal proteins and sperm function during fertilization. *Biol Res* 34:165–78.

Lee, I. C., K. H. Kim, S. H. Kim, H.S. Baek, C. Moon, S. H. Kim, W. K. Yun, K. H. Nam, H. C. Kim and J. C. Kim. 2013. Apoptotic cell death in rat epididymis following epichlorohydrin treatment. *Hum Exp Toxicol* 32:640–46.

Leung, A. Y. H., P. Y. D. Wong, J. R. Yankaskas and R. C. Boucher. 1996. cAMP- but not Ca2(+)-regulated Cl– conductance is lacking in cystic fibrosis mice epididymides and seminal vesicles. *Am J Physiol* 271:C188-C93.

Leung, G. P., K. H. Cheung, C. T. Leung, M. W. Tsang and P. Y. Wong. 2004. Regulation of epididymal principal cell functions by basal cells: Role of transient receptor potential (Trp) proteins and cyclooxygenase-1 (COX-1). *Mol Cell Endocinol* 216:5–13.

Leung, G. P. H., C. M. Tse, S. B. Chew and P. Y. D. Wong. 2001. Expression of multiple Na^+/H^+ exchanger isoforms in cultured epithelial cells from rat efferent duct and cauda epididymidis. *Biol Reprod* 64:482–90.

Leung, P. S., X. Q. Yao, H. C. Chan, L. X. M. Fu and P. Y. D. Wong. 1998. Differential gene expression of angiotensin II receptor subtypes in the epididymides of mature and immature rats. *Life Sci* 62:461–68.

Li, P., H. C. Chan, B. He, S. C. So, Y. W. Chung, Q. Shang, Y. D. Zhang and Y. L. Zhang. 2001. An antimicrobial peptide gene found in the male reproductive system of rats. *Science* 291:1783–85.

Liu, H. W., Y. C. Lin, C. F. Chao, S. Y. Chang and G. Sun. 2000. GP-83 and GP-39, two glycoproteins secreted by human epididymis were conjugated to sperm during maturation. *Mol Hum Reprod* 6:422–28.

Lundwall, A. 2007. A locus on chromosome 20 encompassing genes that are highly expressed in the epididymis. *Asian J Androl* 9:540–44.

Malm, J., O. Sorensen, T. Persson, M. Frohm-Nilsson, B. Johansson, A. Bjartell, H. Lilja, M. Ståhle-Bäckdahl, N. Borregaard and A. Egesten. 2000. The human cationic antimicrobial protein (hCAP-18) is expressed in the epithelium of human epididymis, is present in seminal plasma at high concentrations, and is attached to spermatozoa. *Infect Immun* 68:4297–02.

Markkula-Viitanen, M., V. Nikkanen and A. Talo. 1979. Electrical activity and intraluminal pressure of the cauda epididymidis of the rat. *J Reprod Fertil* 57:431–35.

Mewe, M., C. K. Bauer, J. R. Schwarz and R. Middendorff. 2006. Mechanisms regulating spontaneous contractions in the bovine epididymal duct. *Biol Reprod* 75:651–59.

Miranda, P. V., A. Brandelli and J. G. Tezón. 1995. Characterization of beta-N-acetylglucosaminidase from human epididymis. *Int J Androl* 18:263–70.

Mital, P., B. T. Hinton and J. M. Dufour. 2011. The blood-testis and blood-epididymis barriers are more than just their tight junctions. *Biol of Reprod* 84:851–58.

Nicander, L. 1957. On the regional histology and cytochemistry of the ductus epididymidis in rabbits. *Acta Morphol Neerl Scand* 1:99–118.

Nicander, L. 1958. Studies on the regional histology and cytochemistry of the duct use pididymidis installions, rams and bulls. *Acta Morphol Neerl Scand* 1:337–62.

Nicander, L. and T. D. Glover. 1973. Regional histology and fine structure of the epididymal duct in the golden hamster (*Mesocricetus auratus*). *J Anat* 114:347–64.

Noblanc, A., A. Kocer and J. R. Drevert. 2012. Post-testicular protection of male gametes from oxidative damage. The role of the epididymis. *Med Sci* (Paris) 28:519–25.

Nonogaki, T., Y. Noda, K. Narimoto, M. Shiotani, T. Mori, T. Matsuda and O. Yoshida. 1992. Localization of CuZn-superoxide dismutase in the human male genital organs. *Hum Reprod* 7:81–85.

O'Bryan, M. K., C. Mallidis, B. F. Murphy and H. W. Baker. 1994. Immunohistological localization of clusterin in the male genital tract in humans and marmosets. *Biol Reprod* 3:502–09.

O'Rand, M. G., E. E. Widgren, K. G. Hamil, E. J. Silva and R. T. Richardson. 2011. Epididymal protein targets: A brief history of the development of epididymal protease inhibitor as a contraceptive. *J Androl* 32:698–704.

Orgebin-Crist, M. C. and N. Jahad. 1979. The maturation of rabbit epididymal spermatozoa in organculture: Stimulation by epididymal cytoplasmic extracts. *Biol Reprod* 21:511–15.

Orsi, A. M., P. Pinto e Silva, S. Mello Dias and V. R. de Melo. 1980. Anatomy of the epididymis of the South American opossum (*Didelphis azarae*). *Anat Histol Embryol* 9:164–68.

Palacios, J., J. Regadera, M. Nistal and R. Paniagua. 1991. Apicalmitochondria-rich cells in the human epididymis: A nultra structural, enzymohisto chemical, and immunohisto chemical study. *Anat Rec* 231:82–88.

Palladino, M. A., T. A. Mallonga and M. S. Mishra. 2003. Messenger RNA (mRNA) expression for the antimicrobial peptides beta-defensin-1 and beta-defensin-2 in the male rat reproductive tract: Beta-defensin-1 mRNA in initial segment and caput epididymidis is regulated by androgens and not bacterial lipopolysaccharides. *Biol Reprod* 68:509–15.

Park, H. K., S. H. Paick, H. G. Kim, Y. S. Lho and S. R. Bae. 2014. Induction of contraception by intraepididymal sclerotherapy. *World J Mens Health* 32:83–86.

Pastor-Soler, N. M., K. R. Hallows, C. Smolak, F. Gong, D. Brown and S. Breton. 2008. Alkaline pH- and cAMP-induced V-ATPase membrane accumulation is mediated by protein kinase A in epididymal clear cells. *Am J Physiol Cell Physiol* 294:C488–94.

Păunescu, T. G., W. W. Shum, C. Huynh, L. Lechner, B. Goetze, D. Brown and S. Breton. 2014. High-resolution heliumion microscopy of epididymal epithelial cells and their interaction with spermatozoa. *Mol Hum Reprod* 20:929–37.

Pearl, A. and J. F. Roster. 2014. Lactoferrin expression and secretion in the stallion epididymis. *Reprod Biol* 14:148–154.

Penttinen, J., D. A. Pujianto, P. Sipila, I. Huhtaniemi and M. Poutanen. 2003. Discovery in silico and characterization in vitro of novelgene sexclusively expressed in the mouse epididymis. *Mol Endocrinol* 17:2138–51.

Phelps, B. M., D. E. Koppel, P. Primakoff and D. G. Myles. 1990. Evidence that proteolysis of the surface is an initial step in the mechanism of formation of sperm cell surface domains. *J Cell Biol* 111:1839–47.

Pöllänen, P. and T. G. Cooper. 1994. Immunology of the testicular excurrent ducts. *J Reprod Immunol* 26:167–216.

Potts, R. J., T. M. Jefferies and L. J. Notarianni. 1999. Antioxidant capacity of the epididymis. *Hum Reprod* 14:2513–16.

Priesel, A. 1924. Über das Verhalten von Hoden und Nebenhoden bei angeborenem Fehlen des Ductus deferens, zugleich ein Beitrag zur Frage des Vorkommens von Zwischenzellen im menschlichen Nebenhoden. *Virchows Arch* 249:246–304.

Pujianto, D. A., E. Loanda, P. Sari, Y. H. Midoen and P. Soeharso. 2013. Sperm-associated antigen 11A is expressed exclusively in the principal cells of the mouse caput epididymis in an androgen-dependent manner. *Reprod Biol Endocrinol* 11:1–12.

Pyttel, S., A. Nimptsch, J. Böttger, K. Zschörnig, U. Jakop,J. Wegener,K. Müller, U. Paasch, and J. Schiller. 2014. Changes of murine sperm phospholipid composition during epididymal maturation determined by MALDI-TOF mass spectrometry. *Theriogenology* 82:396–402.

Ramos-Ibeas, P., E. Pericuesta, R. Fernández-González, M. A. Ramírez and A. Gutierrez-Adan. 2013. Most regions of mouse epididymis are able to phagocytose immature germ cells. *Reproduction* 146:481–89.

Reddy, E. S. and P. M. Bhargava. 1979. Seminal plasmin – an antimicrobial protein from bovine seminal plasma which acts in E. coli by specific inhibition of rRNA synthesis. *Nature* 279:725–28.

Reid, B. L. and K. W. Cleland. 1957. The structure and function of the epididymis. *Aust J Zoo* 5: 223.

Ricker, D. D. and T. S. Chang. 1996. Neuronal input from the inferior mesenteric ganglion (IMG) affects sperm transport within the rat cauda epididymis. *Int J Androl* 19:371–76.

Ricker, D. D., S. L. Chamness, B. T. Hinton and T. S. Chang. 1996. Changes in luminal fluid protein composition in the rat cauda epididymis following partial sympathetic denervation. *J Androl* 17:117–26.

Robaire, B. and L. Hermo. 1988. Efferent ducts, epididymis, and vas deferens: Structure, functions and their regulation. In *The physiology of reproduction,* eds. E. Knobil and J. D. Neill, 999–1080. Raven Press, New York.

Robaire, R., B. T. Hinton and M. C. Orgebin-Crist. 2006. The epididymis. In *Physiology of Reproduction*, eds. E. Knobil and J. D. Neill, 3rd edn., 1071–147. Elsevier, New York.

Rochwerger, L. and P. S. Cuasnicu. 1992. Redistribution of a rat sperm epididymal glycoprotein after in vivo and in vitro capacitation. *Mol Reprod Dev* 31:34–41.

Rodríguez, C. M., J. C. Labus and B. T. Hinton. 2002. Organic cation/carnitine transporter, OCTN2, is differentially expressed in the adult rat epididymis. *Biol Reprod* 67:314–19.

Saez, F., G. Frenette and R. Sullivan. 2003. Epididymosomes and prostasomes: Their roles in posttesticular maturation of the sperm cells. *J Androl* 24:149–54.

Sakkas, D., E. Mariethoz, G. Manicardi, D. Bizzaro, P. G. Bianchi and U. Bianchi. 1999. Origin of DNA damage in ejaculated human spermatozoa. *Rev Reprod* 4:31–37.

Scheer, H. and B. Robaire. 1980. Steroid delta 4-5 alpha-reductase and 3 alpha-hydroxysteroid dehydrogenase in the rat epididymis during development. *Endocrinol* 107: 948–53.

Schimming, B. C. and C. A. Vicentini. 2008. Morphological features of the apical region in the principal cells of mongrel dog epididymis. *Int J Morphol* 26:149–53.

Schwaab, V., J. J. Lareyre, P. Vernet, E. Pons, J. Faure, J. P. Dufaure and J. R. Drevet. 1998. Characterization, regulation of the expression and putative roles of two glutathione peroxidase proteins found in the mouse epididymis. *J Reprod Fertil Suppl* 53: 157–62.

Seiler, P., T. G. Cooper and E. Nieschlag. 2000. Sperm number and condition affect the number of basal cells and their expression of macrophage antigen in the murine epididymis. *Int J Androl* 23:65–76.

Seiler, P., T. G. Cooper, C. H. Yeung and E. Nieschlag. 1999. Regional variation in macrophage antigen expression by murine epididymal basal cells and their regulation by testicular factors. *J Androl* 20:738–46.

Seiler, P., I. Wenzel, A. Wagenfeld, C. H. Yeung, E. Nieschlag and T. G. Cooper. 1998. The appearance of basal cells in the developing murine epididymis and their temporal expression of macrophage antigens. *Int J Androl* 21:217–26.

Serre, V. and B. Robaire. 1998. Segment-specific morphological changes in aging Brown Norway rat epididymis. *Biol Reprod* 58:497–513.

Serre, V. and B. Robaire. 1999. Distribution of immune cells in the epididymis of the aging Brown Norway rat is segment-specific and related to the luminal content. *Biol Reprod* 61:705–14.

Sever, D. M. 2010. Ultra structure of the reproductive system of the black swamp snake (*Seminatrix pygaea*). VI. Anterior testicular ducts and their nomenclature. *J Morphol* 271:104–115.

Shum, W. W., T. B. Smith and V. Cortez-Retamozo. 2014. Epithelial basal cells are distinct from dendritic cells and macrophages in the mouse epididymis. *Biol Reprod* 90:90.

Shum, W. W., N. DaSilva, D. Brown and S. Breton. 2009. Regulation of luminal acidification in the male reproductive tract via cell-cell crosstalk. *J Exp Biol* 212:1753–61.

Shum, W. W, N. Da Silva, M. McKee, P. J. Smith, D. Brown and S. Breton. 2008. Transepithelial projections from basal cells are luminal sensors in pseudostratified epithelia. *Cell* 135:1108–17.

Shum, W. W. C., E. Hill, D. Brown and S. Breton. 2013. Plasticity of basal cells during postnatal development in the rat epididymis. *Reproduction* 146:455–69.

Silva, E. J. R., K. G. Hamil, R. T. Richardson and M. G. O'Rand. 2012. Characterization of EPPIN's semenogelin I binding site: A contraceptive drug target. *Biol Reprod* 87:1–8.

Sinha Hikim, A. P. and R. S. Swerdloff. 1999. Hormonal and genetic control of germ cell apoptosis in the testis. Rev Reprod 4:38–47.

Solomon, H. M., P. J. Wier, D. L. Ippolito and T. V. Toscano. 1997. Effect of prazosin on sperm transport in male rats. *Reprod Toxicol* 11:627–31.

Srivastav, A. 2000. Maturation-dependent glycoproteins containing both N- and O-linked oligosaccharides in epididymal sperm plasma membrane of rhesus monkeys (*Macaca mulatta*). *J Reprod Fertil* 119:241–52.

Stanley, A. and M. A. Akbarsha. 1992. Male reproductive toxicity of the anticancer drug vincristine: A review. In *Recent Advances in Life Sciences*, eds. A. K. Saxena, R. Ramamurthi, G. Sree and V. L. Saxena, 122–136. Manu Publications, Kanpur, India.

Strous, G. J. and R. Govers. 1999. The ubiquitin-proteasome system and endocytosis. *J Cell Sci* 112:1417–23.

Sullivan, R., J. Cabellero, O. D'Amours and C. Belleannée. 2015. The sperm journey in the excurrent duct: Functions of microvesicles on sperm maturation and gene expression along the epididymis. *Anim Reprod* 12:88–92.

Sullivan, R. and F. Saez. 2013. Epididymosomes, prostasomes, and liposomes: Their roles in mammalian male reproductive physiology. *Reproduction* 146:21–35.

Sullivan, R., F. Saez, J. Girouard and G. Frenette. 2005. Role of exosomes in sperm maturation during the transit along the male reproductive tract. *Blood Cells Mol Dis* 35:1–10.

Sun, E. L. and C. J. L. Flickinger. 1979. Development of cell types and regional differences in the postnatal rat epididymis. *Am J Anat* 154:27–56.

Suryawanshi, A. R., S. A. Khan, R. K. Gajbhiye, M. Y. Guray and V. V. Khole. 2011. Differential proteomics leads to identification of domain-specific epididymal sperm proteins. *J Androl* 32:240–59.

Sutovsky, P., R. Moreno, S. J. Ramalho-Santo, T. Dominko and W. E. Thompson. 2001. A putative, ubiquitin-dependent mechanism for the recognition and elimination of defective spermatozoa in the mammalian epididymis. *J Cell Sci* 114:1665–75.

Temple-Smith, P. D., S. S. Zheng, T. Kadioglu and G. J. Southwick. 1998. Development and use of surgical procedures to bypass selected regions of the mammalian epididymis: Effects on sperm maturation. *J Reprod Fertil Suppl* 53:183–95.

Tezón, J. G. and J. A. Blaquier. 1981. The organ culture of human epididymal tubules and their response to androgens. *Mol Cell Endocrinol* 21:233–42.

Thimon, V., O. Koukoui, E. Calvo and R. Sullivan. 2007. Region-specific gene expression profiling along the human epididymis. *Mol Hum Reprod* 13:691–704.

Toshimori, K., S. Araki and C. Oura. 1990. Epithelial cells with vacuoles containing 54,000 dalton sialoglyco protein in the mouse epididymal duct. *Arch Histol Cytol* 53:333–38.

Tracy, C. R., W. D. Steers and R. Costabile. 2008. Diagnosis and management of epididymitis. *Urol Clin North Am* 35:101–08.

Tubbs, C. E., J. C. Hall, R. O. Scott, V. P. Clark, T. L. Hermon and C. Bazemore-Walker. 2002. Binding of protein D/E to the surface of rat epididymal sperm before ejaculation and after deposition in the female reproductive tract. *J Androl* 23:512–21.

Tulsiani, D. R., S. K. NagDas, M. D. Skudlarek and M. C. Orgebin-Crist. 1995. Rat sperm plasma membrane mannosidase: Localization and evidence for proteolytic processing during epididymal maturation. *Dev Biol* 167:584–95.

Turner, T. T. 2008. De Graaf's thread: The human epididymis. *J Androl* 29:237–50.

Turner, T. T. and T. A. Riley. 1999. p53 independent, region-specific epithelial apoptosis is induced in the rat epididymis by deprivation of luminal factors. *Mol Reprod Dev* 53:188–97.

Vázquez, M. H., M. A. deLarminat and J. A. Blaquier. 1986. Effect of androgen on androgen receptors in cultured human epididymis. *J Endocrinol* 111:343–48.

Veri, J. P., L. Hermo and B. Robaire. 1993. Immuno cytochemical localization of the Yfsub unit of glutathione S-transferase P shows regional variation in the staining of epithelial cells of the testis, efferent ducts, and epididymis of the male rat. *J Androl* 14: 23–44.

Verma, R. J. 2001. Sperm quiescence in cauda epididymis. *Asian J Androl* 3:181–183.

Von Horsten, H. H., S. S. Johnson, S. K. SanFrancisco, M. C. Hastert, S. M. Whelly and G. A. Cornwall. 2007. Oligomerization and transglutaminase cross-linking of the cystatin CRES in the mouse epididymal lumen: Potential mechanism of extracellular quality control. *J Biol Chem* 282:32912–23.

Wang, L., Q. Yuan, S. Chen, H. Cai, M. Lu, Y. Liu and C. Xu. 2012. Antimicrobial activity and molecular mechanism of the CRES protein. *PLoS One* 7: e48368.

Wang, Y. F. and A. F. Holstein. 1983. Intraepithelial lymphocytes and macrophages in the human epididymis. *Cell Tissue Res* 233:517–21.

Weir, C. P. and B. Robaire. 2007. Spermatozoa have decreased antioxidant enzymatic capacity and increased reactive oxygen species production during aging in the Brown Norway rat. *J Androl* 28:229–40.

White, S. H., W. C. Wimley and M. E. Selsted. 1995. Structure, function, and membrane integration of defensins. *Curr Opin Struct Biol* 5:521–27.

Wong, P. Y. 1998. CFTR gene and male fertility. *Mol Hum Reprod* 4:107–10.

Wong, P. Y. and C. H. Yeung. 1978. Absorptive and secretory functions of the perfused rat cauda epididymidis. *J Physiol* 275:13–26.

Wong, P. Y. D., X. D. Gong, G. P. H. Leung and B. L. Y. Cheuk. 2002. Formation of the epididymal fluid microenvironment. In: *The Epididymis: From Molecules to Clinical Practice*, eds. B. Robaire and B. T. Hinton, 119–30. Kluwer Academic/Plenum Publishers, Dordrecht/New York.

Wong, P. Y. D., A. Y. F. Tsang and W. M. Lee. 1981. Origin of the luminal fluid proteins of the rat epididymis. *Int J Androl* 4:331–41.

Wright, C., S. Milne and H. Leeson. 2014. Sperm DNA damage caused by oxidative stress: Modifiable clinical, lifestyle and nutritional factors in male infertility. *Reprod Biomed Online* 28:684–703.

Yamazaki, K., T. Adachi, K. Sato, Y. Yanagisawa, H. Fukata, N. Seki, C. Mori and M. Komiyama. 2006. Identification and characterization of novel and unknown mouse epididymis-specific genes by complementary DNA micro array technology. *Biol Reprod* 75:462–68.

Yanagimachi, R. 1994. Mammalian fertilization. In *The Physiology of Reproduction*, eds. E. Knobil and J. D. Neill, 189–317. Raven Press, New York.

Yeung, C. H., M. Bergmann and T. G. Cooper. 1991. Non-specific uptake of IgG by rat epididymal tubules in vitro. *Int J Androl* 14:364–73.

Yeung, C. H., T. G. Cooper and E. Nieschlag. 1997. Human epididymal secreted protein CD52 on ejaculated spermatozoa: Correlations with semen characteristics and the effect of its antibody. *Mol Hum Reprod* 3:1045–51.

Yeung, C. H., D. Nashan, C. Sorg, F. Oberpenning, H. Schulze, E. Nieschlag and T. G. Cooper. 1994. Basal cells of the human epididymis – antigenic and ultrastructural similarities to tissue-fixed macrophages. *Biol Reprod* 50:917–26.

Yoshinaga, K., I. Tanii, T. Oh-Oka and K. Toshimori. 2000. Transport and rearrangement of the intra-acrosomal protein acrin1 (MN7) during spermiogenesis in the guinea pig testis. *Anat Rec* 259:131–40.

Zini, A. and P. N. Schlegel. 1997. Identification and characterization of antioxidant enzyme mRNAs in the rat epididymis. *Int J Androl* 20:86–91.

Zuo, W. L., J. H. Huang, J. J. Shan, S. Li, P. Y. Wong and W. L. Zhou. 2010. Functional studies of acid transporter in cultured rat epididymal cell. *Fertil Steril* 93:2744–49.

Chapter 7

Regulation of Growth and Function of Epididymides

Nirmala Singh Yaduvanshi, Shayu N. Deshpande, Vijayakumar Govindaraj and A. Jagannadha Rao

Contents

7.1 The Male Reproductive System 168
7.2 The Epididymis: Structure, Function and Regulation 168
7.2.1 Structure 168
7.2.1.1 Cell Types of the Epithelium 169
7.2.1.2 Luminal Compartment of the Epididymis 170
7.2.2 Functions of the Epididymis 171
7.2.2.1 Fluid Absorption and Secretion 171
7.2.2.2 Endocytosis 172
7.2.2.3 Protection of the Epithelium and Sperm 172
7.2.2.4 Sperm Storage and Spermiophagy 173
7.2.3 Regulation of Epididymal Growth and Function 173
7.2.3.1 Androgens 174
7.2.3.2 Oestrogens 176
7.2.3.3 Role of FSH in Regulation of Epididymal Functions 188
Acknowledgements 190
References 190

7.1 The Male Reproductive System

The mammalian male reproductive tract includes a pair of testes, which consist of seminiferous tubules separated by the interstitial tissue. The testes are connected to their respective epididymides via the rete testes and efferent ductules. The rete testes, efferent ductules and the epididymides together form the excurrent duct (extragonadal duct) system. Within the seminiferous tubules, the germ cells that eventually develop into spermatozoa are associated closely with the nurse cells called Sertoli cells. The Sertoli cells are responsible for providing nutritional support to the germ cells in addition to providing different growth factors and proteins such as inhibin, ceruloplasmin, transferrin, androgen binding protein, etc. for their development. In addition, the cellular tight junctions formed by the adjacent Sertoli cells are responsible for the formation of the blood–testis barrier, which is critical for excluding components of the immune system from developing an immune response against the developing germ cells. The function of the Sertoli cells is mainly under the control of follicle-stimulating hormone (FSH) and androgens. The interstitial tissue, on the other hand, is composed of the steroidogenic cells called Leydig cells, in addition to mast cells and macrophages. Leydig cells under stimulation by luteinising hormone (LH) are responsible for the secretion of testosterone, a hormone indispensable for spermatogenesis and the development of secondary sexual characteristics in the male. The differentiated spermatozoa are transported in testicular fluid from the seminiferous tubules to the epididymis, where they undergo maturation and storage.

7.2 The Epididymis: Structure, Function and Regulation

7.2.1 Structure

The term *epididymis* has a Greek origin, meaning 'adjacent to the testis'. The origin of the extragonadal ducts, including the epididymis can be traced to the Chondrichthyes, wherein these ducts carried out post-testicular maturation of sperm in addition to storing them for a considerable period. These ducts were envisaged to provide a protected environment for the male gametes. In mammals, the efferent ductules and the epididymis develop from the Wolffian ducts and mesonephric tubules (Byskov and Hoyer, 1994). The latter also gives rise to the kidney, and this in large provided clues in understanding the functions of the excurrent ducts. The numbers of efferent ductules vary from 1 in some marsupials to 14–22 in elephants. The ductuli initially lie parallel to one another, and eventually follow a sinuous path, at times anastomosing each other to form the conus vasculosus. Ultimately, the efferent ductules drain into the epididymis either as a single tubule as seen in the rat, mouse and some guinea pigs or enter the epididymis as parallel ducts as observed in large mammals including humans. The epididymis itself shows diverse structural features in different species. In birds, the epididymis is a fairly simple duct while in higher mammals the epididymal duct is an extensively coiled tubule varying from 3–4 m in man to 80 m in horse. Anatomically, the epididymis can be grossly divided into three segments namely the head (caput), body (corpus) and the tail (cauda). Later a fourth segment upstream of the caput was introduced which had a distinctive histological appearance, and was termed as the initial segment. While morphologically these segments are quite distinct in the rodents, this segmentation cannot be readily visualised in the primates. In the rodent the epididymis can be divided into the caput, corpus and cauda (Robaire et al., 2001).

7.2.1.1 Cell Types of the Epithelium

The epididymis can be divided into a luminal compartment and epithelial compartment (Figure 7.1). The epididymal epithelium is of a simple columnar type and is comprised of various cell types (Table 7.1).

The ESC42 (Epididymis Specific Clone 42) protein is expressed mainly in the caput and efferent ducts, but it is also detected in areas of the corpus and cauda and also present in the lumen and bound to the sperm (Liu et al., 2001).

The interactions among the cells and between the cells and extracellular matrix facilitate formation of a luminal environment distinct from the surrounding milieu. In addition, apical tight junctional complexes between the epithelial cells form the blood–epididymis barrier, which not only creates a specialised luminal environment for the sperm but also provides an immunological barrier, protecting sperm from the immune system. The identification of sperm antigens capable of eliciting production of functionally relevant sperm antibodies is a first step toward understanding the mechanism of immunologic infertility. SAGA-1 (sperm agglutination antigen-1) is a human male reproductive tract glycoform of CD52. SAGA-1 is a highly acidic (pI 2.5–3), polymorphic (~15–25 kDa) glycoprotein secreted by the epididymis that is localised over the entire human sperm plasma membrane (Diekman et al., 1997). A glycoform of CD52 is structurally conserved in chimpanzees but not in lower order non-human primates, identifying the chimpanzee as an appropriate non-human primate model to evaluate the immune-contraceptive vaccine development of SAGA-1 (McCauley et al., 2002).

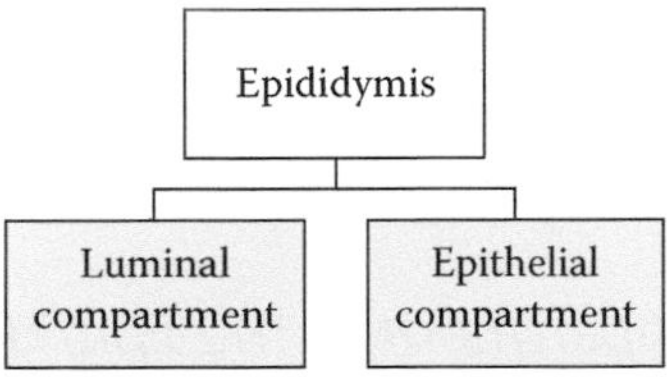

Figure. 7.1 (See colour insert.) Types of epididymal compartments.

Table 7.1 Epididymal Epithelial Cell Types

Cell Types	*Structure*	*Function*
Basal cells	Flat elongated.	Form basement structure of the epithelium and also play a major role in detoxification (Robaire and Viger, 1995).
Principal cells	Presence of tall microvilli, which form a distinct brush border.	Constitute 65%–80% of the epithelium, transport and secretion of small organic molecules, protein synthesis and secretion and absorption of both fluid and particulate matter (Robaire and Viger, 1995).
Narrow cells	Present mostly in the initial segment and are characterised by the presence of several endocytic vesicles.	The interactions among the cells and between the cells and extracellular matrix facilitate formation of a luminal environment distinct from the surrounding milieu.
Clear cells	Large endocytic cells.	Involve in endocytosis of cytoplasmic droplets, released by the sperm as they traverse the epididymal duct.
Halo cells	Halo cells are found throughout the epididymis and in various positions within the epithelium but not spanning it (Robaire and Viger, 1995).	The halo cells have been described as lymphocytes or monocytes. These cells perhaps function as the primary immune cells of the epididymis (Robaire and Viger, 1995).

The epididymal epithelium in the rat undergoes differentiation during 16–44 days of age at the end of which the various cell types can be discerned within the epididymis. Rat epididymal cells become fully differentiated by 49 days and several lines of evidence suggest that these cells exist from then on in a terminally differentiated state. Another striking change observed after day 44–49 is the appearance of sperm within the epididymal lumen (Rodriguez et al., 2002).

7.2.1.2 Luminal Compartment of the Epididymis

The milieu surrounding the sperm is perhaps the most complex fluid found in any exocrine gland. The complexity arises from progressive changes in the luminal composition along the epididymal duct. This specialised composition is maintained in part by the secretion and reabsorption of components by the epithelium, but also by the blood–epididymal barrier. The fluid of the epididymal lumen is hyper-osmolar, due to the presence of small inorganic ions such as calcium, potassium, sodium and small organic molecules such as phospholipids and carnitine. In addition, the luminal fluid also contains steroids such as androgens and oestrogens, and proteins synthesised by the epididymis such as clusterin, human epididymal protein, oxytocin, mannosidase, etc. some of which coat the sperm membrane during maturation (Dacheux and Dacheux, 2002).

7.2.2 Functions of the Epididymis

The three regions of the epididymis not only exhibit morphological differences (in certain species) but are also functionally distinct, and hence understanding their specific functions would shed light on the formation of the unique epididymal environment. The epididymis not only plays a crucial role in maturation of spermatozoa but in addition has an important role in absorbing the testicular fluid in which the sperm are immersed, as a result of which sperm and the 'maturation factors' get concentrated. The epididymis also serves to protect and store sperm, and these functions are described briefly next. Figure 7.2 outlines the function of epididymis.

7.2.2.1 Fluid Absorption and Secretion

Electrolyte and fluid transport is an important function of the epididymis because it affects the concentration of the luminal components. A balance between the absorptive (lumen to cell) and secretory (cell to lumen) processes exerts a fine control over the net movement of water across the epididymal duct. Interestingly, similar to the kidney, the epithelia of the efferent ductules and the epididymis are rich in ion exchangers and water channel proteins, enabling them to maintain fluid homeostasis in the tract. Sperm enter the efferent ductules along with seminiferous fluid that is secreted by the Sertoli cells at a rate of 30–50 μL/hr (Hinton and Setchell, 1993). The time required for spermatozoa to travel the entire length of the efferent ductules is about 45 min in rats. During this period, about 74%–96%

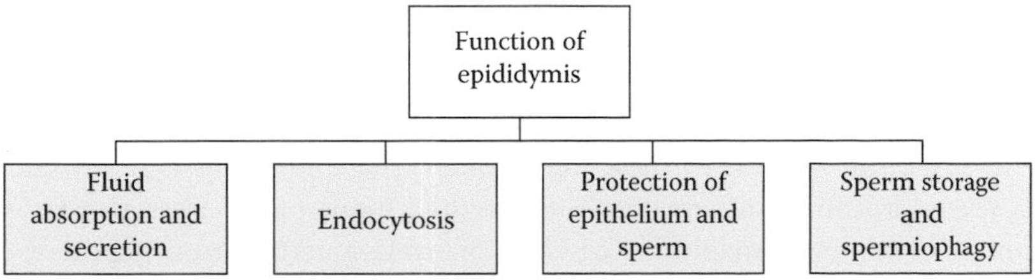

Figure 7.2 (See colour insert.) Function of epididymis.

of fluid is reabsorbed (Hess, 2000). As the sperm then traverse the epididymis, the remaining fluid is reabsorbed, mainly in the head or the caput region of the epididymis. Much of the understanding about the mechanism of fluid absorption comes from the studies in efferent ductules (Ilio and Hess, 1994; Clulow et al., 1997). The movement of water is dependent upon the ionic gradient in the epithelium that is maintained by several ion exchangers such as Na-K ATPase, Na-H exchanger (NHE) and carbonic anhydrases (CA). The passive movement of water is then facilitated via the water channel proteins called aquaporins (AQP) present both on the luminal and basal faces of the epithelium. Secretion, on the other hand, functions as a counterbalance to the absorptive process and is driven by stimulation of epididymis by bradykinins and peptide hormones (Wong and Huang, 1990). Stimulation of the epithelium causes an increase in intracellular cAMP that leads to opening of the apical membrane channels permeable to chloride and bicarbonate ions (Wen and Wong, 1988). The ions then move from the cell into the lumen of the duct and water movement flows in the direction of the ionic gradient via the AQPs. In vitro incubation of efferent ductules with ICI 182780(ER antagonist) resulted in a two fold increase in tubular diameter, indicating affected fluid reabsorption capacity (Shayu et al., 2005).

7.2.2.2 Endocytosis

Endocytosis involves the receptor-mediated uptake of substances such as androgen-binding protein, transferrin, α-2 macroglobulin, etc. by the cells of the epididymal epithelium. The uptake of these substances is followed by lysosomal degradation and recycling of the receptors to the cell surface. Interestingly, a differential expression of lysosomal enzymes has been observed among the various epididymal regions. This suggests that the three regions are capable of processing different substances and this could be one of the mechanisms, by which these regions maintain a distinct environment for the sperm.

7.2.2.3 Protection of the Epithelium and Sperm

The epididymis plays an important role in protecting itself and the sperm residing in its tract. Protection is provided against oxidative stress (De Lamirande et al., 1997; Aitken, 1999) and microbial attack (Hall et al., 2002). Sperm membrane rich in polyunsaturated fatty acids is vulnerable to oxidative damage and in addition, sperm are also an intrinsic source of reactive oxygen species that are essential for capacitation and chromatin condensation (Agarwal et al., 2014). To counteract the deleterious effects of these free radicals, the epididymis is richly endowed with several anti-oxidant enzymes such as glutathione peroxidases also known as gamma-glutamylcysteinylglycine or GSH, gamma glutamyl transpeptidases, glutathione S-transferases and superoxide dismutases that guard the sperm and the epithelium. Besides providing protection from harmful endogenous molecules, the epididymal tract and sperm have to be protected from an external microbial attack. As the sperm themselves are highly immunogenic, innate immunity is the front-line defence mechanism in the epididymis.

7.2.2.4 Sperm Storage and Spermiophagy

Sperm in the epididymis are stored in the cauda until required, and during this period, the epididymis partakes an active role in eliminating defective sperm. The time duration of sperm transit from the caput to cauda varies from 8 to 14 days in most mammals (Rowley et al., 1970), while spermatozoal storage in the cauda ranges from several days to weeks depending on the species (Bedford, 1975). The combination of the milieu and lower temperature in the epididymis are the major contributors for sperm survival. The epididymis also performs a quality control of the sperm, preventing misshapen, genetically abnormal sperm from entering the ejaculate. Defective mammalian sperm during their passage in the epididymis acquire ubiquitinated epitopes that are surface-exposed and these sperm are phagocytosed by the epithelial cells, by a process termed spermiophagy (Sutovsky et al., 2000). Characterisation of LCN6 of novel lipocalin which has an expression in the epididymis and location on the sperm surface suggested a role in male fertility (Hamil et al., 2003). Defective sperm can also be degraded by proteinases, DNAses and hydrolases released by the epithelium. Although many of the functions of the epididymis are performed by the entire epididymal duct, some regions are specialised to perform specific functions. For example, fluid absorption is carried out by the entire epididymis at a basal level. The caput region is thought to be more important for absorbing the remnant fluid coming from the efferent ductules, and the cauda region is equipped to store sperm in addition to maintaining quality control of sperm.

Besides the intrinsic gene expression patterns, secretory activity also varies across the three regions. While about 60% of the total protein synthesised in the caput is secreted in the lumen, the corpus and cauda secrete only between 20% and 40% of protein, indicating a higher secretory activity in the caput (Vreeburg et al., 1990; Turner et al., 1994). In addition to protein expression and secretion in a spatial and temporal fashion, epididymal gene regulation by hormones and growth factors also shows regional differences. These together contribute to the overall functional differences observed among the epididymal segments.

7.2.3 Regulation of Epididymal Growth and Function

The regulation of epididymal growth and differentiation is controlled by several hormones and growth factors. Over the years several reviews have appeared detailing the role of androgens. In the present review emphasis is placed on the role of hormones such as FSH and oestrogen in regulation of growth and function of epididymides. The role of these hormones and growth factors in epididymal growth and function is briefly described.

7.2.3.1 Androgens

Pioneering studies by Benoit (1925) demonstrated that a testicular factor was responsible for maintaining epididymal structure and function, and this factor was identified to be testosterone by Butenandt (1931). Since then, numerous papers have elaborated the role of androgens on morphological, biochemical and molecular aspects of epididymis (Wang et al., 2009; Robaire and Hamzeh, 2011). Testosterone and dihydrotestosterone (DHT) are the two key androgens involved in epididymal regulation. DHT is formed from testosterone by the action of 5α-reductase, an enzyme present in the epididymis. DHT is several times more potent than testosterone, and is the primary hormone that regulates epididymal functions (Orgebin-Crist et al., 1975). Since DHT is the active metabolite in most tissues, high conversion of testosterone to DHT via the 5α-reductase pathway and low metabolism of DHT via the 3α,β-hydroxysteroid dehydrogenase pathway are necessary for optimal androgen action (Nieschlag et al., 2012). The testis serves as the major source of androgens for the epididymis. The concentration of testosterone and DHT in the peripheral circulation and the epididymis is presented in Table 7.2.

7.2.3.1.1 Receptor and Binding Proteins

The action of androgens is mediated via androgen receptors (ARs), which belong to the nuclear receptor family. ARs are present throughout the epididymal tract in all species studied. Among the cell types, stromal cells and principal cells are known to show intense AR immunoreactivity in the nucleus, while staining in the halo and clear cells is negligible. Androgens themselves seem to be sufficient to maintain AR levels in the epididymis (Zhu et al., 2000; Chang et al., 2013). Besides AR, androgen-binding protein (ABP) is another androgen responsive molecule in the epididymis. ABP is synthesised by Sertoli cells and about 80% is secreted into testicular fluid, which is then transported to the epididymis via efferent ductules. ABP functions as a carrier molecule of androgens and in this respect has high affinity for

Table 7.2 Testosterone Concentrations (ng/mL) in the Serum, Rete-Testis Fluid (RTF), Caput and Cauda Epididymal Fluid of Rat

Fluid	*T*	*DHT*
Serum	1–2	—
RTF	17.8 ± 2.1	2.5 ± 0.3
Caput	5.4 ± 0.8	5.87 ± 6.5
Cauda	7.7 ± 1.2	4.4 ± 1.1

Source: From Turner, T. 1984. *Journal of Reproduction and Fertility*, 72, 509–514. With permission.

testosterone (K_a: $0.5 \times 10^9\ M^{-1}$) and DHT (K_a: $1.25 \times 10^9\ M^{-1}$). Within the rat epididymis, the concentration of ABP declines from caput to cauda, which correlates with the concentration of DHT. ABP bound to testosterone is taken up by the epithelial cells by a receptor mediated process, wherein the testosterone released in the cell is converted to DHT by 5α-reductase. Interestingly two forms of 5α-reductase are present in the epididymis, which are products of two separate genes that are differentially regulated by androgens. Between the two 5α-reductases, type 1 enzyme is considered to be the functionally dominant form in the epididymis (Ezer and Robaire, 2002).

7.2.3.1.2 Androgen Action

One of the classical approaches used to study androgen-dependent effects on a target tissue in the male has been orchidectomy. Other approaches involve using androgen receptor antagonists such as flutamide, bicalutamide, etc. that block androgen action. Using these approaches, it has been possible to study the targets of androgen action in the epididymis. Following orchidectomy there is drastic decrease in epididymal weight and morphological changes in cells are observed. The blood–epithelial barrier is also affected as components of gap junctions and tight junctions of the epithelium are regulated by androgens. In addition, the secretory function of the principal cells is severely compromised. Withdrawal of androgens also leads to a wave of apoptosis in the epididymis, which is detectable up to day 5 post-castration in the rat (Fan and Robaire, 1998). Functionally, androgen withdrawal also severely affects transport of molecules and ions across the epithelium (Wong and Yeung, 1977). Expression of several genes and proteins is known to be affected by lack of androgens such as oxidative stress proteins (Chauvin and Griswold, 2004).

A variety of approaches like microarray, proteomics and knockout models have been employed to identify the androgen-regulated genes in the epididymis using rodent as well as non-human primate models. An elegant study by Chauvin and Griswold (2004) using microarray approach documented the region-specific androgen-regulated genes in the different regions of immune epididymis. They have investigated over 12,000 transcripts and identified genes regulated by DHT within the three sub-regions of the epididymis. The maximum numbers of androgen-regulated genes were found in caput region when compared to corpus and cauda. One of the androgen-regulated transcripts was EST A1592789 which has 96% homology to the rat amino acid transporter SN2 and it should be noted that the concentration of several amino acids varies with different regions of the epididymis.

Myomesin 2 is important for structure of the sacromeric myofibrils and muscle contraction. It is possible that androgen regulation of this gene is important for contractility of the corpus to move spermatozoa through the epididymis to the cauda region. The androgen-regulated cauda-specific genes are kallikerins which are serine proteases which are important for proper storage of spermatozoa.

Proteomic analysis of different regions of rat epididymis by two-dimensional electrophoresis revealed 28 differentially expressed proteins altered from caput to cauda (Yuan et al., 2006). It has been suggested that inducible carbonyl reductase (iCR), a major androgen-regulated enzyme in the antioxidative system, shows primary and cell-type specific distribution in the ductal cauda region.

7.2.3.1.3 Androgen-Regulated Genes

The ability of the epididymis to perform its diverse functions stems from its regionalised gene and protein expression patterns. In addition to the rodent studies, differences in the gene expression patterns of the caput and cauda regions of the bonnet monkey epididymis were compared using the technique of differential display reverse transcriptase polymerase chain reaction (DDRT-PCR) (Shayu et al., 2006a, 2007). A transcript showing homology to human whey acidic protein 10 (hWFDC10A) was highly expressed in the monkey caput region. A peptide P2 was designed spanning a region of the monkey WFDC10A (mWFDC10A), which could inhibit the growth of gram-negative bacterial strains of Escherichia coli. P2 could permeabilise the bacterial cell membrane but was unable to permeabilise mammalian cells as evidenced by the lack of hemolysis upon incubation with the peptide. Expression of genes such as mWFDC10A may be essential in providing the first line of defence against microbial infections to the epididymal tract and thus rendering protection to the male gametes sheltered within the epididymis.

Another epididymis-specific androgen-regulated protein is Eppin and it was found that monkeys (*Macaca radiata*) immunised against Eppin, a testis/epididymis-specific protein, were infertile. Seven out of nine males (78%) developed high titers to Eppin, and all of these high-titer monkeys were infertile. Five out of seven (71%) high – anti-Eppin titer males recovered fertility when immunisation was stopped (O'Rand et al., 2004).

In conclusion, androgens are definitely very important for numerous epithelial functions that are related to maintenance of the epididymis, regulation of its secretory functions and controlling synthesis of molecules involved in sperm maturation. Yet it is also becoming apparent that several other factors such as oestrogens, retinoids, vitamin D, melatonin and growth factors secreted by the testes are also involved in epididymal regulation. It is evident that a plethora of information is available on the role of androgens in regulation of epididymal function. Considering this the emphasis in this chapter is on the involvement of oestrogen in regulation of epididymal function, particularly in non-human primates.

7.2.3.2 Oestrogens

The conventional view of estradiol as the 'female' hormone has undergone a drastic change in recent times. An increased interest in the role of oestrogens in the male stemmed from studies demonstrating impaired fertility in mice lacking the

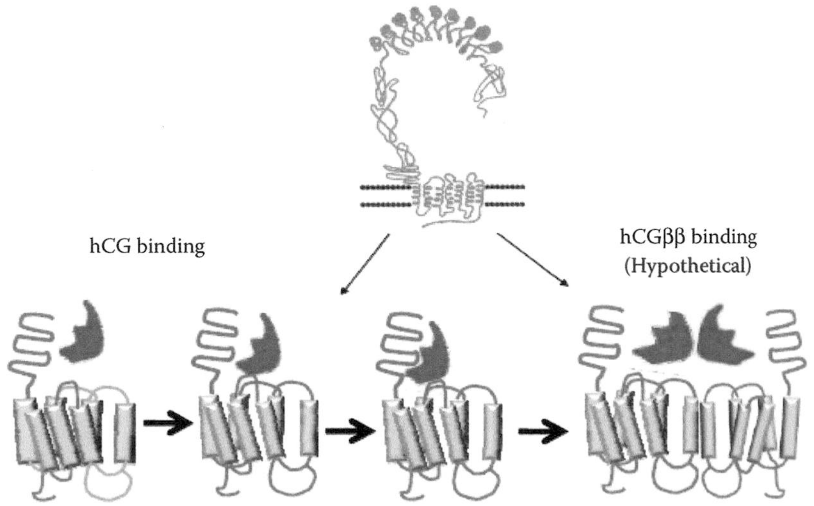

Figure 3.2 A cartoon suggesting the way hCGββ could bind gonadal LH receptor.

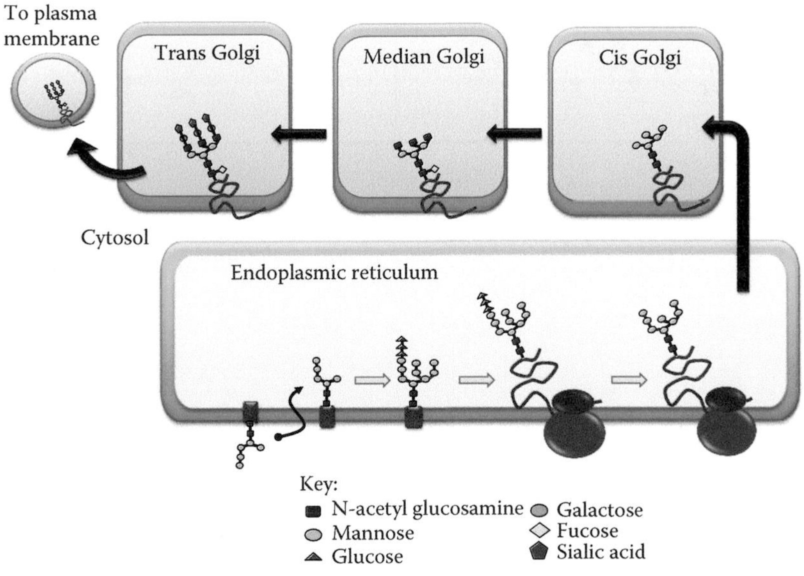

Figure 3.3 Dolichol linked glycosylation of proteins.

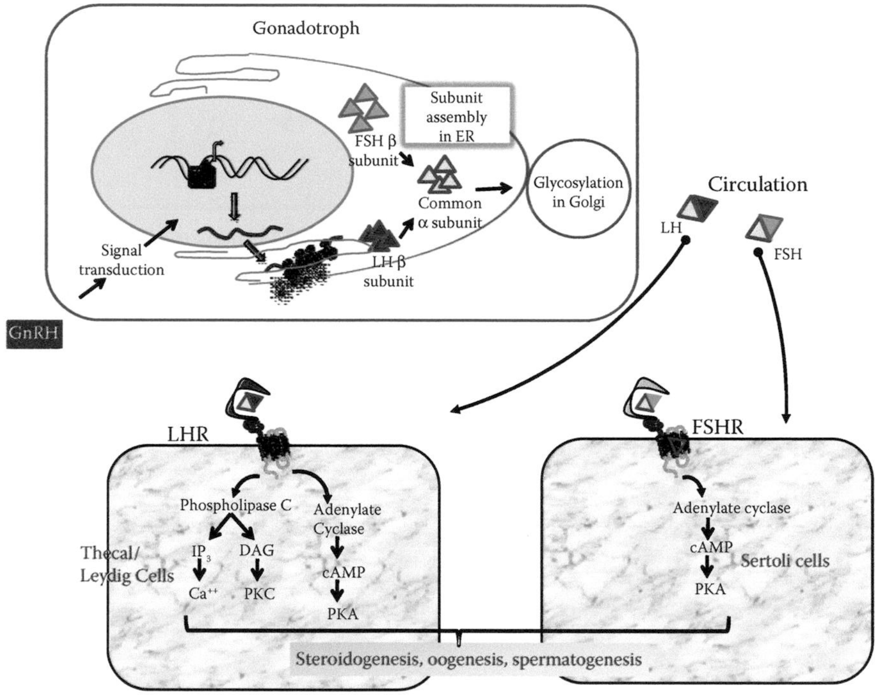

Figure 3.6 Signal transduction pathway of gonadotropins.

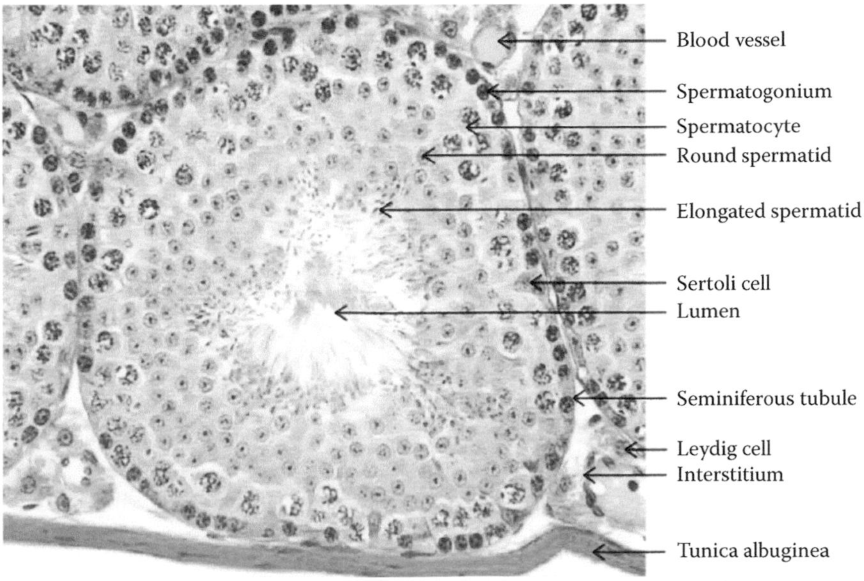

Figure 4.2 **The cross-section of mouse testis showing interstitial and tubular compartments. The interstitium is comprised of blood vessels, Leydig cells and lymph space. The seminiferous tubule shows various germ cells, Sertoli cells and lumen.**

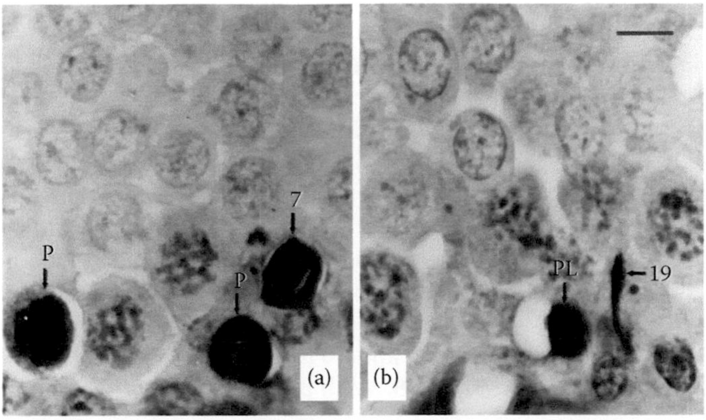

Figure 5.2 ***In situ*** **3′-end labelling of DNA strand breaks in apoptotic germ cells in glutaraldehyde-fixed, paraffin-embedded testicular sections. Methyl green was used as a counterstain. (a and b) Portions of stage VII tubules from a rat treated with GnRH-A for 5 days exhibiting apoptotic preleptotene (PL) and pachytene (P) spermatocytes, and step 7 (7) and 19 (19) spermatids. Scale bar, 10 μm.**

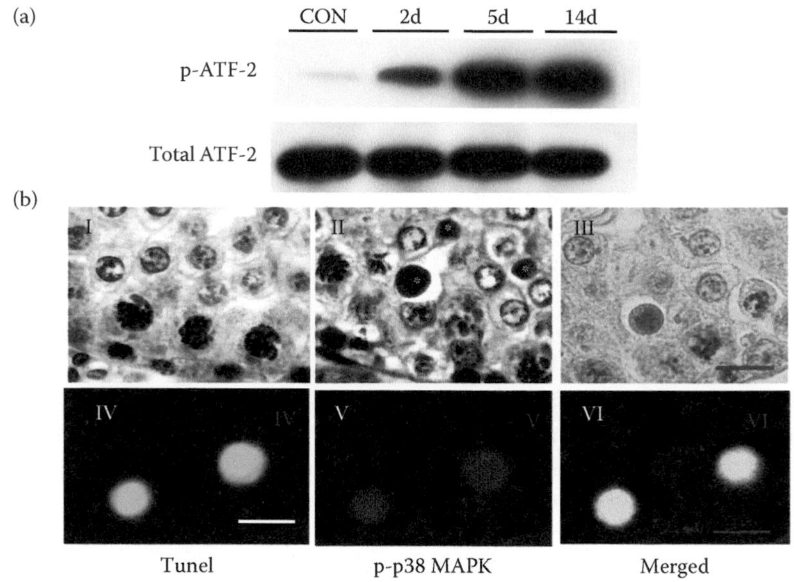

Figure 5.4 **Activation of p38 MAPK in rat testes after GnRH-A treatment. (a) Analysis of p38 MAPK activation by Western blotting using phospho-ATF-2 (Thr 71) antibody in testicular lysates after GnRH-A treatment. Total ATF-2 in the immunoblot is shown as a loading control. (b) p38 MAPK activation visualised by immunocytochemistry and confocal microscopy. Portions of stage VII tubules from control (panel I) and rats treated with GnRH-A for 5 days (panel II) show a strong phospho-p38 MAPK immunoreactivity in the condensed nuclei of apoptotic germ cells (asterisk) after hormone withdrawal. A testicular section from a rat treated with GnRH-A for 5 days incubated with rabbit IgG (negative control) shows no such immunostaining in a stage VII tubule (panel III). Panels IV–VI, confocal images show TUNEL (green), active p38 MAPK (red) and co-localisation of TUNEL and active p38 MAPK (yellow) in apoptotic germ cells triggered by hormone deprivation. Scale bar, 15 μm (panels I–III) and 10 μm (panels IV–VI). (From Vera Y et al. 2006. *Mol Endocrinol* 20: 1597–1609. With permission.)**

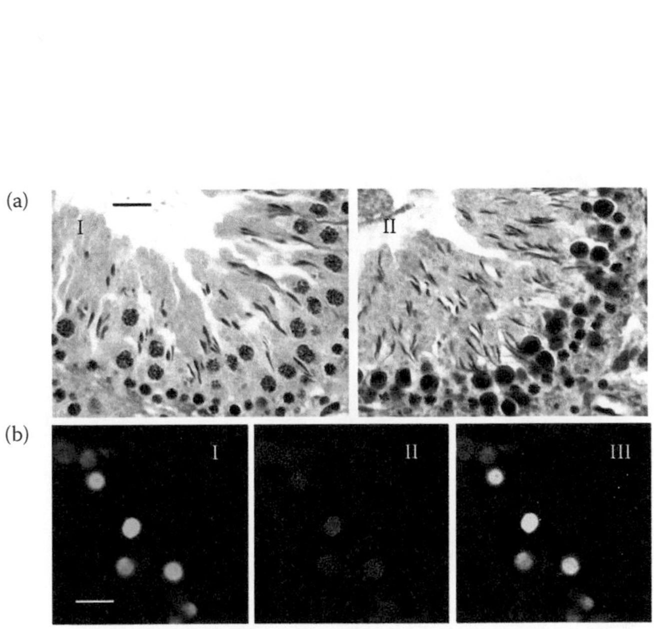

Figure 5.5 Testicular hyperthermia results in serine phosphorylation of BCL-2 in germ cells. (a) Portions of stage XII tubules from control (panel I) and a rat that had been exposed to short-term local testicular heating (panel II) show serine phosphorylation of BCL-2 only in heat-susceptible late pachytene spermatocytes 6 h after heating. Scale bar, 25 μm. (b) (Panels I–III), confocal images of late pachytene spermatocytes at stage XII from a heat-treated rat show TUNEL (green), phospho-BCL-2 (red), and colocalisation of TUNEL and phospho-BCL-2 (yellow) in apoptotic germ cells 6 h after heat treatment. Scale bar, 15 μm.

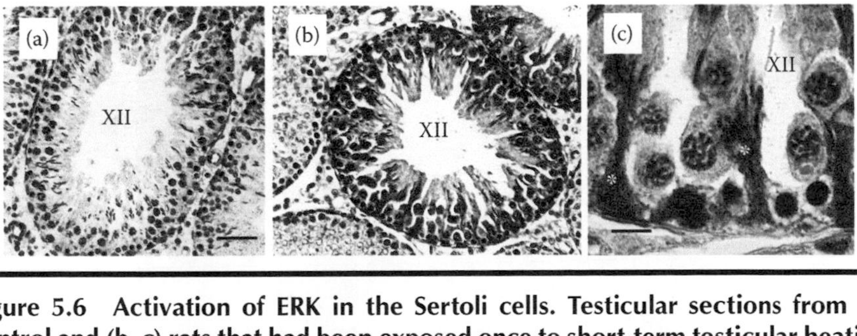

Figure 5.6 Activation of ERK in the Sertoli cells. Testicular sections from (a) control and (b, c) rats that had been exposed once to short-term testicular heating show activation of ERK in Sertoli cells at stage XII (a heat sensitive stage) within half hour of heating. Scale bar, (a, b) 50 μm and (c) 15 μm.

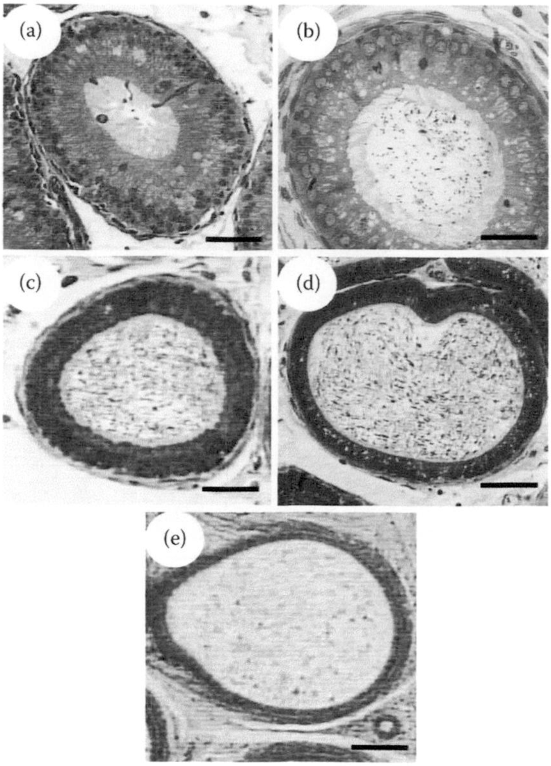

Figure 6.2 **T.S. of different segments of rat epididymis. (a) Initial segment, (b) intermediate zone, (c) caput, (d) corpus, (e) cauda. Scale bar (a) 20 μm, (b, c, d) 18 μm, (e) 15 μm.**

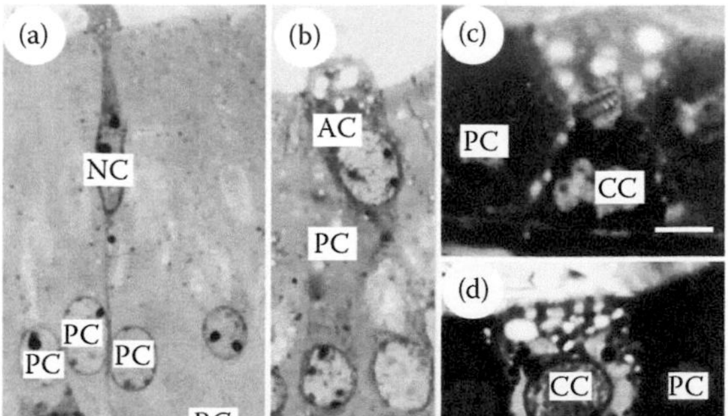

Figure 6.3 **Epithelial organisation of mouse epididymis. LM photomicrographs. (a) Initial segment, (b) intermediate zone, (c) proximal cauda, (d) distal cauda. AC, apical cell; BC, basal cell; CC, clear cell; NC, narrow cell; PC, principal cell. Scale bar 4 μm.**

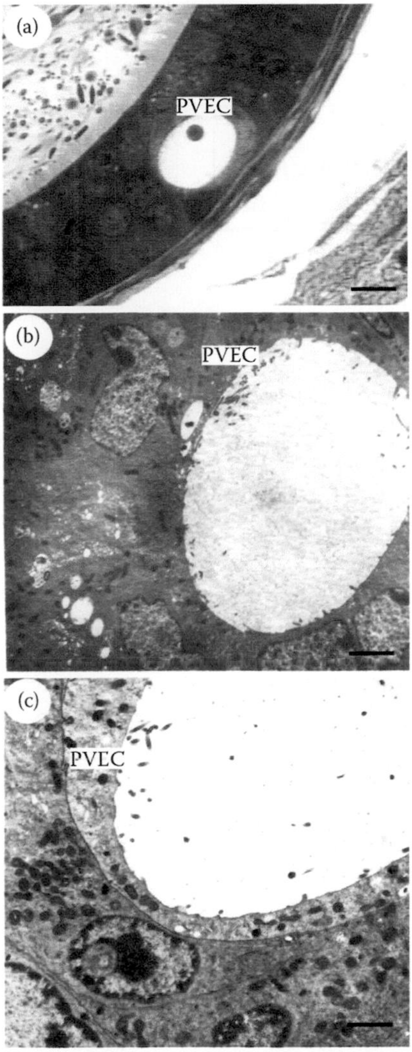

Figure 6.10 (a) LM and (b, c) TEM of pale vacuolated epithelial cell (PVEC) in rat epididymal epithelium.

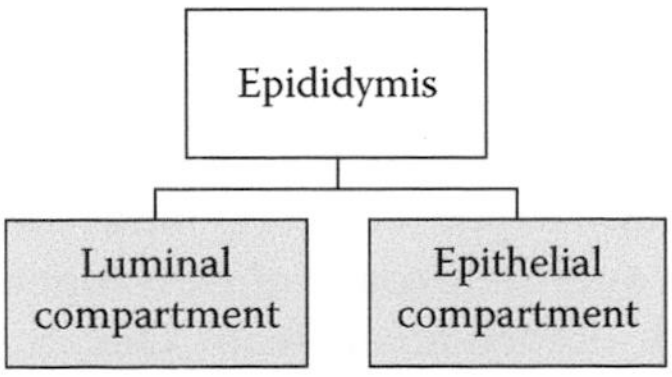

Figure. 7.1 Types of epididymal compartments.

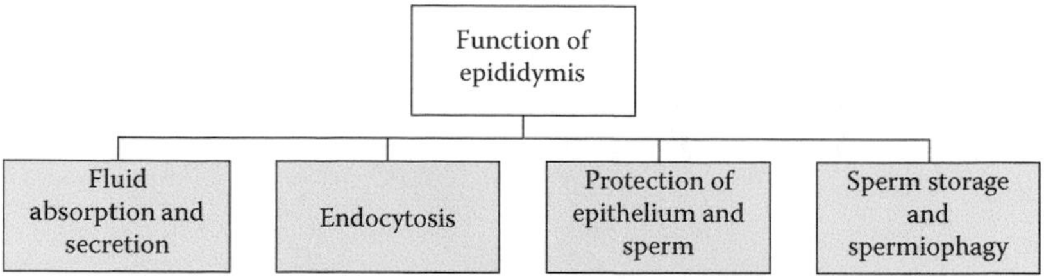

Figure 7.2 Function of epididymis.

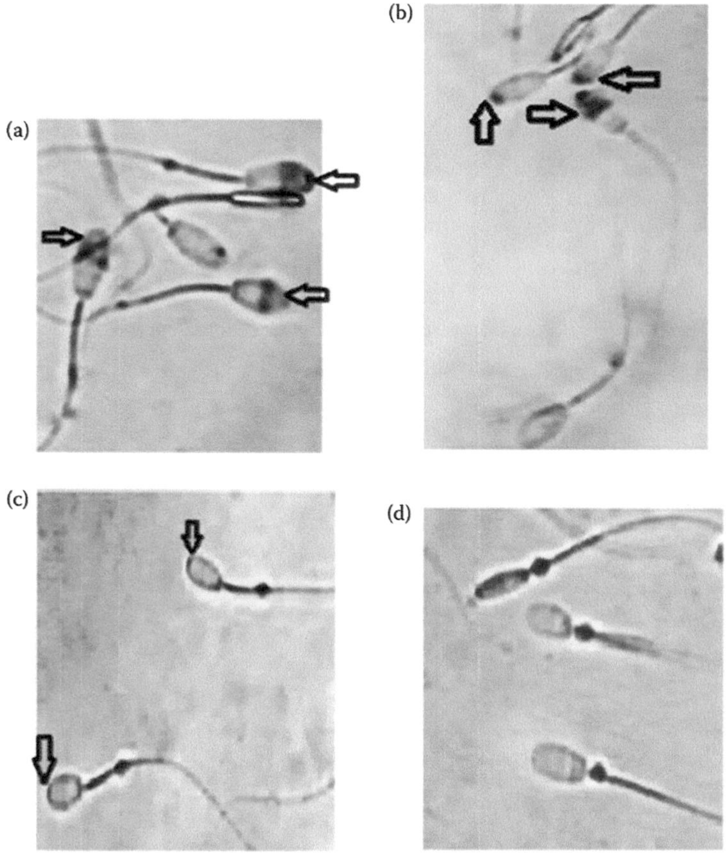

Figure 8.21 Binding pattern of the FMSF-HRP conjugate on maturing spermatozoa. Figures are representative for only binding pattern, not frequency. (a) Caudal cells, (b) distal-corpus cells, (c) mid-corpus cells, (d) caput cells. Arrow indicates towards spermatozoa that developed any sort of coloration on the sperm anterior head. (Reproduced from Dey, S. et al. 2014. *Biochemistry and Cell Biology* 92:43–52.)

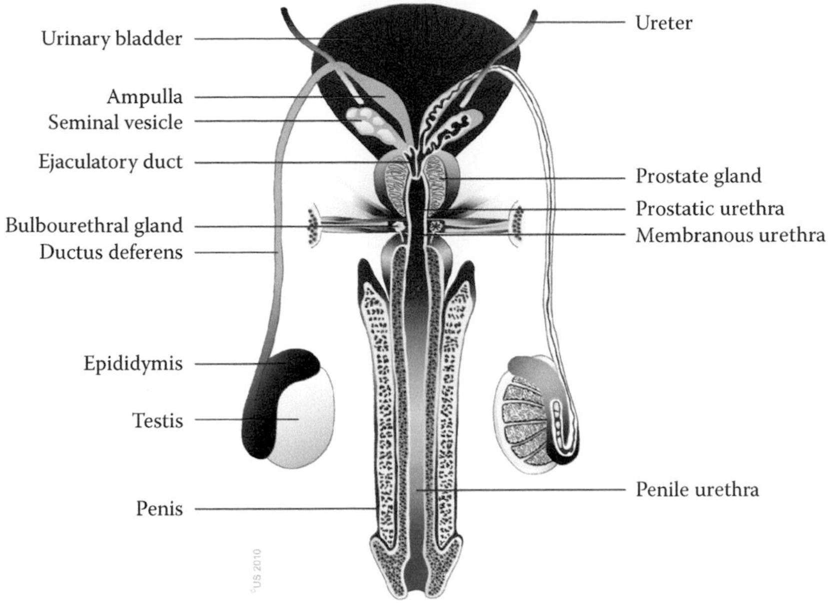

Figure 9.1 A posterior view of the adult human male reproductive system to show the location and structure of the major accessory sex glands.

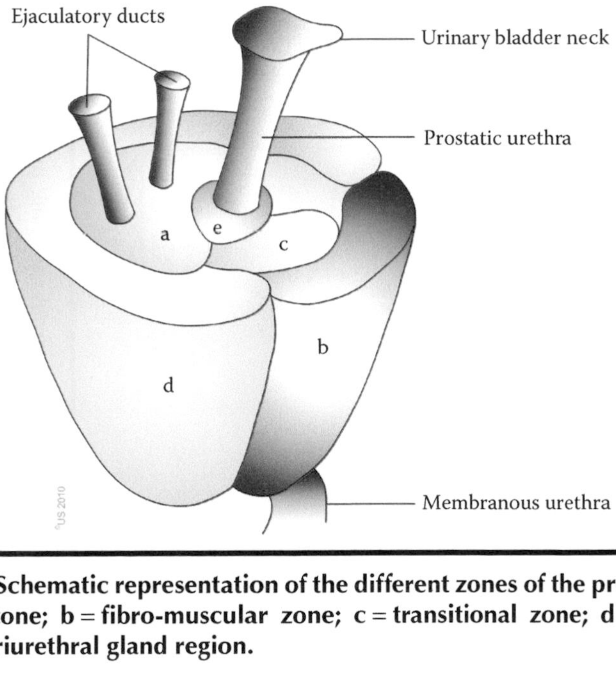

Figure 9.2 Schematic representation of the different zones of the prostate gland. a = central zone; b = fibro-muscular zone; c = transitional zone; d = peripheral zone; e = periurethral gland region.

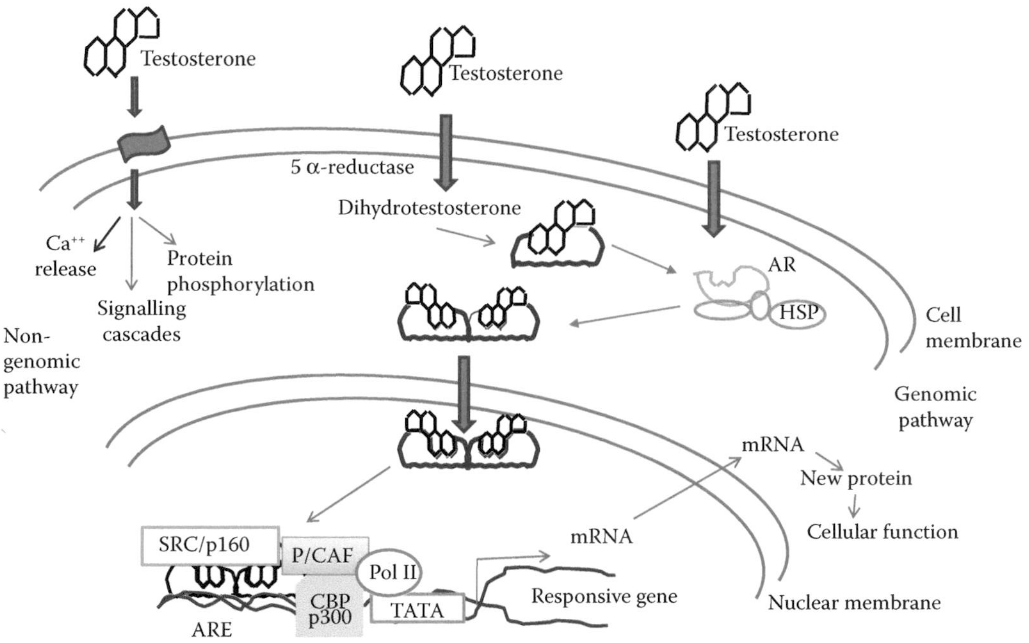

Figure 11.1 **Schematic diagram outlining molecular mechanism of androgen action. In the genomic pathway, testosterone (T) or its 5α-reduced metabolite DHT binds androgen receptor (AR) inducing the dissociation of heat shock proteins (HSP) and dimerisation of the receptor through amino- and carboxy-terminal interactions. Subsequently, the hormone-receptor complex is translocated into the nucleus, interacts with androgen responsive element (ARE) of the responsive gene and transcriptional machinery, and recruits a host of coregulators. Finally, this leads to alteration in transcriptional activity, protein synthesis and cellular functions. In non-genomic pathway, androgens act at the level of cell membrane and interact with uncharacterised receptor proteins to influence the release of calcium, phosphorylation of proteins and other signalling cascades.**

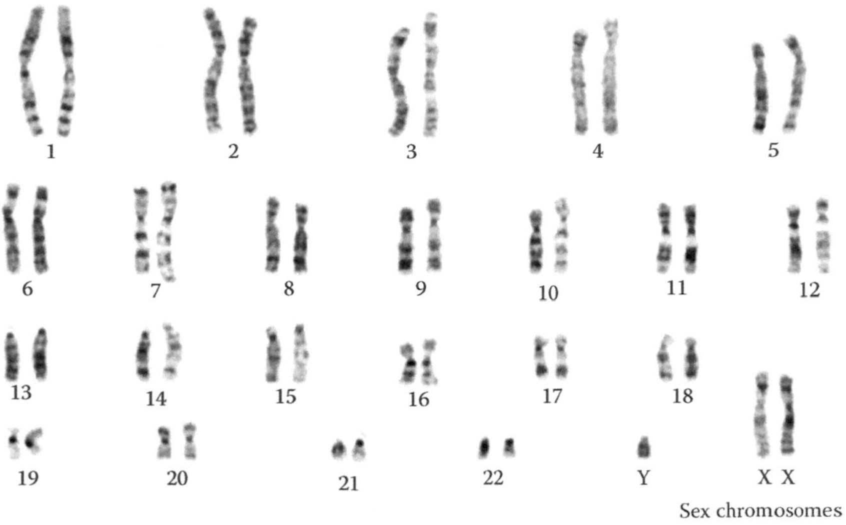

Figure 12.1 Karyotype of a Klinefelter patient 47, XXY.

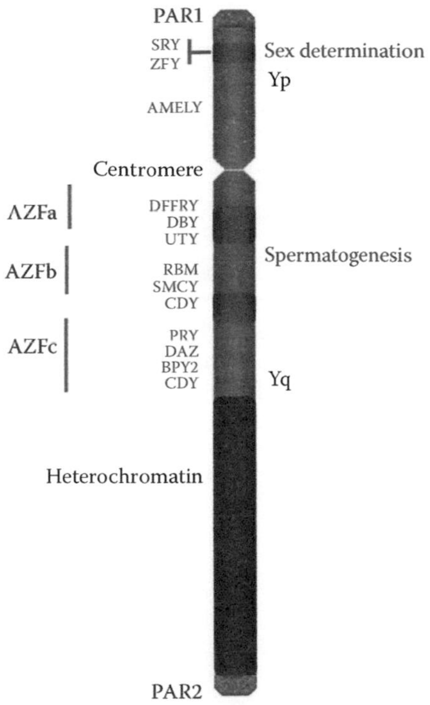

Figure 12.2 Schematic representation of the human Y chromosome.

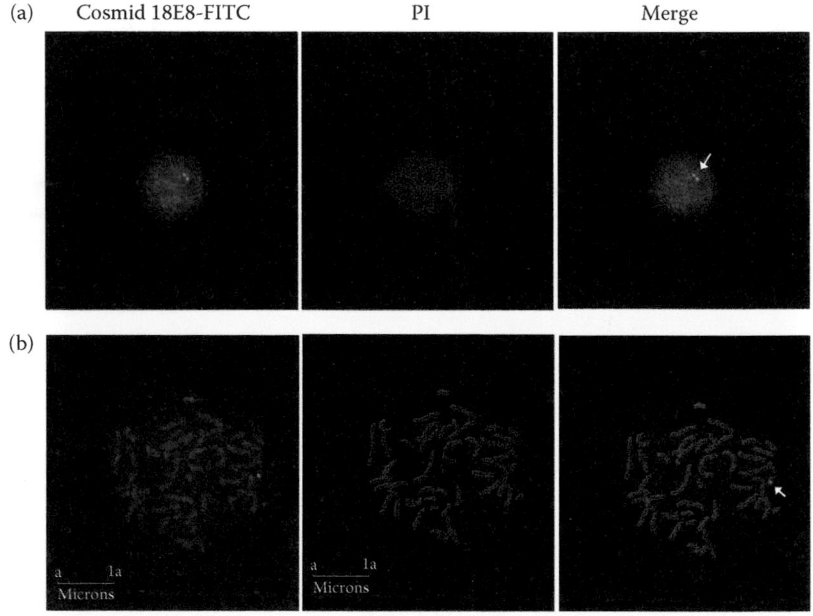

Figure 12.3 Detection of *DAZ* gene cluster by FISH. Confocal laser scanned images of representative (a) interphase nucleus and (b) metaphase chromosomes from a fertile, control individual hybridised with cosmid 18E8, showing two dots (corresponding to two *DAZ* clusters, each containing two *DAZ* genes).

receptor for oestrogen (Lubahn et al., 1993; Korach, 1994) and triggered a need to understand the role of oestrogen in male reproductive development and fertility.

Oestrogen is biosynthesised by a microsomal enzyme complex, called aromatase. This enzyme utilises testosterone or androstenedione as substrates to form the highly potent 17β-estradiol (estradiol) or the weakly oestrogenic product, estrone respectively (Thompson and Siiteri, 1974). Oestrogen is a generic term that includes three steroids: estradiol, estrone and estriol. While estradiol and estrone are formed by aromatisation of androgens, estriol is a catabolic product of oestrogen. Among these, estradiol is the most potent and physiologically relevant steroid and is therefore commonly used interchangeably (even in this chapter) with the term oestrogen. Testicular oestrogen is perhaps the main source of oestrogen for the efferent ductules and the epididymis (Hess, 2003), although recent reports suggest the presence of the enzyme aromatase in the extragonadal ducts.

Oestrogen concentrations are low in the peripheral blood and range from 2 to 180 pg/mL depending upon the species (Table 7.3). Also, oestrogen concentrations are typically higher in the testicular vein and lymph than in general circulation. High concentration of oestrogen has also been reported in semen, which is perhaps due to local oestrogen synthesis by aromatase expressed within the sperm (Nitta et al., 1993).

Table 7.3 Oestrogen Concentrations in the Male

Source	*Concentration (pg/mL)*	*Species*
Peripheral blood	2–46 40–145 19–46	Rat Monkey Man
Testis venous blood	17.5 104–200 926	Rat Monkey Man
RTF	249 14–149 —	Rat Monkey Man
Semen	— — 14–162	Rat Monkey Man

Source: From Springer Science+Business Media: *The Epididymis: From Molecules to Clinical Practice. A Comprehensive Survey of the Efferent Ducts, the Epididymis and the Vas Deferens,* The role of estrogens in the endocrine and paracrine regulation of the efferent ductules, epididymis and vas deferens, 2002, Hess, R. A., Zhou, Q. and Nie, R.

7.2.3.2.1 Oestrogen Receptor Localisation

There is considerable tissue specificity in the expression of ERα and ERβ. ERα is the predominant form expressed in the uterus, liver, adipose, pituitary and hypothalamus, while ERβ is the major form in the ovary, cerebellum, cerebral cortex, etc. Within the male reproductive tract, ERβ is expressed in the Sertoli cells and germ cells of the testis, and in almost all the accessory organs such as the efferent ductules, the entire epididymal tract, prostate and vas deferens (O'Donnell et al., 2001). Interestingly, expression of ERβ is observed in the adult mouse Leydig cells but not in the rat Leydig cells, indicating species-specific differences in expression pattern. ERβ is the predominant and perhaps the only ER expressed in the germ cells, although one report showed the presence of ERα in the rat spermatocytes and spermatids (Pelletier, 2002). ERα has a more specific localisation pattern and is found to be expressed within the Leydig cells of testis, the efferent ductules, epididymis, and vas deferens (O'Donnell et al., 2001). Among these tissues, very high expression was observed in the efferent ductules, in almost all species studied until date (Hess et al., 2002; Joseph et al., 2011).

7.2.3.2.2 Effect of Oestrogenic Compounds on Epididymis

While the expression of the ERs has been established decisively in the efferent ductules and the epididymis, it is only during the last two decades that the precise functions of oestrogen in these tissues have begun to unfold. The generation of the knockout mice for ERs and aromatase has provided an impetus to the study of the role of oestrogen in the male. Classically the role of oestrogen in development of the excurrent ducts came from studies by Arai et al. (1982) and Wilson et al. (1986) who observed that neonatal exposure to oestrogenic compounds such as diethylstilbestrol (DES) led to the formation of rete testis neoplasms and induction in epididymal granulomas. It has now been shown that *in utero* exposure to DES induces an early appearance of ERs and results in a failure to express AR (Atanassova et al., 2001; McKinnell et al., 2001) in addition to other abnormalities such as testicular atrophy, under-developed epididymis and sperm granulomas that have a direct bearing on male development and fertility (Fisher et al., 1999). Similar to DES, environmental oestrogens such as bisphenol A and octylphenol affect the testis and excurrent ducts while the plasticiser di(n-butyl)phthalate when given neonatally results in an absence of the epididymis from a high percentage of the offspring (Mylchreest et al., 1998; Fisher et al., 1999). In addition, oestrogen alone can cause a decrease in sperm transit time through the epididymis, an effect mimicked by neonatal treatment of tetrachlorodibenzo-p-dioxin, an environmental oestrogen. This reduction in sperm passage time through the epididymis would directly affect sperm maturation and therefore have repercussions on its fertilising ability. While these studies showed that oestrogens can have a myriad of effects in male reproduction, these results need to be interpreted with care as oestrogen or oestrogenic compounds are known to cause a negative feedback inhibition of hormones in the hypothalamus–pituitary–gonadal (hpg) axis, thus also affecting testosterone levels. The potential for androgen and oestrogen co-regulation of these ducts is also likely, given the presence of both AR and ER within the same tissue. Nevertheless, these results provided the first evidence that oestrogen could have a role in epididymal development.

7.2.3.2.3 ERαKO

Among the oestrogen-related knockouts, ERαKO was the first to be generated and strikingly, the males were found to be infertile (Lubahn et al., 1993). These males had normal testes until puberty, but began to show a transient increase in testis weight from 32 to 81 days of age followed by a continual decrease in weight, ending with atrophy by 150 days. Interestingly, in the ERαKO mice while FSH levels remained unchanged, serum LH was slightly elevated and testosterone levels were increased twofold compared to wild-type mice (Lindzey et al., 1998).

A detailed histology of the efferent ductules in the ERαKO revealed several abnormalities in the epithelium. Besides the notable loss in endocytic organelles, a flattening of the nucleus and a loss in the microvillus border was observed. The contribution of the epididymis to infertility in the ERαKO male is presently unknown. However, considering the high expression of ERs in the epididymis and defects in sperm motility and fertilising ability of ERαKO sperm, it was likely that the epididymal functions would be negatively affected by a loss in ERα.

7.2.3.2.4 ERβKO

In contrast to these defects in the ERαKO males, the ERβKO males were completely fertile (Krege et al., 1998). Testicular histology appeared normal and preliminary studies suggested abnormal morphology in only some epithelial cells. Males showed normal sexual behaviour and sperm count (Ogawa et al., 1999) indicating that ERβ loss was less disruptive than ERα. The lack of any spermatogenic phenotype in the ERβKO was surprising because ERβ appears to be the only ER in the germ cells. It was also interesting to note that there were no gross abnormalities in the efferent ductules and the epididymis, although these ducts are rich in ERβ expression (Hess, 2000). ERβ expression in the efferent ductules and epididymis of ERαKO was relatively unchanged although the subcellular localisation differed in comparison to wild type (Rosenfeld et al., 1998), thereby indicating that ERα-mediated effects play a predominant regulatory role in functioning of the excurrent ducts. This fact was reiterated in the double knockouts of ERα and ERβ, which were infertile and the phenotype resembled that of ERαKO (Couse and Korach, 1999). The loss in fertility was due to fluid build-up as seen in ERαKO mice. Overall, among the two receptors, ERα-mediated actions seemed to be of primary importance in the excurrent ducts.

7.2.3.2.5 ArKO

Loss in the enzyme aromatase (ArKO) responsible for oestrogen biosynthesis, showed an age-dependent disruption in spermatogenesis. Testicular histology was normal in mice at 14 weeks and mice were initially fertile, but at 7 months of age mice showed variable degrees of fertility (Fisher et al., 1998). Strikingly, the disruption in spermatogenesis was not due to fluid build-up in the excurrent ducts, which appeared normal, indicating that their function was not markedly impaired. Although, the sperm from ArKO were able to fertilise oocytes, sperm count was reduced and there was a drop in sperm motility (Toda et al., 2001). While the lack of any apparent effects on the excurrent ducts is difficult to reconcile in light of the importance of oestrogen-mediated action in these tissues, it is possible that the ArKO mice may have had access to maternal oestrogens during development. It is also likely that the ERs in these tissues could have been activated in an oestrogen-independent manner (O'Donnell et al., 2001).

7.2.3.2.6 ER Antagonists

Another tool that has provided vital information regarding oestrogen action is the use of oestrogen receptor antagonists. A vast majority of antagonists developed in the late 1950s were derivatives of triphenylethylene such as MER-25, clomifene and tamoxifen. These anti-oestrogens were non-steroidal compounds that functioned as competitive oestrogen receptor antagonists. While these compounds could effectively combat breast cancer, their usage inadvertently resulted in uterine cancer (Macgregor and Jordan, 1998). It is now known that promoter (AF1 or AF2 dependent) usage and the overall balance of the relative expression levels of coactivators and corepressors, can be an important determinant of tissue specificity of these compounds (Macgregor and Jordan, 1998). Later a 'pure' steroidal anti-oestrogen, ICI 182780 (ICI), was developed which could inhibit both AF1- and AF2-dependent activities, thereby probably functioning as a complete antagonist (Howell et al., 2000). ICI treatment in rodents showed ERαKO-like effects in the efferent ductules and provided valuable information regarding oestrogen action in the excurrent ducts. Together both the knockout models and antagonists have resulted (Shayu et al., 2006b) in a better understanding of the role of oestrogen in the male. It was found that chronic administration of ICI in adult bonnet monkey resulted in upregulation of the expression of ERα and ERβ in caput region of epididymis both at mRNA and protein levels (Shayu et al., 2005).

7.2.3.2.7 Oestrogen-Regulated Genes

The development of oestrogen receptor knockout mice established the importance of oestrogen in regulation of epididymal function. It also paved the way for identification of several oestrogen-regulated genes in the murine model. These genes were primarily involved in regulation of fluid reabsorption and oestrogen was found to regulate the expression of Na^+/H^+ exchange and the rate of $^{22}Na^+$ transport. Immunochemical staining for NHE3 carbonic anhydrase 11 and aquaporin was decreased in ERαKO mice efferent ductules. Recently, using an affirmetrix M430-2.0 microarray approach oestrogen-regulated genes were identified in the proximal male mouse reproductive tract. These include transcripts which include proteins associated with lipids metabolism, transcription and steroid metabolism. In contrast to the information available with rodent models, relatively much is less known about role of oestrogen in regulation of epididymal function in non-human primates. Recently, studies were carried out to identify the oestrogen -regulated genes in different regions of epididymis of the adult male bonnet monkeys, which were administered ICI, a specific oestrogen receptor antagonist via an Alzet pump (Shayu et al., 2007).

7.2.3.2.7.1 RT-PCR Analysis for ERα, AQP-1 and Na +-K+ ATPase-α1 in the Three Regions of the Epididymis As a first step the presence of ERα was ascertained in the bonnet monkey epididymis, RNA from the caput, corpus and cauda was subjected to RT-PCR analysis using specific primers. A high level of ERα message was detected in the caput, corpus and cauda regions. The three regions did not differ appreciably in their level of expression. It has been well documented that the efferent ductules of rodents express components like AQP-1 and Na +-K+ ATPase, which are important for fluid absorption. The aforementioned components of the fluid absorption machinery were present in the bonnet monkey epididymis and further their distribution within the three regions was also tested. Analysis of the steady-state mRNA levels of these genes revealed specific signals for each of these components in all the three regions of the monkey epididymis, with varying levels of expression. A twofold higher expression of AQP-1 in the cauda compared to the caput was seen. The level of expression of Na +-K+ ATPase-α1 was fourfold higher in the corpus and cauda compared to the caput.

7.2.3.2.7.2 Western Blots Analyses for ERα and ERβ in the Three Regions of the Bonnet Monkey Epididymis Specific signals for ERα (65 kDa) and ERβ (a prominent band of 56 kDa) in the caput, corpus and cauda regions were seen. Comparative analysis revealed no significant difference in the intensity of expression of the two-receptor subtypes (at the protein level) in the three regions of the monkey epididymis.

7.2.3.2.7.3 Immunolocalisation of ERα and ERβ in the Three Regions of the Bonnet Monkey Epididymis Intense staining for ERα was observed in all the three regions that is, the caput, corpus and cauda of the bonnet monkey epididymis. Staining was localised mainly to the nuclei of the epithelial cells lining the lumen. However, weak staining was also observed in the surrounding peri-tubular smooth muscle nuclei and in the vascular smooth muscle nuclei. The three regions of the epididymis stained intensely for ERβ. Intense staining was observed in the peritubular and vascular smooth muscle nuclei. Staining was also observed in the tubular epithelial cell nuclei of caput, corpus and cauda regions. The cytoplasm in these cells was also weakly stained. This staining is in sharp contrast to the ERα staining which was mostly restricted to the tubular epithelial cell nuclei.

7.2.3.2.7.4 Effect of ICI Treatment on the Bonnet Monkey Serum Testosterone Levels Studies revealed that there was no change in serum testosterone levels in the 30-day, 60-day or 180-day ICI-treated monkeys. Nocturnal surge of serum testosterone also remained unaltered. This is in accordance with previous reports that ICI does not cross the blood–brain barrier and hence does not alter the hormonal profile (Shayu et al., 2007).

7.2.3.2.7.5 Expression Profile of ERβ, AQP-1 and Na +-K+ ATPase-α1 in the Caput Region of 30-Day and 60-Day ICI-Treated Monkey: RT-PCR Analysis In order to assess the effect of ICI treatment on the steady-state mRNA levels of ERα, AQP-1 and Na +-K+ ATPase-α1, RNA from the vehicle and ICI treated-caput regions was subjected to RTPCR analysis using specific primers. A significant fourfold increase was found for ERα in the 30-day ICI-treated samples compared to the vehicle control. A significant increase in the ERα message in the 60-day ICI-treated samples was observed. A twofold reduction in the message for AQP-1 was observed in the caput of both 30- and 60-day ICI treated monkeys compared to the corresponding vehicle controls. In sharp contrast, no significant change in the expression of Na +-K+ ATPase-α1 between the vehicle control and ICI-treated samples was seen. These results indicated that ICI treatment significantly affects ERα and AQP-1 mRNA levels in the caput region of the epididymis.

7.2.3.2.7.6 ERα Expression in the Caput Region of Vehicle and ICI-Treated Bonnet Monkey Epididymis: Iimmunohistochemical Analysis The increase in the ERα mRNA observed in the ICI-treated caput samples was confirmed at the protein level, ERα was immunolocalised using a well-characterised antibody. Interestingly, no change was appreciable in the ERα signal in the 30-day ICI-treated caput samples (compared to vehicle control), although a significant increase in the ERα mRNA at this time-point was observed. However, a striking increase in the ERα expression was seen in the 60-day ICI-treated caput samples (compared to the corresponding vehicle control), reflecting the increase in the mRNA levels.

7.2.3.2.7.7 Effect of ICI on the Fluid Reabsorption Capacity of Efferent Ductules Additional evidence for a role for oestrogen in fluid reabsorption was obtained by in vitro incubation of efferent ductules from ICI-treated monkeys fluid reabsorption. Efferent ductules were chosen for the study, as previous reports have shown that these ducts absorb the majority of the testicular fluid. Following 48 h of the total incubation duration with ICI, an almost twofold increase in the efferent ductule diameter compared to control ductules incubated without ICI was seen. These results support the observations made with rodents by other workers, who showed that efferent ductules are very sensitive to lack of oestrogen. Since this experiment was carried out with efferent ductules from normal monkeys, it ruled out any possible involvement of testicular factors being modulated by in vivo ICI treatment and indirectly regulating fluid absorption in efferent ductules (Shayu et al., 2007).

7.2.3.2.7.8 Effect of ICI on Sperm Count Short-term (30-day period) and long-term (180-day period) ICI treatment had no adverse effect on the sperm count in all the monkeys tested at the given dose and duration of treatment. All the samples analysed had more than 90% viable sperms as tested by eosin-nigrosin stain.

7.2.3.2.7.9 Effect of ICI on Sperm Motility Parameters Nevertheless, it is pertinent to note that in the 180-day ICI-treated monkeys, a drastic reduction in progressive motility was observed. Also a decrease in the beat frequency and lateral amplitude of the sperms was observed in the same animals.

7.2.3.2.7.10 Identification of Oestrogen-Regulated Genes In order to identify oestrogen-regulated genes, the approach of differential display reverse transcription polymerase chain reaction (DDRT-PCR) and microarray analysis was employed. DD-RT-PCR was carried out comparing the caput regions of vehicle-treated and 30-day or 90-day ICI-treated bonnet monkeys. Among the differentially expressed transcripts identified in the vehicle- and 30-day ICI-caput regions, keratin 19 (K19) the intermediate filament, family of cytoskeletal proteins was found to be down-regulated in the 30 day ICI-caput. The trend was confirmed by Northern blot analysis in the monkey caput regions of vehicle- and 30-day ICI-treated bonnet monkey. In addition, K19 mRNA level of the caput was reduced relative to vehicle controls in the rat, when assessed by semi-quantitative RTPCR analysis (Shayu et al., 2007).

Another transcript that was found by DDRT-PCR analysis showed to be regulated by oestrogen was phosphotidylethanolamine N-methyltransferase (PEMT). PEMT catalyses three sequential methylation reactions converting phosphotidylethanolamine (PE) to phosphatidylcholine (PC). While PEMT expression is mainly found in the liver, its expression has also been observed in the brain, testis and heart. PC is a major component of mammalian cellular membranes and its appropriate concentration is necessary for the fluidity of the membrane. In addition, oxidation of PC provides an energy source to sperm. Expression of PEMT was found to be up-regulated in the 90-day ICI-caput in the bonnet monkey; however, interestingly ICI treatment in the rat did not elicit a similar response. The role of PEMT in epididymal function needs to be studied further in light of its effects on sperm membrane fluidity and energy metabolism.

Expression of genes such as PEMT and URO that were up-regulated in the ICI-treated monkey caput and ERαKO-caput respectively, were not affected by 8-day ICI treatment in the rat, indicating that some changes require a long-term loss of ER-mediated signalling (Table 7.4). In addition to fluid absorption, oestrogen may have a role in regulating expression of genes involved in lipid metabolism, maintenance of cell structure, which may or may not be related to fluid absorption. The importance of oestrogen in the sperm maturation was established in a study using adult male bonnet monkeys which were chronically administered tamoxifen.

No significant difference was observed in the sperm count of control monkeys and tamoxifen-treated monkeys either during the treatment or recovery period which lasted for more than 6 months. Similar results were obtained with regard to serum testosterone levels. In contrast, it was observed that within 20 days of the start of treatment with tamoxifen, sperm motility decreased in all the treated monkeys. Spermatozoa lacked forward motility and exhibited sluggish motility which is normally shown by spermatozoa with coiled tails. This was observed beyond 90 days, up until day 225, the period up to which motility was monitored. In addition, for a limited period (up to 90 days), sperm quality was examined by checking spermatozoa for morphological defects such as coiled tails, head defects and neck defects. It was observed that with progression of treatment, normal spermatozoa decreased from 84% to 29% and the percentage of abnormal spermatozoa increased from an initial 16% to 71%.

In addition to the identification of oestrogen-regulated genes in monkey epididymis the changes in the protein profile in the cauda region of the epididymis following ICI treatment in the monkeys was also investigated by 2D PAGE. Although six differentially expressed protein spots could be seen, the identity of only one protein could be established. This protein was consistently down-regulated in repeated experiments using three different vehicle-treated and 90-day ICI-treated cauda samples. Based on database search analysis, the identity of this differentially expressed protein was established as WNT4 (Deshpande et al., 2009).

7.2.3.2.8 E2 Synthesis in the Epididymis

Considering the fact that we have been able to demonstrate that E2 has an important role in regulation of function of epididymis including maturation of sperms, it was of interest to investigate the source of oestrogen. Detailed studies revealed that an active oestrogen biosynthesising system is present in the epididymis. Both caput and cauda are capable of synthesising oestrogen and P450Aromatase mRNA expression was demonstrated in both the caput and cauda regions of epididymis. Analysis of RNA from caput and cauda of animals of 40 days, which are devoid of spermatozoa, established that the expression is due to presence in epididymis and not due to spermatozoal contribution. In vitro studies also established that epididymal aromatase is functional in that androstene was converted into oestrogen in a concentration dependent manner. Though it is well known that FSH regulates

Table 7.4 Down-Regulation Caused by Oestrogen Blockade versus Up-Regulation

Gene	*Function*	*ERαKO*	*Regulation*	*Result*
AQP and CAII	Involve in fluid absorption by generating H+	Yes	Down-regulation in relative to wild type	Effect Na transport NHE3
Phospholipase A2V (PLA2V)	Localisation of PLA2V in the epididymis coincides with the localisation of cyclo-oxygenases, which are implicated in regulating fluid secretion. Yes	Yes	Down-regulation	Found to be down-regulated in caput epididymides of ERαKO and ICI-treated males.
Uromodulin (URO)	URO or Tamm-Horsfall protein is a mucoprotein found in urine (Tamm and Horsfall, 1950; Muchmore and Decker, 1985). Function in protecting the urinary tract from bacterial infections. In addition, the major distribution of URO in the thick ascending duct of the kidney suggests a role in ion transport and water permeability.	Yes	Up-regulation	Increased URO expression would impede fluid concentration, thereby augmenting the effect offset by reduced ion exchangers in the ERαKO. Microarray analysis revealed a very high expression of URO in the ERαKO caput. While URO expression showed a 3–3.5-fold up-regulation in the ERαKO-caput, interestingly there was no change in its expression in the ICI-caput.

aromatase in Sertoli cells and granulosa cells, it had no effect on epididymal aromatase. However, epididymal aromatase was found to be regulated by LH, and LHR mRNA was demonstrated both in cauda and caput and expression was more in cauda (Shayu and Rao, 2006; Shayu et al., 2007).

7.2.3.2.8.1 P450AROM and 17-HSDI mRNA Are Expressed in the Caput and Cauda Regions of the Epididymis P450AROM mRNA was expressed in both the caput and cauda regions of the epididymis. A single band of 290 base pair was obtained by RT-PCR analysis. Interestingly, the level of P450AROM expression did not differ between the two regions. The enzyme aromatase can utilise either testosterone or androstenedione as substrates. In addition to P450AROM, the mRNA for 17-HSDI could also be detected in the caput and cauda. Ontological analysis of P450AROM expression revealed increase in the mRNA level with age, which is likely due to contribution from spermatozoa which express P450AROM (Janulis et al., 1998). Hence, for all studies involving P450AROM expression or activity, animals of age 40 days were employed, since at this age epididymides are bereft of spermatozoa.

7.2.3.2.8.2 Caput and Cauda Regions Are Capable of E2 Biosynthesis In order to assess whether the epididymal aromatase was functional, preliminary studies were carried out using caudal minces. A dose and time-dependent increase in E2 formation was observed upon incubation of cauda with increasing concentration of androstenedione. Subsequently, E2 production in the presence of the androgenic substrates was measured in the caput and cauda regions. Although there was no significant difference in the mRNA expression between the two regions, basal E2 levels (in presence of vehicle control) were higher in the caput compared to cauda. Incubation with the substrate androstenedione or testosterone resulted in at least a twofold increase in E2 level above the basal values in the caput and cauda regions. In each region, the net E2 formed using either androstenedione or testosterone as substrates was similar. Importantly, neither caput nor cauda were able to utilise DHT for aromatisation.

7.2.3.2.8.3 Other Factors The contribution of other factors such as retinoids or growth hormones such as TGFβ, fibroblast growth factor (FGF) etc. needs a mention as these molecules are also necessary for epididymal functioning. Several proteins in the epididymis are regulated by these growth factors; FGF is important for gamma glutamyl transpeptidase expression (Lan et al., 1998), while TGFβ is an important paracrine factor for the epididymis.

7.2.3.2.8.4 LHR mRNA is Differentially Expressed in the Caput and Cauda Regions The expression of LHR was analysed in the epididymis by semi-quantitative RT-PCR analysis. Interestingly, a differential level of LHR expression was observed in the two regions of the epididymis; expression was higher in the cauda when compared to the caput. As anticipated, abundant expression of LHR was seen in the Leydig cells employed as positive control for the RT-PCR reaction. The LHR PCR products from both the Leydig cells and cauda were found to be identical upon sequencing.

7.2.3.2.8.5 Differential Binding of 125I hCG to Caput and Caudal Membrane Extracts and cAMP Production upon hCG Stimulation

To assess whether the LHR in the epididymis was capable of binding its cognate ligand, binding experiments were performed using hCG as the ligand. In these studies, hCG was used in place of LH due to its greater stability and high affinity for rat LHR (Thomas and Segaloff, 1994; Bernard et al., 1998). Binding of hCG was at least two- to threefold higher in the cauda compared to caput. The binding results correlated well with the mRNA level of LHR in these tissues. Binding of hCG to the testicular interstitial cell extract, employed as positive control was very high while in contrast, liver membrane extract used as negative control, showed minimal binding. The ability of the epididymal LHR to transduce cellular signalling was assessed by estimating the level of intracellular cAMP generated in the caput and cauda upon stimulation with 100 ng/mL hCG. In the cauda, hCG stimulation resulted in a 1.5–2-fold increase in cAMP level compared to control (vehicle stimulated), while in the caput region no appreciable increase in cAMP could be detected. Leydig cells showed a seven- to eightfold increase in cAMP upon hCG stimulation while liver tissue employed as negative control, did not show any cAMP production upon stimulation. Importantly, a dose-dependent increase in cAMP was observed in caudal minces stimulated with increasing concentration of hCG (50–200 ng/mL). The binding of hCG to LHR in cauda and cAMP analysis indicated that the LHR in cauda was capable of binding to the cognate ligand and generation of the downstream signal molecule, cAMP. These results together suggested that the LHR was functional in the cauda.

7.2.3.2.9 Regulation of P450AROM mRNA by Androgens and LH

7.2.3.2.9.1 P450AROM mRNA Expression Is Modulated by Androgens In Vivo and In Vitro

The role of androgens in regulating P450AROM expression in the caput and cauda regions of castrated and castrated + testosterone supplemented rats was analysed by semi-quantitative RT-PCR analysis. P450AROM mRNA levels in the caput and cauda were almost undetectable post-castration, compared to sham-operated control animals. However, there was a recovery of P450AROM expression in the castrated group by 70% after testosterone treatment suggesting that testicular androgens play an essential role in maintaining P450AROM expression. The androgen dependence of P450AROM mRNA expression was also emphasised by flutamide treatment, as P450AROM mRNA expression was severely reduced in the caput and cauda of flutamide-treated animals. Finally, the effect of androgens on P450AROM mRNA level was studied in vitro. A non-aromatisable androgen, DHT was employed in place of testosterone, since testosterone can be further metabolised to either DHT or E2. In order to analyse the effect of androgen without the potential interference from any other newly formed metabolites, caput or caudal minces were incubated with DHT and P450AROM mRNA expression was analysed. A twofold increase in P450AROM mRNA level could be observed in the DHT-treated caput and cauda. Together both the in vivo and in vitro studies highlighted androgens as vital modulators of P450AROM expression.

7.2.3.2.9.2 P450AROM mRNA Expression Is Regulated by hCG In Vitro To determine if LHR-mediated function had any role in regulating P450AROM mRNA expression, caudal minces were incubated with 100 ng/mL hCG for 6 h and P450AROM mRNA level was analysed at the end of the incubation period. It was observed that compared to control, there was a twofold increase in the P450AROM mRNA levels in caudal minces incubated in vitro with hCG. This increase was mimicked by the addition of forskolin, an adenylyl cyclase activator, under identical incubation conditions. This clearly indicated that hCG was capable of inducing P450AROM mRNA in the cauda.

7.2.3.2.9.3 hCG Enhances E2 Synthesis in the Cauda Similar to the increase in P450AROM expression, hCG could stimulate aromatase activity in the cauda. The addition of testosterone and hCG in the cauda resulted in higher E2 synthesis as compared to testosterone alone. Addition of androstenedione and hCG to caudal minces also resulted in enhanced E2 formation; however, hCG could not stimulate aromatase activity in the caput.

7.2.3.3 Role of FSH in Regulation of Epididymal Functions

7.2.3.3.1 Proliferation of Cauda Cells by FSH and Testosterone

Deprival of endogenous FSH in immature rats and adult rats resulted in reduction of steroidogenic capacity and LH responsiveness of Leydig cells (Sriraman and Jagannadha Rao, 2004). Since it is known that FSH stimulates proliferation of Sertoli cells (Prahalada et al., 1975) the possibility of such an effect by FSH in cauda cells was also examined. Initially, the maximum proliferative stage of cauda was determined by Western blot analysis for PCNA (proliferating cell nuclear antigen). It was observed that the level of expression of PCNA was maximal on day 10 and decreased by days 20 and 40. The expression of PCNA was very low in 80-day-old rat cauda, indicating that cauda cells are highly proliferative during the early neonatal period. BrdU incorporation analysis revealed that proliferation was high in the 10-day-old rat cauda cells compared to the 80-day-old rat cauda cells. Following addition of FSH (0.25 μg/mL, 0.5 μg/mL and 1 μg/mL) a concentration-dependent increase in the proliferation of 10-day-old rat cauda cells was observed. However, no stimulation in proliferation of 80-day-old rat cauda cells was observed even with 1 μg/mL of FSH.

The role of FSH in the regulation of proliferation of immature rat cauda was validated by in vivo studies. FSH was neutralised in the immature rats from day 7 until day 13 of age, and, as expected, deprival of FSH resulted in a decrease in the expression of PCNA in the cauda epididymis. It was observed that following neutralisation of endogenous FSH, there was a decrease in the incorporation of BrdU by immature rat cauda cells. This agrees with the decrease in PCNA, and these results establish that proliferation was decreased in cauda following neutralisation of FSH.

As it is well known that epididymis is an androgen-regulated tissue (Tsolkas et al., 1967; Brooks, 1981) it was of interest to evaluate the involvement of androgens in the proliferation of immature rat cauda cells. Analysis for expression of mRNA for androgen receptor in the immature and adult rat cauda cells by RT-PCR revealed that the expression was quite low in the 10-day-old rat cauda cells compared to the 80-day-old rat cauda cells. It was also observed that neither hCG nor testosterone was able to stimulate the proliferation of cauda cells from 10-day-old rats.

The absence of spermatozoa was noted in all three regions of epididymis in the FSH-deprived animals. As expected, FSH deprivation had a severe effect on testicular development. The size of the seminiferous tubules as well as the lumen was much smaller following FSH antiserum administration. Since Sertoli cell function was compromised, spermatogenesis was affected, resulting in a decrease in the number of spermatogonia.

In the case of rats, the effect of FSH deprivation was observed as early as 13 days. A decrease in the size of tubule and change in the histology of the epithelial cells was observed following short-term deprivation of endogenous FSH for only 5 days in immature rats. An increase in the stromal cells was also observed following FSH deprivation. A decrease in the size of the seminiferous tubules in the FSH-antiserum-treated rats confirmed that endogenous FSH had been neutralised (Dahia et al., 2008). Furthermore, it should be noted that at this age the lumen of the seminiferous tubules was not open in either the FSH-antiserum-treated rats or the NMS (normal monkey serum)-treated controls, suggesting that the effect in the epididymis was not due to the absence of testicular factors as a result of FSH deprivation. The involvement of FSH in regulation of function of epididymis was further established by the demonstration of the presence of FSHR in rat and monkey epididymis (Moudgal, 1969; Dahia and Rao, 2006).

In summary, recent literature has provided a deeper insight into the importance of epididymis, its role in the fascinating process of sperm maturation, and the complexity of hormonal regulation that control these events. The declining male fertility and the possibility of environmental oestrogens affecting male reproduction provide an impetus to examine the molecular mechanisms involved in the hormonal control of epididymal function, with a special emphasis on oestrogen.

Acknowledgements

Financial support for the work presented in this review was provided by Institute of Science, Bangalore, Department of Science and Technology, Department of Biotechnology, and Indian Council of Medical Research, Government of India; INDO-US programme, CONRAD/Mellon Foundation and Population Council, United States. A.J. Rao is thankful to the Department of Science and Technology for award of Rajaramanna fellowship and to Mrs. Shailaja for secretarial assistance.

References

Agarwal, A., Virk, G., Ong, C. and Du Plessis, S. S. 2014. Effect of oxidative stress on male reproduction. *The World Journal of Men's Health,* 32, 1–17.

Aitken, R. J. 1999. The Amoroso Lecture. The human spermatozoon – a cell in crisis? *Journal of Reproduction and Fertility,* 115, 1–7.

Arai, Y., Mori, T., Suzuki, Y. and Bern, H. A. 1982. Long-term effects of perinatal exposure to sex steroids and diethylstilbestrol on the reproductive system of male mammals. *International Review of Cytology,* 84, 235–268.

Atanassova, N., McKinnell, C., Williams, K., Turner, K., Fisher, J., Saunders, P., Millar, M. and Sharpe, R. 2001. Age-, cell- and region-specific immunoexpression of estrogen receptor α (but not estrogen receptor β) during postnatal development of the epididymis and vas deferens of the rat and disruption of this pattern by neonatal treatment with diethylstilbestrol 1. *Endocrinology,* 142, 874–886.

Bedford, J. 1975. Maturation transport and fate of spermatozoa in the epididymis. In: Greep, R. O. and Hamilton, D. W., eds. *Handbook of Physiology*, Section 7: *Endocrinology,* Vol. 5, *Male Reproductive System.* Washington, D.C., American Physiological Society, 303–317.

Benoit, J. 1925. *Recherches anatomiques, cytologiques et histophysiologiques sur les voies excrétrices du testicule, chez les mammifères. Contribution à l'étude de quelques problèmes de cytologie générale relatifs à la cellule glandulaire (planches I-VII), par le docteur Jacques Benoit, licencié ès sciences, préparateur à la Faculté de médecine de Strasbourg. (Travail de l'Institut d'histologie.),* Impr. alsacienne. (Anatomy, cytology and histophysiology of urinary tract and testis in mammals. Contribution to the study of problems in cytology by Dr. Jacques Benoit, Bachelor of Science, Preparator, Institute of Histology, Faculty of Medicine, University of Strasbourg (Printed in French).)

Bernard, M., Myers, R. and Moyle, W. 1998. Lutropins appear to contact two independent sites in the extracellular domain of their receptors. *Biochemical Journal,* 335, 611–617.

Brooks, D. E. 1981. Metabolic activity in the epididymis and its regulation by androgens. *Physiological Reviews,* 61, 515–555.

Butenandt, A. 1931. Über die chemische untersuchung der sexualhormone. *Angewandte Chemie,* 44, 905–908.

Byskov, A. and Hoyer, P. 1994. Embryology of mammalian gonads and ducts. *The Physiology of Reproduction,* 1, 487–540.

Chang, C., Lee, S. O., Wang, R.-S., Yeh, S. and Chang, T.-M. 2013. Androgen receptor (AR) physiological roles in male and female reproductive systems: Lessons learned from ARKO mice lacking AR in selective cells. *Biology of Reproduction,* 89, 21.

Chauvin, T. R. and Griswold, M. D. 2004. Androgen-regulated genes in the murine epididymis. *Biology of Reproduction,* 71, 560–569.

Clulow, J., Jones, R., Hansen, L. and Man, S. 1997. Fluid and electrolyte reabsorption in the ductuli efferentes testis. *Journal of Reproduction and Fertility* (Suppl.), 53, 1–14.

Couse, J. F. and Korach, K. S. 1999. Estrogen receptor null mice: What have we learned and where will they lead us? *Endocrine Reviews,* 20, 358–417.

Dacheux, J.-L. and Dacheux, F. 2002. Protein secretion in the epididymis. In: Robaire, B. and Barry, T. Eds. *The Epididymis: From Molecules to Clinical Practice.* Springer: Kluwer Academic/Plenum Publishers, New York, pp. 297–316.

Dahia, C. L., Petrusz, P., Hall, S. H. and Rao, A. J. 2008. Effect of deprivation of endogenous follicle stimulating hormone on rat epididymis: A histological evaluation. *Reproductive Biomedicine Online,* 17, 331–337.

Dahia, C. L. and Rao, A. J. 2006. Demonstration of follicle-stimulating hormone receptor in cauda epididymis of rat. *Biology of Reproduction,* 75, 98–106.

De Lamirande, E., Leclerc, P. and Gagnon, C. 1997. Capacitation as a regulatory event that primes spermatozoa for the acrosome reaction and fertilization. *Molecular Human Reproduction,* 3, 175–194.

Deshpande, S. N., Vijayakumar, G. and Rao, A. J. 2009. Oestrogenic regulation and differential expression of *WNT4* in the bonnet monkey and rodent epididymis. *Reproductive Biomedicine Online,* 18, 555–561.

Diekman, A. B., Westbrook-CASE, V. A., Naaby-Hansen, S., Klotz, K. L., Flickinger, C. J. and Herr, J. C. 1997. Biochemical characterization of sperm agglutination antigen-1, a human sperm surface antigen implicated in gamete interactions. *Biology of Reproduction,* 57, 1136–1144.

Ezer, N. and Robaire, B. 2002. Androgenic regulation of the structure and functions of the epididymis. In: Robaire, B. and Barry, T. Eds. *The Epididymis: From Molecules to Clinical Practice.* Springer: Kluwer Academic/Plenum Publishers, New York.

Fan, X. and Robaire, B. 1998. Orchidectomy induces a wave of apoptotic cell death in the epididymis 1. *Endocrinology,* 139, 2128–2136.

Fisher, C. R., Graves, K. H., Parlow, A. F. and Simpson, E. R. 1998. Characterization of mice deficient in aromatase (ArKO) because of targeted disruption of the cyp19 gene. *Proceedings of the National Academy of Sciences,* 95, 6965–6970.

Fisher, J. S., Turner, K. J., Brown, D. and Sharpe, R. M. 1999. Effect of neonatal exposure to estrogenic compounds on development of the excurrent ducts of the rat testis through puberty to adulthood. *Environmental Health Perspectives,* 107, 397.

Hall, S. H., Hamil, K. G. and French, F. S. 2002. Host defense proteins of the male reproductive tract. *Journal of Andrology,* 23, 585–597.

Hamil, K. G., Liu, Q., Sivashanmugam, P., Anbalagan, M., Yenugu, S., Soundararajan, R., Grossman, G., Rao, A., Birse, C. E. and Ruben, S. M. 2003. LCN6, a novel human epididymal lipocalin. *Reproductive Biology and Endocrinology,* 1, 112.

Hess, R. A. 2000. Oestrogen in fluid transport in efferent ducts of the male reproductive tract. *Reviews of Reproduction,* 5, 84–92.

Hess, R. A. 2003. Estrogen in the adult male reproductive tract: A review. *Reproductive Biology and Endocrinology,* 1, 52.

Hess, R. A., Zhou, Q. and Nie, R. 2002. The role of estrogens in the endocrine and paracrine regulation of the efferent ductules, epididymis and vas deferens. In: *The Epididymis: From Molecules to Clinical Practice: A Comprehensive Survey of the Efferent Ducts, the Epididymis and the Vas Deferens.* Springer, New York.

Hinton, B. and Setchell, B. 1993. Fluid secretion and movement. *The Sertoli Cell,* 12, 249–267.

Howell, A., Osborne, C. K., Morris, C. and Wakeling, A. E. 2000. ICI 182,780 (Faslodex™): Development of a novel, 'pure' antiestrogen. *Cancer,* 89, 817–825.

Ilio, K. Y. and Hess, R. A. 1994. Structure and function of the ductuli efferentes: A review. *Microscopy Research and Technique,* 29, 432–467.

Janulis, L., Bahr, J. M., Hess, R. A., Janssen, S., Osawa, Y. and Bunick, D. 1998. Rat testicular germ cells and epididymal sperm contain active P450 aromatase. *Journal of Andrology,* 19, 65–71.

Joseph, A., Shur, B. D. and Hess, R. A. 2011. Estrogen, efferent ductules, and the epididymis. *Biology of Reproduction,* 84, 207–217.

Korach, K. S. 1994. Insights from the study of animals lacking functional estrogen receptor. *Science,* 266, 1524–1527.

Krege, J. H., Hodgin, J. B., Couse, J. F., Enmark, E., Warner, M., Mahler, J. F., Sar, M., Korach, K. S., Gustafsson, J.-Å. and Smithies, O. 1998. Generation and reproductive phenotypes of mice lacking estrogen receptor β. *Proceedings of the National Academy of Sciences,* 95, 15677–15682.

Lan, Z.-J., Labus, J. C. and Hinton, B. T. 1998. Regulation of gamma-glutamyl transpeptidase catalytic activity and protein level in the initial segment of the rat epididymis by testicular factors: Role of basic fibroblast growth factor. *Biology of Reproduction,* 58, 197–206.

Lindzey, J., Wetsel, W. C., Couse, J. F., Stoker, T., Cooper, R. and Korach, K. S. 1998. Effects of castration and chronic steroid treatments on hypothalamic gonadotropin-releasing hormone content and pituitary gonadotropins in male wild-type and estrogen receptor-α knockout mice. *Endocrinology,* 139, 4092–4101.

Liu, Q., Hamil, K. G., Sivashanmugam, P., Grossman, G., Soundararajan, R., Rao, A. J., Richardson, R. T., Zhang, Y.-L., O'rand, M. G. and Petrusz, P. 2001. Primate epididymis-specific proteins: Characterization of ESC42, a novel protein containing a trefoil-like motif in monkey and human. *Endocrinology,* 142, 4529–4539.

Lubahn, D. B., Moyer, J. S., Golding, T. S., Couse, J. F., Korach, K. S. and Smithies, O. 1993. Alteration of reproductive function but not prenatal sexual development after insertional disruption of the mouse estrogen receptor gene. *Proceedings of the National Academy of Sciences,* 90, 11162–11166.

Macgregor, J. I. and Jordan, V. C. 1998. Basic guide to the mechanisms of antiestrogen action. *Pharmacological Reviews,* 50, 151–196.

Mccauley, T. C., Kurth, B. E., Norton, E. J., Klotz, K. L., Westbrook, V. A., Rao, A. J., Herr, J. C. and Diekman, A. B. 2002. Analysis of a human sperm CD52 glycoform in primates: Identification of an animal model for immunocontraceptive vaccine development. *Biology of Reproduction,* 66, 1681–1688.

McKinnell, C., Atanassova, N., Williams, K., Fisher, J., Walker, M., Turner, K., Saunders, P. and Sharpe, R. 2001. Suppression of androgen action and the induction of gross abnormalities of the reproductive tract in male rats treated neonatally with diethylstilbestrol. *Journal of Andrology,* 22, 323–338.

Moudgal, N. R., Madhwa RAJ, H. G., Rao, A. J. and Sairam, M. R. 1969. Need of luteinising hormone for maintaining early pregnancy in rat. *Indian Journal of Experimental Biology,* 7, 2.

Muchmore, A. V. and Decker, J. M. 1985. Uromodulin: A unique 85-kilodalton immunosuppressive glycoprotein isolated from urine of pregnant women. *Science,* 229, 479–481.

Mylchreest, E., Cattley, R. C. and Foster, P. M. 1998. Male reproductive tract malformations in rats following gestational and lactational exposure to di (n-butyl) phthalate: An antiandrogenic mechanism? *Toxicological Sciences,* 43, 47–60.

Nieschlag, E., Behre, H. M. and Nieschlag, S. 2012. *Testosterone: Action, Deficiency, Substitution*. New York: Cambridge University Press.

Nitta, H., Bunick, D., Hess, R., Janulis, L., Newton, S. C., Millette, C. F., Osawa, Y., Shizuta, Y., Toda, K. and Bahr, J. M. 1993. Germ cells of the mouse testis express P450 aromatase. *Endocrinology,* 132, 1396–1401.

O'Donnell, L., Robertson, K. M., Jones, M. E. and Simpson, E. R. 2001. Estrogen and spermatogenesis 1. *Endocrine Reviews,* 22, 289–318.

Ogawa, S., Chan, J., Chester, A. E., Gustafsson, J.-Å., Korach, K. S. and Pfaff, D. W. 1999. Survival of reproductive behaviors in estrogen receptor β gene-deficient (βERKO) male and female mice. *Proceedings of the National Academy of Sciences,* 96, 12887–12892.

O'Rand, M., Widgren, E., Sivashanmugam, P., Richardson, R., Hall, S., French, F., Vandevoort, C., Ramachandra, S., Ramesh, V. and RAO, A. J. 2004. Reversible immunocontraception in male monkeys immunized with eppin. *Science,* 306, 1189–1190.

Orgebin-Crist, M., Danzo, B. and Davies, J. 1975. Endocrine control of the development and maintenance of sperm fertilizing ability in the epididymis. In: Greep, R.O. and Hamilton, D.W., eds., *Handbook of Physiology, Section 7: Endocrinology, Vol. 5, Male Reproductive System*. Washington, D.C.: American Physiological Society, 319–338.

Pelletier, G. 2002. Effects of estradiol on prostate epithelial cells in the castrated rat. *Journal of Histochemistry and Cytochemistry,* 50, 1517–1523.

Prahalada, S., Venkatramaiah, M., Jagannadha Rao, A. and Moudgal, N. 1975. Termination of pregnancy in macaques (*Macaca radiata*) using monkey antiserum to ovine luteinizing hormone. *Contraception,* 12, 137–147.

Robaire, B. and Hamzeh, M. 2011. Androgen action in the epididymis. *Journal of Andrology,* 32, 592–599.

Robaire, B., Hinton, B. and Orgebin-Crist, M.-C. 2001. *The Epididymis: From Molecules to Clinical Practice: From Molecules to Clinical Practice: A Comprehensive Survey of the Efferent Ducts, the Epididymis and the Vas Deferens.* New York: Springer.

Robaire, B. and Viger, R. S. 1995. Regulation of epididymal epithelial cell functions. *Biology of Reproduction,* 52, 226–236.

Rodriguez, C. M., Kirby, J. L. and Hinton, B. T. 2002. The development of the epididymis. *The Epididymis: From Molecules to Clinical Practice: A Comprehensive Survey of the Efferent Ducts, the Epididymis and the Vas Deferens.* New York: Springer.

Rosenfeld, C. S., Ganjam, V. K., Taylor, J. A., Yuan, X., Stiehr, J. R., Hardy, M. P. and Lubahn, D. B. 1998. Transcription and translation of estrogen receptor-β in the male reproductive tract of estrogen receptor-α knock-out and wild-type mice 1. *Endocrinology,* 139, 2982–2987.

Rowley, M., Teshima, F. and Heller, C. 1970. Duration of transit of spermatozoa through the human male ductular system. *Fertility and Sterility,* 21, 390–396.

Shayu, D., Chennakesava, C. and RAO, A. 2006a. Differential expression and antibacterial activity of WFDC10A in the monkey epididymis. *Molecular and Cellular Endocrinology,* 259, 50–56.

Shayu, D., Hardy, M. P. and Rao, A. J. 2007. Delineating the role of estrogen in regulating epididymal gene expression. *Society of Reproduction and Fertility Supplement,* 63, 31–43.

Shayu, D., Kesava, C., Soundarajan, R. and Rao, A. J. 2005. Effects of ICI 182780 on estrogen receptor expression, fluid absorption and sperm motility in the epididymis of the bonnet monkey. *Reproductive Biology and Endocrinology,* 3, 10.

Shayu, D. and Rao, A. 2006. Expression of functional aromatase in the epididymis: Role of androgens and LH in modulation of expression and activity. *Molecular and Cellular Endocrinology,* 249, 40–50.

Shayu, D., Sriraman, V. and Rao, A. 2006b. Estrogen receptor antagonists and inhibitors of aromatase: Use in reproduction research. *Journal of the Indian Institute of Science,* 86, 1–13.

Sriraman, V. and Jagannadha Rao, A. 2004. Evaluation of the role of FSH in regulation of Leydig cell function during different stages of its differentiation. *Molecular and Cellular Endocrinology,* 224, 73–82.

Sutovsky, P., Moreno, R. D., Ramalho-Santos, J., Dominko, T., Simerly, C. and Schatten, G. 2000. Ubiquitinated sperm mitochondria, selective proteolysis, and the regulation of mitochondrial inheritance in mammalian embryos. *Biology of Reproduction,* 63, 582–590.

Tamm, I. and Horsfall, F. L. 1950. Characterization and separation of an inhibitor of viral hemagglutination present in urine. *Experimental Biology and Medicine,* 74, 108–114.

Thomas, D. M. and Segaloff, D. 1994. Hormone-binding properties and glycosylation pattern of a recombinant form of the extracellular domain of the luteinizing hormone/chorionic gonadotropin receptor expressed in mammalian cells. *Endocrinology,* 135, 1902–1912.

Thompson, E. A. and Siiteri, P. K. 1974. The involvement of human placental microsomal cytochrome P-450 in aromatization. *Journal of Biological Chemistry,* 249, 5373–5378.

Toda, K., Okada, T., Takeda, K., Akira, S., Saibara, T., Shiraishi, M., Onishi, S. and Shizuta, Y. 2001. Oestrogen at the neonatal stage is critical for the reproductive ability of male mice as revealed by supplementation with 17beta-oestradiol to aromatase gene (Cyp19) knockout mice. *Journal of Endocrinology,* 168, 455–463.

Tsolkas, P., Fahrmann, W. and Schuchardt E. 1967. The epididymal epithelium of the rat following hypophysectomy and androgen substitution. *Experientia,* 23, 61–63.

Turner, T. 1984. Resorption versus secretion in the rat epididymis. *Journal of Reproduction and Fertility,* 72, 509–514.

Turner, T., Avery, E. and Sawchuk, T. 1994. Assessment of protein synthesis and secretion by rat seminiferous and epididymal tubules in vivo. *International Journal of Andrology,* 17, 205–213.

Vreeburg, J., Holland, M., Cornwall, G. and Orgebin-Crist, M. 1990. Secretion and transport of mouse epididymal proteins after injection of 35S-methionine. *Biology of Reproduction,* 43, 113–120.

Wang, R.-S., Yeh, S., Tzeng, C.-R. and Chang, C. 2009. Androgen receptor roles in spermatogenesis and fertility: Lessons from testicular cell-specific androgen receptor knockout mice. *Endocrine Reviews,* 30, 119–132.

Wen, R. Q. and Wong, P. Y. 1988. Reserpine treatment increases viscosity of fluid in the epididymis of rats. Biol Reprod., 38(5), 969–974.

Wilson, T., Therrien, A. and Harkness, J. 1986. Sperm counts and reproductive tract lesions in male Syrian hamsters exposed in utero to diethylstilbestrol. *Laboratory Animal Science,* 36, 41–44.

Wong, P. and Huang, S. 1990. Secretory agonists stimulate a rise in intracellular cyclic AMP but not Ca2+ and inositol phosphates in cultured rat epididymal epithelium. *Experimental Physiology,* 75, 321–337.

Wong, P. and Yeung, C. 1977. Hormonal regulation of fluid reabsorption in isolated rat cauda epididymidis. *Endocrinology,* 101, 1391–1397.

Yuan, H., Liu, A., Zhang, L., Zhou, H., Wang, Y., Zhang, H., Wang, G., Zeng, R., Zhang, Y. and Chen, Z. 2006. Proteomic profiling of regionalized proteins in rat epididymis indicates consistency between specialized distribution and protein functions. *Journal of Proteome Research,* 5, 299–307.

Zhu, L.-J., Hardy, M. P., Inigo, I. V., Huhtaniemi, I., Bardin, C. W. and Moo-Young, A. J. 2000. Effects of androgen on androgen receptor expression in rat testicular and epididymal cells: A quantitative immunohistochemical study. *Biology of Reproduction,* 63, 368–376.

Chapter 8

Role of Sperm Surface Molecules in Motility Regulation

Gopal C. Majumder, Sudipta Saha, Kaushik Das, Debjani Nath, Arunima Maiti, Souvik Dey, Debarun Roy, Chinmoy Sankar Dey, Sutapa Mitra, Ajay Rana, Jitamanyu Chakrabarty, Sujoy Das, Arpita Bhoumik, Saswati Banerjee, Mahitosh Mandal, Bijay Shankar Jaiswal, Prasanta Ghosh, Abhi Das, Debdas Bhattacharyya and Sandhya Rekha Dungdung

Contents

8.1 Introduction ... 198
8.2 Sperm Motility Assays ... 199
8.3 Isolation of Sperm Plasma Membrane ... 200
8.4 Membrane Lipids, Asymmetry and Fluidity ... 202
8.5 Sperm Ecto-Protein Kinases and Their Protein Substrates ... 205
  8.5.1 Ecto-Cyclic AMP-Dependent Protein Kinase (RC) ... 205
  8.5.2 Ecto-CIK and Cell-Surface Protein Phosphorylation Mechanism ... 206
  8.5.3 MPS: Major Protein Substrate of Ecto-CIK ... 209
  8.5.4 Sperm Surface Protein Dephosphorylation Mechanism ... 214
8.6 Sperm Surface Lectins and Their Receptors ... 219
  8.6.1 A Novel D-Galactose-Specific Lectin on Sperm Surface ... 220
  8.6.2 D-Galactose-Specific Lectin Receptor on Sperm Surface ... 221

8.6.3 Novel Copper-Dependent Sialic Acid-Specific Lectin on Sperm Surface 222
8.7 Sperm Motility Stimulating Proteins 223
8.7.1 Forward Motility Stimulating Factor (FMSF) 223
8.7.2 Motility Initiating Protein 228
8.8 Membrane-Bound Sperm Motility Inhibitor 232
8.9 Ecto-ATPase 233
8.10 Other Studies 233
8.11 Conclusion 234
Acknowledgements 235
References 236

8.1 Introduction

Sperm cell is the male gamete and it is haploid in nature. This unique microscopic motile cell performs an important function in biology: fertilisation of ova. A typical mammalian sperm consists of a head, neck, middle piece and tail. The inner core of the sperm flagella contains microtubules that serve as the basic infrastructure for the ATP-dependent bending of the sperm tail. The flagellar beat kinematics, sperm morphology and surface properties are responsible for the rate of forward progression which is essential for the *in vivo* fertilisation process (Katz et al. 1989). Mammalian testicular spermatozoa are immotile and infertile and these cells undergo a maturation process in epididymis before they acquire the capacity for forward progression and fertility. Finally upon ejaculation into the female reproductive tract, the male gametes undergo capacitation and acrosomal reaction that are essential for their fertility potential. The three major parts of mammalian epididymis are the caput, corpus and cauda (Figure 8.1). The immature testicular spermatozoa acquire testosterone-dependent maturity as they pass through different parts of epididymis and finally the mature sperm cells are stored in the terminal part (cauda) of the organ (Prasad et al. 1970; Orgebin-Crist and Tichenor 1972; Hoskins et al. 1978; Glander 1984). During the epididymal transit, spermatozoa undergo a variety of biochemical alterations (Cooper 1986; Majumder et al. 1990). There is a marked increase in the intrasperm level of cyclic AMP (cAMP) and pH during the epididymal sperm maturation suggesting thereby that elevated intrasperm levels of cAMP and pH have an important role in the *in vivo* initiation of sperm forward progression (Hoskins et al. 1978; Lee et al. 1983; Brokaw 1987). However, the molecular basis of the initiation of flagellar motility in epididymis and its subsequent regulation are not well understood (for reviews, see Hoskins et al. 1978; Tash and Means 1983; Majumder et al. 1999, 2012; Gagnon and de Lamirande 2006). Recent study by Bhoumik et al. (2014) showed that extracellular calcium has a biphasic role in caprine motility regulation, optimal calcium concentration being 10 μM.

Cell surface molecules regulate the functions of the mammalian cells by modulating cell–cell interactions, effector–receptor interactions, membrane permeability,

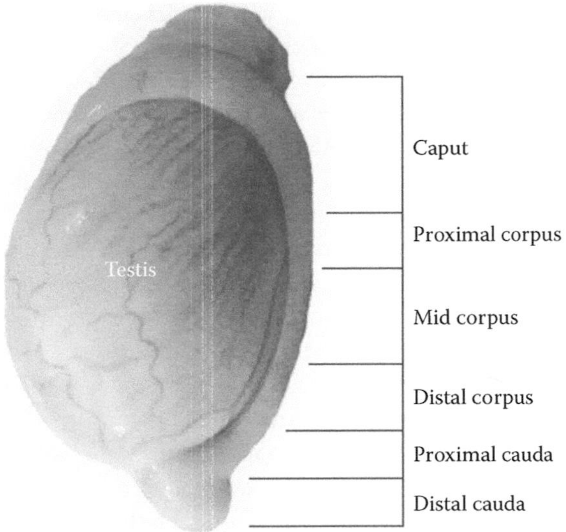

Figure 8.1 A photograph of caprine epididymis. The epididymis is closely apposed tubule to the testicles that fuses with the vasa efferentia at its testicular end and the vas deferens at its distal end. It can be roughly divided into three parts: the enlarged proximal end known as the head or caput, the slender middle region or the corpus and the pointed tail end or cauda.

membrane fluidity, transmembrane signalling etc. (Yamada et al. 1980; Vestweber and Blanks 1999). Cell surface of sperm like the other mammalian cells is believed to play a vital role in the regulation of sperm capacitation, acrosomal reaction and fertilisation (Yanagimachi 1994; Ghosh and Datta 2003; de Lamirande and O'Flaherty 2008; Yoshida et al. 2008). Early studies implicating the role of external cell surface molecules in the regulation of epididymal sperm maturation and forward progression have already been reviewed (Johnson 1975; Hammerstedt and Parks 1987; Majumder et al. 1990). This chapter reviews primarily the forward motility initiating and regulating molecules located on sperm external cell surface with special reference to the caprine sperm model.

8.2 Sperm Motility Assays

As all the studies that will be discussed here are based on sperm motility, it is necessary to have a glimpse of various motility assay methods. The microscopic method is the most widely used subjective method for sperm motility analysis (Acott and Hoskins 1978; Mandal et al. 2006). Subsequently, more objective methods were developed such as the light scattering method, laser beam method and multiple exposure photographic method (Kamidono et al. 1983; Freund and Oliveira 1987;

Schieferstein et al. 1998; Iguer-Ouada and Verstegen 2001). The turbidimetric method uses a spectrophotometer for the estimation of sperm motility, where the sperm cells passing across the light path are analysed in terms of change in absorbance or optical density (Sokoloski et al. 1977; Majumder and Chakrabarti 1984). Computer-aided semen analysis (CASA) based on microscopic video photographic method is widely used for estimating sperm horizontal velocity from the horizontal plane of a glass slide or haemocytometer or Makler Chamber (Devi and Shivaji 1994; Perez-Sanchez et al. 1996). But, even the parameters from CASA are not well correlated with the fertility potential of spermatozoa. Further, all the available techniques consider the 'horizontal' velocity only and there is not a single instrument available for measuring the 'vertical' velocity of spermatozoa.

Using caprine sperm as the model, a unique computer-based spectrophotometric system has been developed for the first time to determine the average 'vertical' velocity of motile cells (Saha et al. 2007). Sperm cells were layered at the bottom of the cuvette containing buffer solution and exposed to the spectrophotometric light path at different heights (2, 4, 6 and 8 mm) to track the vertically moving spermatozoa. The vertical movement was materialised with the development of an electromechanical up-down movement devise for the cuvette accomplished with the help of a cuvette holder-stepper motor–computer assembly. The entire system was controlled by the necessary motion control, data acquisition and data processing software developed for cuvette movement and data analysis. Spermatozoa that swim up against the gravity are recorded at different heights of the cuvette at 545 nm. Average vertical velocity is calculated (with the help of the software) from the values of distances travelled by sperm cells at specified time intervals (Figure 8.2). Undertaking upward movement against gravity is much tougher as compared to horizontal movement. Consequently average vertical velocity is expected to be a much better identifying parameter for assessing semen and other motile cell quality. The novel instrument has the potential for immense application in infertility clinics, sperm banks, animal breeding centres, centres for conservation of endangered species, research laboratories etc. Because of the remarkable applied potential, patent applications have already been filed in India (Paul et al. 2004) and abroad (Paul et al. 2005).

8.3 Isolation of Sperm Plasma Membrane

The sucrose gradient method is most widely used for the isolation of mammalian cell membranes (Lunstra et al. 1974). But, this method is very cumbersome and it requires the use of an ultracentrifuge. We have developed an improved method for isolation of caprine sperm plasma membrane (Rana and Majumder 1987) by modifying the procedure of Ivanov and Profirov (1981). Briefly the standardised method consists of hypotonic shock of intact sperm cells with 1.25 mM EDTA to dissociate the plasma membrane and dispersion of these cells to a two-phase

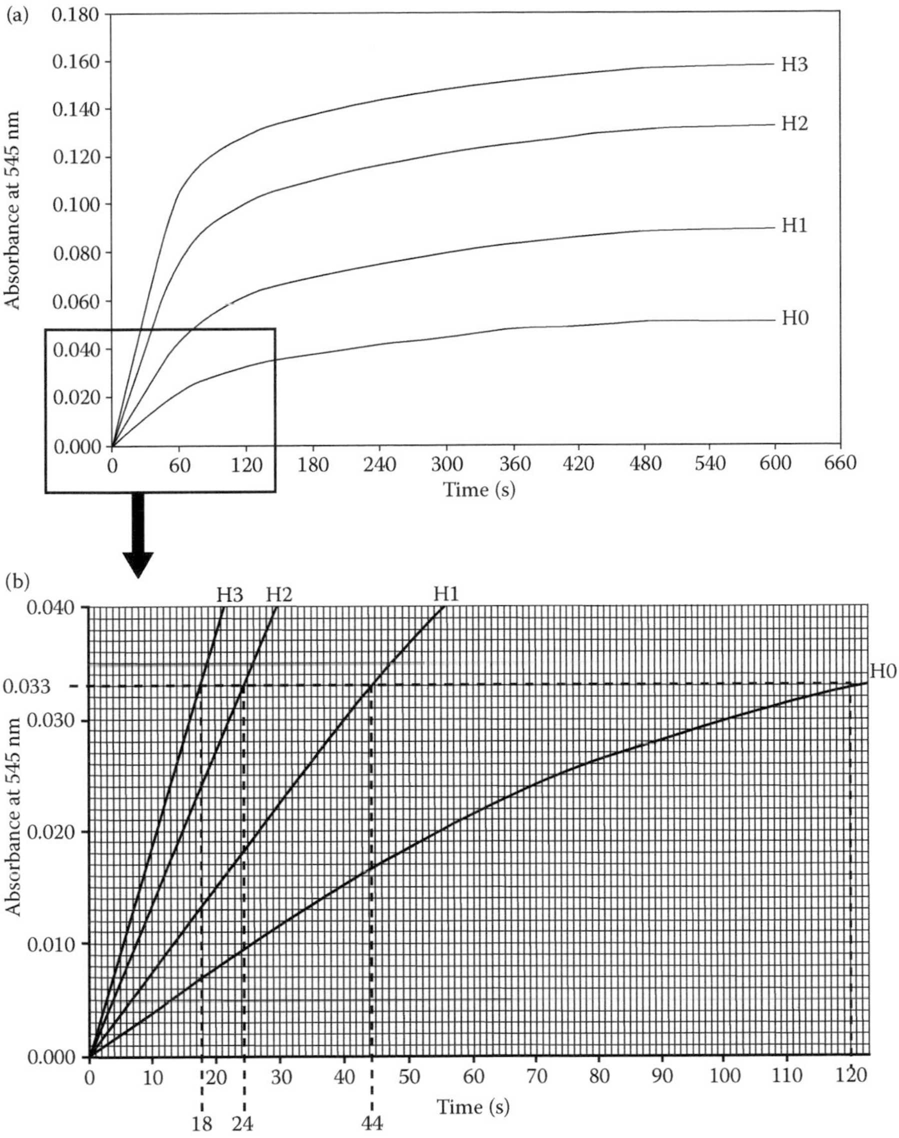

Figure 8.2 **(a) The entire absorbance vs. time plot showing the curves at different heights. (b) Highlighted part of this figure as shown in the rectangular box has been magnified to calculate the time required by the sperm cells to reach specified heights of the cuvette. H0, H1, H2 and H3 represent 2, 4, 6 and 8 mm of cuvette height respectively. (Reproduced from Saha, S. et al. 2007. *Cytometry Part A* 71A:308–316.)**

polymer system consisting of 5.5% 252 kDa dextran and 4.2% 20 kDa polyethylene glycol followed by centrifugation at 9700 × g for 30 min when the two phases get separated and the membrane fraction gets sedimented at the interphase. The yield of plasma membrane by this method was 35%–40% as judged by recovery of membrane bound marker enzymes, alkaline phosphatase and 5′-nucleotidase. The isolated membrane showed a high degree of purity as evidenced by phase contrast and electron microscopic studies and analysis of marker enzymes: phosphatase, 5′-nucleotidase, acrosin, cytochrome oxidase and glucose-6-phosphatase. This method is simpler, more rapid and economical than sucrose gradient (Lunstra et al. 1974) and nitrogen cavitation (Gillis et al. 1978; Bongalhardo et al. 2002) methods. The yield and purity are better than sucrose gradient method and at least comparable with the nitrogen cavitation methods.

8.4 Membrane Lipids, Asymmetry and Fluidity

During epididymal maturation there was a significant decrease in total lipids, phospholipids (PL) and glycolipids (GL) of caprine sperm membrane (Rana et al. 1991). Phosphatidylcholine (PC), phosphatidylethanolamine (PE) and sphingomyelin (SPH) were the constituents of the PL fraction (Table 8.1). Among the phospholipids PE suffered maximum decrease (65% w/w) during sperm maturation. On the contrary neutral lipid (NL) fraction enhanced (about 50% w/w) during the transit from caput to cauda. Sterol and steryl esters, the major constituents of NL, enhanced (about 60% and 200% w/w, respectively) during maturation while other membrane-bound neutral lipids decreased. Since cholesterol (CH) was the major

Table 8.1 Composition of Various Lipid Classes in Maturing Sperm Plasma Membranes

Lipids	*Caput*	*Corpus*	*Cauda*
Total lipids[a]	1.1	0.9	0.8
Neutral lipid[b]	154.0	164.0	221.0
Glycolipid[b]	99.0	43.2	19.2
Phospholipid[b]	847.0	693.0	568.0

Source: Reproduced from Rana, A.P.S. et al. 1991. *Biochimica et Biophysica Acta* 1061:185–196.

Note: The data shown are means of three experiments that are within ±10.0% of the mean.

[a] Expressed as mg of lipid/mg of membrane protein.
[b] Expressed as μg/mg of membrane protein.

component of the sterol fraction, the aforementioned changes led to appreciable enhancement in the cholesterol/phospholipid (CH/PL) ratio. Phosphatidylinositol (PI) has also been detected in the caprine sperm plasma membrane (Chakrabarty et al. 2007) although little is known about its role in epididymal sperm maturation. Linear correlation exists between DHA concentration and normal motile sperm cells (Nissen and Kreysel 1983; Gulaya et al. 2001). The DHA concentration in human semen gradually decreased for normo-, terato-, astheno-, oligo- and azoospermic persons. The oxidation of phospholipid-bound DHA has been shown to be one of the major factors that limit the motile life span of sperm *in vitro* (Ollero et al. 2000). DHA increased from 5.1% of total fatty acids in infants and juveniles to 18.1% in post-pubertal young adults of rhesus monkey (Connor et al. 1997). However, as sperm cells transit from caput to cauda there is a net decrease in DHA content in sperm (Evans and Setchell 1979; Ollero et al. 2000). We also observed similar fate of DHA during maturation. In general, it has been observed that the amount of PL of sperm plasma membrane decrease during epididymal maturation. The changes in different PL fractions are not uniform and are species specific. For example, PC increases during maturation for human (Haidl and Opper 1997) but deceases in case of rat (Aveldaño et al. 1992).

The lipid composition of inner and outer membrane leaflets is not same (Rothman and Lenard 1977). Treating the intact spermatozoa with phospholypase C and trinitrobenzene sulphonate we have shown the asymmetric distribution of phospholipids in maturing sperm plasma membrane (Rana et al. 1993). During maturation the phospholipids asymmetry and distribution of the saturated and unsaturated fatty acids in the phospholipid fractions in both membrane leaflets undergo profound alteration (Table 8.2). The outer membrane was rich in PC and SPH whereas PE dominated the inner leaflet. The ratio of PE/PC in the inner membrane is similar for mature cauda and immature caput sperms but it is less for maturing corpus portion. The profound increase in the amount of PC in the outer leaflet of sperm plasma membrane implicates that it is rich in ether-linked PC as we have found that sperm membrane PC is rich in ether lipids (Rana et al. 1991). Ether linkage retards the action of lipase on phospholipids and therefore contributes to the stability of plasma membrane (Rana et al. 1993). The distribution of phospholipids in the inner/outer leaflets of sperm membrane is also species specific because this distribution in case of ejaculated ram sperm (Hinkovska et al. 1986) differs from that of goat. The major difference lies in localisation of PE in the outer leaflet in case of ram. Müller et al. (1994) reported an asymmetric transversal lipid distribution with aminophospholipids preferentially located in the inner leaflet and choline-containing phospholipids in the outer leaflet. A number of studies on membrane lipid asymmetry have been carried out in relation to capacitation and acrosome reactions of sperm cells when translocation of PS from inner to outer leaflet is observed (Martin et al. 2005).

During maturation an increase in the ratio of saturated/unsaturated fatty acids was observed. This, along with enhanced CH/PL ratio, are instrumental in the

Table 8.2 Phospholipid Compositions of Sperm PM Leaflets Obtained by Phospholipase C Treatments of the Sperms Derived from Corpus, Caput and Cauda

		Membrane Phospholipids			
Sperm Source	*Leaflets*	*PE*	*PC*	*SPH*	*PE/PC*
Caput-epididymis	Whole	352.8	378.0	109.2	0.93
	Inner	307.4	151.2	42.0	2.0
	Outer	45.4	226.8	67.2	0.20
Corpus-epididymis	Whole	208.8	403.2	108.0	0.52
	Inner	170.6	180.0	43.2	0.95
	Outer	38.2	223.2	64.8	0.17
Cauda-epididymis	Whole	209.0	300.0	91.0	0.69
	Inner	178,0	93.0	29.0	1.91
	Outer	30.0	207.0	61.2	0.14

Source: Reproduced from Rana, A.P.S. et al. 1993. *Biochimica et Biophysica Acta* 1210:1–7.

Note: Results were expressed as μg/mg of membrane protein. The data shown are means of three experiments that are ±10.0% of the mean.

observed decrease of caprine sperm membrane fluidity during the epididymal maturation process (Rana and Majumder 1990), strengthening thereby the postulation of Stubbs and Smith (1984). The lipid coefficient increased significantly during epididymal maturation of ram, mouse, goat, dog and monkey spermatozoa (Christova et al. 2004). However, in boar the diffusion coefficient value does not change significantly between epididymal regions (James et al. 1999). Microviscosity analysis of isolated plasma membrane from caput, corpus and cauda epididymal sperm of goat using 1,6-diphenyl-1,3,5-hexatriene and Arrhenius plot of the data showed two distinct lipid phase transitions in the temperature zones 19°C–25°C and 35°C–37°C (Table 8.3). The phase transition of the mature male gamete at 36°C–37°C may have a great impact on the subsequent events of the sperm life cycle since the mature sperm cells are stored in the epididymis at a few degrees below the body temperature and experience higher temperature when ejaculated into the female reproductive tract. The phase transition temperatures of phospholipids isolated from sperm membrane of boar, bull, stallion and rooster are 24.0°C, 25.4°C, 20.7°C and 24.5°C respectively. The phase transition temperatures of glycolipids for the first three species were 36.2°C, 42.8°C and 33.4°C, respectively (Parks and Lynch 1992). For ram the transition temperature was found to be 23°C

Table 8.3 Lipid Phase Transition Temperatures of the Caput, Corpus and Cauda Epididymal Sperm Plasma Membranes

	Low Phase Transition Temperature (°C)	*High Phase Transition Temperature (°C)*
Caput	19–20	36–37
Corpus	21–22	35–36
Cauda	24–25	36–37

(Holt and North 1985). All the data implicate significant alteration of the sperm membrane structure during epididymal sperm maturation.

8.5 Sperm Ecto-Protein Kinases and Their Protein Substrates

8.5.1 Ecto-Cyclic AMP-Dependent Protein Kinase (RC)

Since the appearance of the first two reports on the localisation of a protein kinase (ecto-kinase) on the external surface of mammalian cells (Mastro and Rosengurt 1976; Schlaeger and Kohler 1976), many papers have been published demonstrating various types of ecto-kinase and cell surface phosphoproteins in a variety of cell types (for review, Redegeld et al. 1999; Nath et al. 2008; Majumder et al. 2012). Studies from our laboratory have reported for the first time the occurrence of a cyclic AMP (cAMP)-dependent protein kinase on the external surface of sperm using rat epididymal spermatozoa as a model (Majumder 1978, 1981). Subsequently several investigations provided evidence supporting the externally displayed localisation of a cAMP-dependent ecto-protein kinase in spermatozoa derived from rat (Atherton et al. 1985) and human (Schoff et al. 1982; Pariset et al. 1983). The sperm ecto-kinase consists of a catalytic subunit (C) and a regulatory subunit (R) that binds with high affinity to cAMP. These kinases are thus referred to as ecto-RCs. The goat sperm ecto-cyclic AMP-dependent kinase is incapable of phosphorylating the sperm surface ecto-phosphoproteins (Dey and Majumder 1990b). But as shown in Table 8.4, the ecto-RC is capable of phosphorylating proteins located in the extra-cellular fluids: epididymal plasma and seminal plasma (Majumder et al. 1990). The presence of a cAMP-binding protein (R) on the external surface of goat epididymal spermatozoa has been demonstrated (Dey and Majumder 1987a). Highly purified plasma membrane (PM) isolated from the goat cauda-epididymal spermatozoa, was found to possess RC activity. As shown in Figure 8.3, the sperm PM-bound RC solubilised in Triton X-100, resolved into type I and II isoenzymes by DEAE-cellulose chromatography. The postulated surface topography of these kinases has been

Table 8.4 Substrate Specificity of Intact Sperm Ecto-Kinase for Phosphorylation of Epididymal and Seminal Plasma Proteins

Protein Source	*Protein Concentration (mg/mL)*	*Protein Kinase Activity (Units)*	
		Control	*+cAMP*
Epididymal plasma	2.5	0.28	0.62
	5.0	0.83	1.91
Seminal plasma	2.5	0.37	1.40
	5.0	0.47	2.61

Source: Reproduced from Majumder, G.C. et al. 1990. *Archives of Andrology* 24:287–303.

Note: Protein kinase activity of goat cauda epididymal intact sperm was estimated as described (Dey and Majumder 1990b) with some modifications. Instead of histones, specified proteins were used as substrate. The amount of radioactivity in each assay was approximately 2×10^6 count/min. The system without protein substrate served as a blank. Epididymal plasma was extracted from goat cauda epididymis (Barua and Majumder 1987). Seminal plasma was obtained from human semen by removing sperm by centrifugation. Both reproductive fluids were heated at 100°C for 1 min to inactivate endogenous enzymes and then were subjected to centrifugation to remove denatured proteins. The clear fluids were dialysed against medium A (Haldar and Majumder 1986) before being used in the protein kinase assays.

shown in Figure 8.4. Ecto-RC is an inactive complex of the kinase. Extracellular cAMP binds to R and thereby dissociates C which is now active to cause phosphorylation of the exogenous proteins in presence of ATP. There is little type-II kinase in the immature sperm and it appears specifically during motility initiation in the mature cauda sperm (Dey and Majumder 1990a). Type-II RC is thus maturation specific. Forward-motile sperm cells are enriched in both the kinases (Dey and Majumder 1990b; Majumder et al. 1990) (Table 8.5). However, the exact role of the ecto-RC in the induction and regulation of sperm motility is largely unknown.

8.5.2 Ecto-CIK and Cell-Surface Protein Phosphorylation Mechanism

A cAMP-independent protein kinase has been demonstrated on the outer surface of goat cauda-epididymal spermatozoa (Haldar and Majumder 1986). Several lines of evidence demonstrated that the observed ecto-kinase is not due to leaky and/or damaged spermatozoa or to the epididymal plasma contamination. The ecto-protein kinase causes phosphorylation of the serine and threonine residues

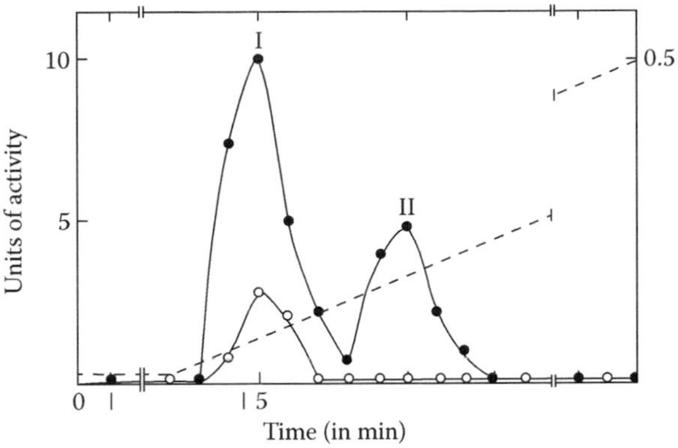

Figure 8.3 Effect of treatment of intact spermatozoa with the surface probe DSS on the profile of PM-bound type I and II RCs. Washed spermatozoa (500×10^6 cells/mL) were incubated in medium A at 37°C for 5 min without or with freshly prepared diazonium salt of sulphanilic acid (DSS) (2 mM). The cells were then sedimented by centrifugation at $500 \times g$ for 5 min and the cell pellets were washed once with medium A. Membranes were isolated from these cell preparations by the procedure described earlier. DEAE-cellulose chromatography of Triton X-100 solubilised RC obtained from membrane with (o) or without (•) DSS treatment and the kinase assays (+ cAMP) of the resulting fractions were then performed. Two milligrams of the membrane protein were loaded in the column and the recovery of the kinase activity was ~50%. (Reproduced from Dey, C.S., Majumder, G.C. 1990. *Biochemistry and Cell Biology* 68: 459–470.)

of multiple endogenous proteins localised on the sperm outer surface. The intact-sperm bound ectoenzyme is also capable of phosphorylating exogenous proteins such as casein, phosvitin, histone and protamine (Haldar et al. 1986). Our studies on the goat sperm model have shown that the ecto-cyclic AMP-independent protein kinase (ecto-CIK) of the intact cells causes phosphorylation of the endogenous membrane-bound phosphoproteins. The effects of several phospholipids and related compounds on intact sperm ecto-ClK were investigated (Dey and Majumder 1987b). Of all the phospholipids, phosphatidyl inositol showed maximal inhibitory effect (Table 8.6). Other phospholipids such as phosphatidyl choline and serine had no detectable effect on the kinase whereas phosphatidyl ethanolamine (125 μg/mL) inhibited slightly (approx. 25%) the activity of the enzyme. The ecto-CIK has been partially purified and characterised (Mitra et al. 1994). It is activated by Mg^{2+}. The ecto-CIK is not activated by any of the cyclic nucleotides, calcium-calmodulin and phosphatidylserine. It is not casein kinase. The properties of the kinase appear to be different from all the reported protein kinases. The ecto-protein phosphorylation profile of intact goat sperm by the endogenous ecto-CIK as well as CIK activity undergo marked alteration during epididymal maturation (Mitra

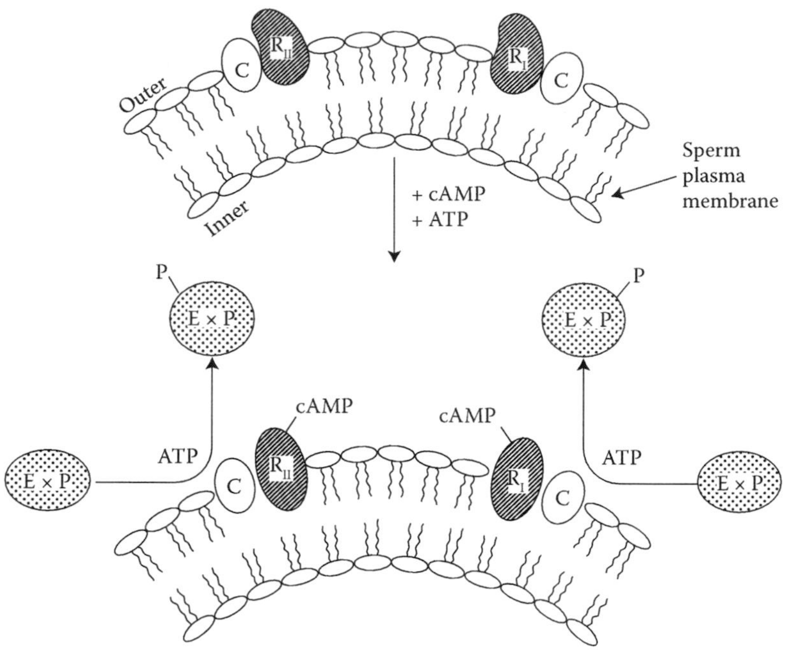

Figure 8.4 Schematic representation of the surface topography of the sub-units of the sperm plasma membrane bound type I (R_IC) and type II ($R_{II}C$) cAMP-dependent protein kinase and the phosphorylation of the extracellular phosphoprotein. ExP: exogenous proteins. (Reproduced from Dey, C.S., Majumder, G.C. 1990. *Biochemistry and Cell Biology* 68: 459–470.)

and Majumder 1991) thereby implicating that ecto-CIK may play an important role in the expression of flagellar forward progression during the epididymal sperm maturation event. Maturation-related alteration of the cell surface phosphoproteins by the endogenous protein kinase has also been demonstrated in the hamster sperm model (Devi et al. 1997).

Subsequently the ecto-protein kinase (ecto-CIK) has been purified for the first time to apparent homogeneity (using the caprine sperm model) and characterised (Nath et al. 2008). The isolated kinase is a dimer possessing two subunits: 63 and 55 kDa (Figure 8.5). The CIK is a strongly basic protein. The enzyme showed maximal activity at pH 8.5–9.5. Table 8.7 shows the general properties of the purified protein kinase. CIK is activated maximally by Mg^{2+}. It is a unique membrane protein-specific kinase, which specialises for phosphorylating the serine and threonine residues of the outer cell-surface phosphoproteins. The ecto-protein kinase is localised in the acrosomal cap area of the external surface of the mature sperm head as demonstrated by indirect immunofluorescence studies (Figure 8.6). Indirect immunofluorescence study demonstrates that CIK is primarily localised in

Table 8.5 Specific Activity of Ecto-Enzymes in Intact Forward Motile Spermatozoa

Ecto-Enzyme	*No. of Experiments*	*Specific Activity of Ecto-Enzymes(units/10^7 cells)*	
		Composite Cells (Mean ± SEM)	*Forward Motile Cells (Mean ± SEM)*
Cyclic AMP-dependent protein kinase (Ecto-RC)	5	11.7 ± 1.8	52.4 ± 7.9*
Cyclic AMP-independent protein kinase (Ecto-CIK)	13	5.0 ± 2.3	11.0 ± 5.8*
Phosphoprotein phosphatase (Ecto-PPase)	11	0.52 ± .06	1.84 ± 0.28*

Source: Reproduced from Majumder, G.C. et al. 1988. *Indian Journal of Biochemistry and Biophysics* 25:215–218.

Note: Vigorously forward motile spermatozoa were separated from non-motile and weakly motile cells of 'composite' spermatozoa based on their capacity to move upward against the gravity. The ecto-enzymic activities of these cells were estimated as described before (*$p < 0.05$; when compared with the composite cells).

the plasma membrane of the sperm (Figure 8.7). Although earlier investigators have provided several lines of evidences for the occurrence of ecto-protein kinases in a variety of mammalian cells (Mastro and Rosengurt 1976; Schlaeger and Kohler 1976; Majumder 1981; Haldar and Majumder 1986; Dey and Majumder 1990b), the aforementioned finding (Nath et al. 2008) provides confirmatory evidence for the localisation of an ecto-protein kinase on a cell surface. Recent study, using ELISA based on ecto-CIK antibody demonstrates that ecto-CIK level is remarkably higher in the sperm membrane than in the cytosol. The epididymal sperm maturational event as well as sperm vertical velocity is associated with a significant increase in the ecto-CIK level (Nath et al. 2012).

8.5.3 MPS: Major Protein Substrate of Ecto-CIK

^{32}P-labelled membrane proteins phosphorylated by endogenous ecto-CIK of intact cauda-epididymal spermatozoa were solubilised with 1% Triton X-100, and then subjected to various fractionation procedures. We have purified to apparent homogeneity and characterised the major physiological protein substrate (MPS) of ecto-CIK (Maiti et al. 2004). It is a 100 kDa membrane-bound monomeric protein. Three isoforms of MPS have been found with pI of 6.37, 6.05 and 5.14 and

Table 8.6 Effect of Different Phosphlipids and Related Compounds on the Protein Kinase Activity of Intact Spermatozoa

Additions (μg/mL)		*Protein Kinase Activity (Units)*
Control		15.4
Phosphatidyl inositol	25	6.5
	125	2.6
Phosphatidyl choline	25	15.1
	125	14.5
Phosphatidyl ethanolamine	25	14.1
	125	11.8
Phosphatidyl serine	125	16.0
Phosphatidic acid	25	16.0
	125	16.0
Diacyl glycerol	25	16.7
Myo-inositol	25	16.6
	125	16.1
Myo-inositol-2-mono phosphate	25	16.1
	125	18.4

Source: Reproduced from Dey, C.S., Majumder, G.C. 1987. *Biochemical Biophysical Research Communication* 146:422–429.

Note: Assays were carried out under the standard assay conditions, except where the specified conditions were made.

all these isoforms served as the specific high affinity protein substrates of ecto-CIK (Figure 8.8). The data show that CIK has remarkably high substrate specificity for the PM-bound phosphoprotein: MPS. Like the ecto-CIK, MPS is also located on the tip of sperm head overlaying the acrosomal cap.

Since the phospholipid bilayer of the plasma membrane has a hydrophilic exterior and hydrophobic interior, polar molecules including MPS are unable to freely pass through the membrane. Cell electroporation is a novel biophysical method of cell membrane permeabilisation (Neumann et al. 1982; Tsong

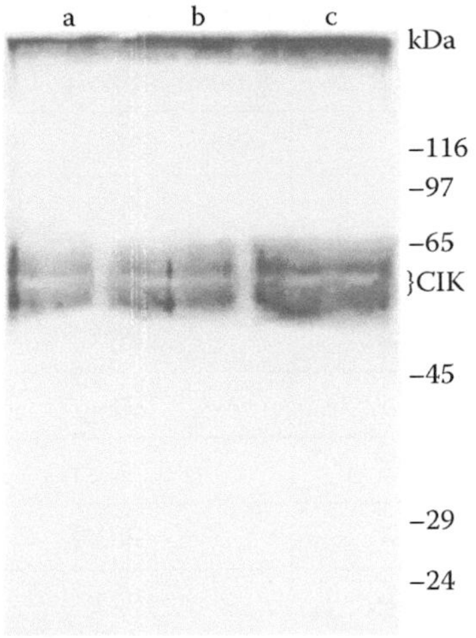

Figure 8.5 SDS-PAGE of CIK using 10% polyacrylamide gel. Markers were β-galactosidase (116 kDa), phosphorylase b (97 kDa), bovine serum albumin (66 kDa), ovalbumin (45 kDa), carbonic anhydrase (29 kDa) and trypsin inhibitor (20 kDa). Purified CIK 5 µg (lane a), 10 µg (lane b) and 25 µg (lane c) were loaded in three successive lanes. (Reproduced from Nath, D., Maiti, A., Majumder, G.C. 2008. *Biochimica et Biophysica Acta* 1778:153–165.)

1991). The advantage of this method has been taken to permit entry of ^{32}P- PMS into sperm cytosol, with a view to investigate its role in sperm flagellar forward progression (Maiti et al. 2008). Optimal electroporation condition developed for caprine sperm comprised exposure of 0.2 mL of sperm cells (2×10^8) to external electric field of intensity 1.5 KV/cm, capacitation at 4°C and post-pulse incubation at 37°C for 1 h. It was noticed that motility and forward motility increased markedly, with increase of MPS incorporation. At maximum MPS incorporation, increments in motility and forward motility were also maximal (Figure 8.9). The results of this investigation based on the novel cell electroporation technique demonstrate conclusively that MPS, the sperm membrane-bound phosphoprotein, serves as an activator of sperm forward motility. To determine the role of membrane protein-phosphorylation on forward motility, the degree of phosphorylation was estimated in forward motile cells (Nath et al. 2008). Forward motile cells were separated by swim up technique. Using the ELISA technique it has been found that the amount of MPS is approximately twofold higher in forward motile cells compared to the rest of spermatozoa (non-forward motile cell population) (Figure 8.10). It has further been observed that the rate of phosphorylation of MPS is nearly fivefold higher in the vertically motile sperm as compared to non-motile cauda sperm cells (Figure 8.11). ELISA study demonstrates that like CIK, MPS is also predominantly localised in the plasma membrane of the sperm

Table 8.7 General Properties of the Ecto-CIK

Assay System	*Kinase Activity (Units)*	*Kinase Activity (%)*
Complete	210	100
$-Mg^{2+}$	4.2	0.47
$-Mg^{2+}$ +EDTA (5 mM)	0.12	0
$-Mg^{2+}$ + Co (5 mM)	20	9.5
$-Mg^{2+}$ +Co (20 mM)	91	42.8
$-Mg^{2+}$ + Mn (2 mM)	5.21	2.48
$-Mg^{2+}$ + Mn (20 mM)	20.03	9.5
$+Zn^{2+}$ (5 mM)	184.0	87.6
$+Zn^{2+}$ (10 mM)	120.0	57.14
+NaF (10 mM)	205	97.61
Na_3VO4 (100 μM)	200	97
EGTA (200 μM)	181	86.1
Ca^{2+} (100 μM)	199.25	94.8
Ca^{2+} (500 μM)	205.01	97.6
Ca^{2+} (1 mM)	200	95.2
Ca^{2+} (100 μM) + Calmodulin (5 μg)	192	91.4
+cAMP (5 μM)	190	88
+cAMP (10 μM)	182	86.1
+cGMP (5 μM)	200	97
+cGMP (10 μM)	205	97.61

Source: Reproduced from Nath, D., Maiti, A., Majumder, G.C. 2008. *Biochimica et Biophysica Acta* 1778:153–165.

Note: Standard assay conditions were used except for the alterations indicated. The data are representative of five such experiments.

(Figure 8.7). Our recent study shows that the cell-surface protein kinase and its protein substrate also have important roles in the membrane fusion component of acrosome reaction (Maiti et al. 2009). It is thus clear that ecto-CIK through its substrate protein, MPS, plays a vital role in the regulation of sperm forward progression and velocity as well as acrosomal reaction.

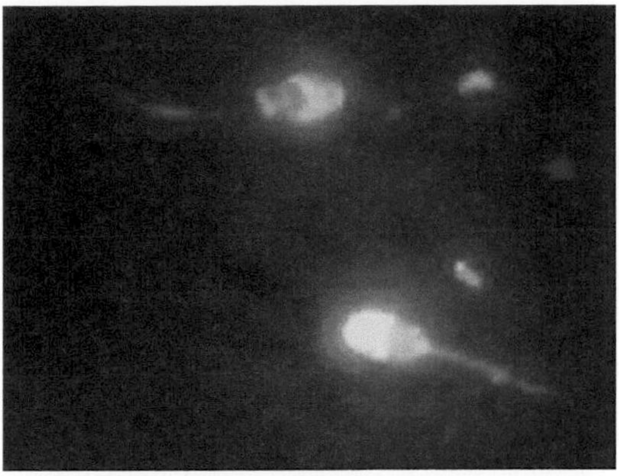

Figure 8.6 Immunofluorescence of goat epididymal mature spermatozoa. Spermatozoa were isolated from cauda part of the epididymis. Cells were incubated with CIK-antibody, followed by FITC-labelled goat anti-rabbit IgG. Spermatozoa were examined by fluorescence microscope at 1000× magnification. Preimmune rabbit sera treated control cells did not show any detectable fluorescence. (Reproduced from Maiti, A., Mishra, K.P., Majumder, G.C. 2008. *Molecular Reproduction and Development* 75:1185–1195.)

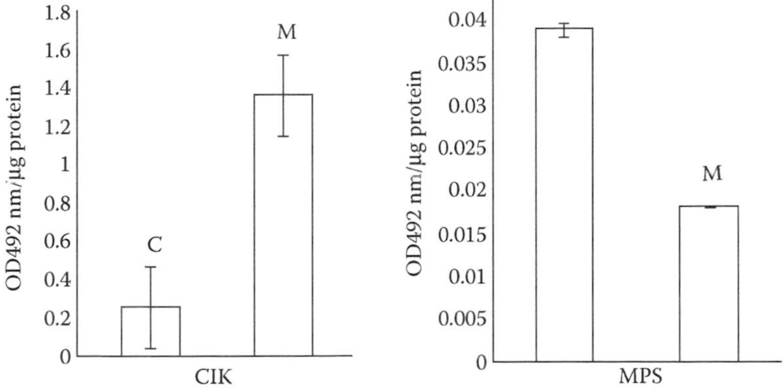

Figure 8.7 Immunoassay of CIK in caprine mature sperm cytosol and plasma membrane: Distribution of CIK in the cauda sperm fractions was determined by ELISA under standard assay condition. The data shown are Mean ± SEM of 5 experiments. 'C' denotes cytosol and 'M' denotes membrane. (Adapted from Nath, D., Maiti, A., Majumder, G.C. 2008. *Biochimica et Biophysica Acta* 1778:153–165.)

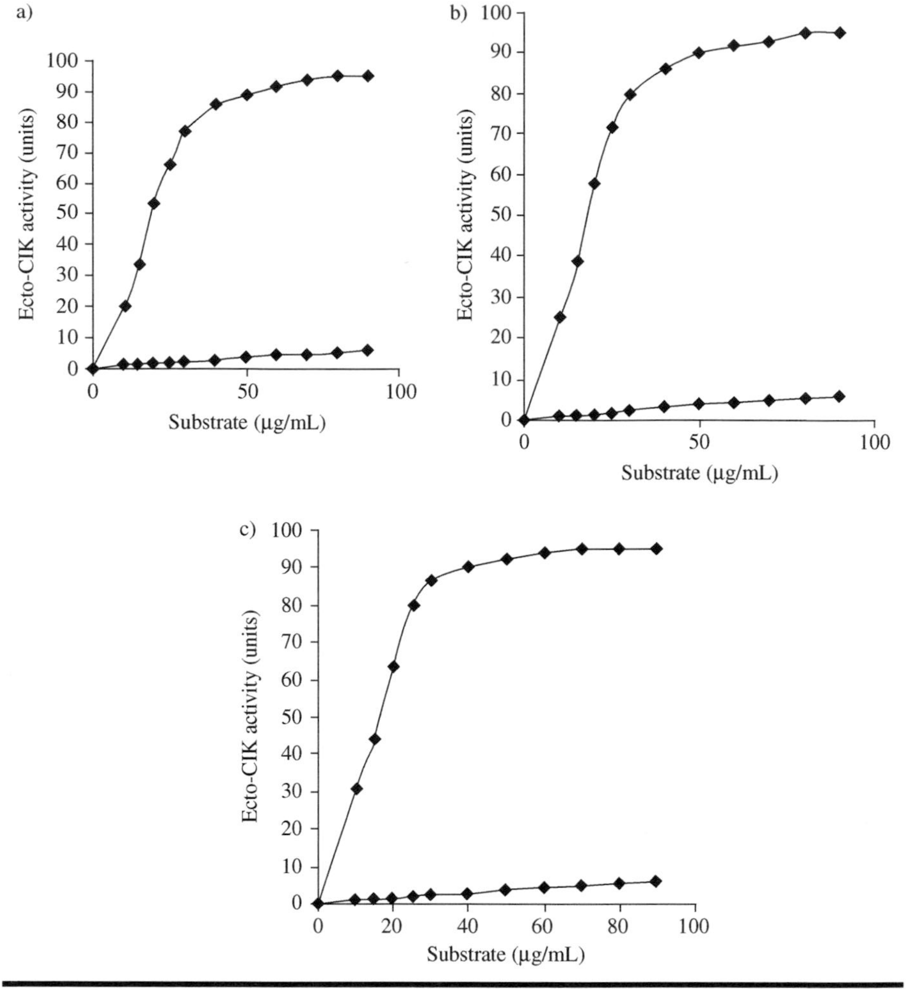

Figure 8.8 Substrate specificity of purified ecto-CIK. Varying concentrations of (-♦-): MPS (all three isoforms: a, b and c) and (-•-): casein have been used to assay CIK-activity. (Reproduced from Maiti, A., Mishra, K.P., Majumder, G.C. 2004. *Journal of Cellular Biochemistry* 92:164–177.)

8.5.4 Sperm Surface Protein Dephosphorylation Mechanism

Limited studies have been carried out on the occurrence of phosphoprotein phosphatase (ecto-PPase) on outer cell surface. Makan (1979) provided evidence for the first time for the localisation of an ecto-PPase in normal and transformed 3T3 fibroblasts. Localisation of an ecto-PPase on the caprine sperm outer surface has been demonstrated. The ecto-PPase causes dephosphorylation of exogenous

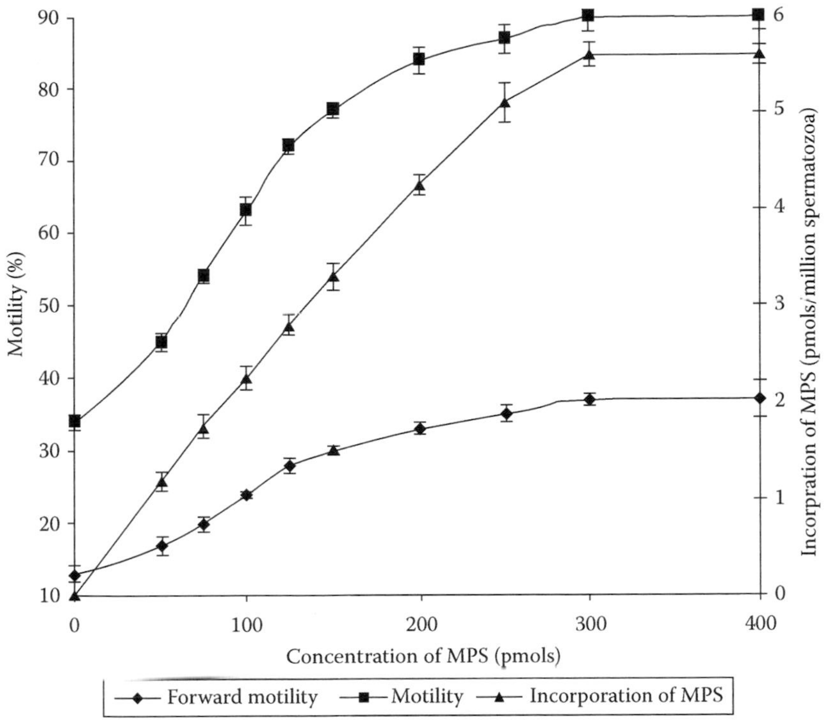

Figure 8.9 Effect of MPS incorporation on motility (dose course). Spermatozoa (2×10^8 cells/mL) were incubated with ^{32}P-labelled MPS in absence of external electric field. Spermatozoa were electroporated under standard assay conditions in presence of variable amounts of MPS (0–300 pmol). Amount of incorporation of [^{32}P]-MPS was assayed by precipitating the cells with 10% TCA and estimating the radioactivity in the cells in a liquid scintillation counter. Motility and forward motility were determined by the microscopic method. These data are representative of five separate experiments. (Reproduced from Maiti, A., Mishra, K.P., Majumder, G.C. 2008. *Molecular Reproduction and Development* 75:1185–1195.)

proteins such as histone, casein, phosvitin and protamine (Barua et al. 1985; Barua and Majumder 1987). Figure 8.12 shows the time course of the sperm ecto-protein phosphorylation by the intact-cell CIK and the dephosphorylation of the membrane bound ^{32}P-labelled proteins when the kinase reaction was arrested with excess of non-radioactive ATP (or EDTA that chelates Mg^{2+} essential for the kinase reaction). The results elicit that the cell surface ^{32}P-proteins undergo rapid turnover thereby showing that these surface proteins may serve as important regulatory proteins. The dephosphorylated proteins were found to undergo rephosphorylation under appropriate condition such as upon the addition of excess $MgCl_2$ in the EDTA treated system (Barua et al. 1990). This study demonstrates for the first time

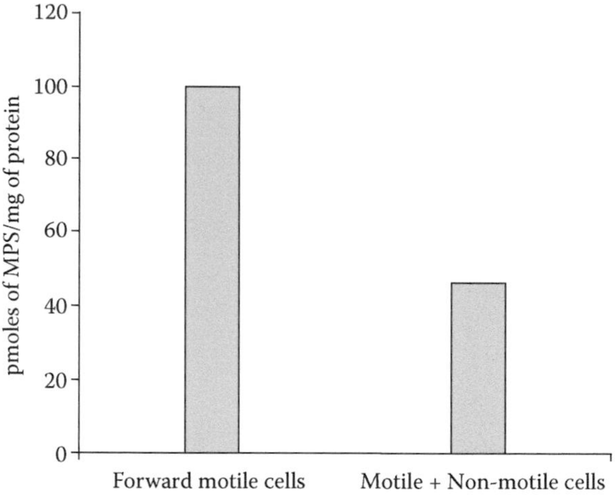

Figure 8.10 Immunodetection of MPS in forward motile and other (motile + non-motile) cells. Cells were separated by swim up technique and plasma membranes were isolated, solubilised in Triton X-100 and centrifuged. The supernatant was used in ELISA. Presence and concentration of MPS was detected with MPS antibody. The data are representative of three such experiments. (Reproduced from Maiti, A., Mishra, K.P., Majumder, G.C. 2008. *Molecular Reproduction and Development* 75:1185–1195.)

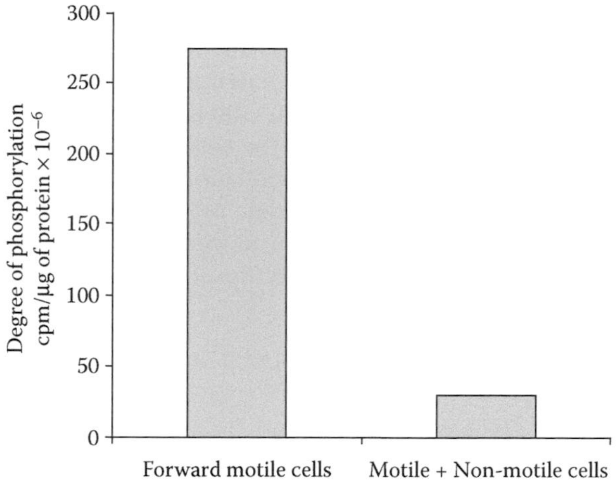

Figure 8.11 Degree of surface protein-phosphorylation in forward motile cells and motile + non-motile cells. Cells were separated by swim up technique and plasma membranes were isolated. The degree of phosphorylation was estimated under standard assay conditions. The data are representative of six such experiments. (Reproduced from Nath, D., Maiti, A., Majumder, G.C. 2008. *Biochimica et Biophysica Acta* 1778:153–165.)

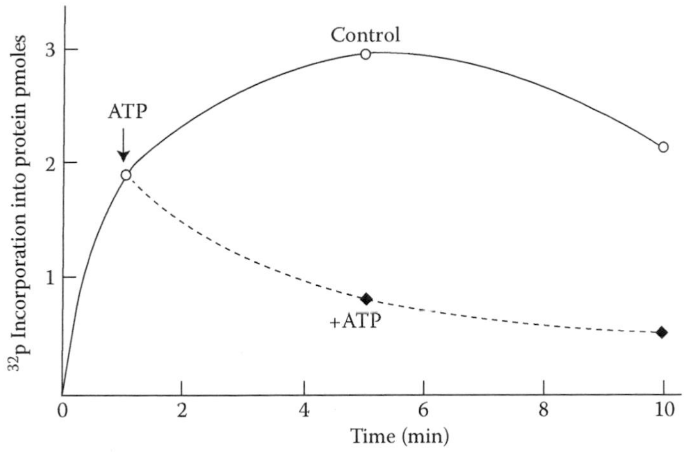

Figure 8.12 Turnover of sperm ecto-phosphoproteins. For the phosphorylation of sperm ecto-phosphoproteins, the intact cells were incubated with γ-[^{32}P] ATP under standard assay conditions for specified periods. To another set of tubes a large excess (5 mM) of unlabelled ATP was added after 1 min of incubation to arrest further incorporation of ^{32}P into protein, and the tubes were incubated again for specified periods. The reaction was stopped with trichloroacetic acid, and cells were processed to assay for radioactivity in protein. (Reproduced from Haldar, S., Majumder, G.C. 1986. *Biochimica et Biophysica Acta* 887:291–303.)

the occurrence of a coupled-enzyme system (consisting of ecto-CIK and PPase) on the mammalian outer cell surface (using caprine sperm as the cell model) that regulates phosphorylation/dephosphorylation of the specific phosphoproteins of the cellular ecto-domain. As mentioned earlier the ecto-CIK phosphorylates sperm ecto proteins. The ecto-PPase localised on the outer surface of spermatozoa dephosphorylates sperm outer-surface phosphoproteins (Barua and Majumder 1990; Barua et al. 1990).

The sperm ecto-PPase has been solubilised by alkali treatment, partially purified and characterised (Barua et al. 1999). The ecto-PPase has subsequently been purified to apparent homogeneity (Barua et al. 2001). This study reports for the first time not only purification but also confirmation of an ecto-PPase in a mammalian cell type. The purified enzyme is a 36 kDa molecule with an isoelectric point of 5.95. The isolated PPase is a serine/threonine type phosphatase and it is not dependent on any metal ions for its activity (Barua et al. 1999). The ecto-PPase is a glycoprotein as it binds to the Concanavalin A-Sepharose and the bound enzyme can be eluted with α-methyl mannoside. The observation that PPase antibody conjugated to fluorescein isothiocyanate (FITC) binds to the sperm surface shows that the PPase isolated from sperm plasma membrane is localised on the sperm external surface.

Using the intact sperm as well as isolated membrane models, we have extensively analysed for the first time maturational profile of ecto-PPase at various stages of epididymal sperm transit (Barua et al. 2001). The sperm maturation process is associated with an initial significant increase in ecto-PPase activity of the intact sperm followed by sharp fall in its activity (Figure 8.13). The results show that the ecto-PPase undergoes marked modulation during epididymal transit of the male gametes. Immunoreactivity was observed all over the surface of caput-sperm but was restricted primarily to the anterior-tip of the head in the corpus-sperm and to the posterior part of the head in cauda-sperm cells. The maturation-dependent

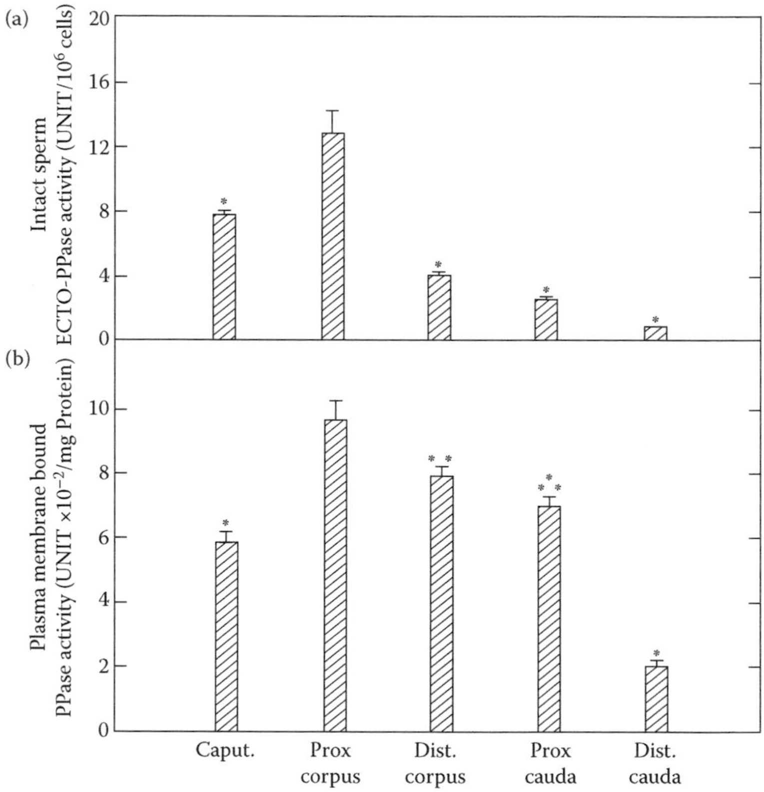

Figure 8.13 (a) The profile of phosphoprotein phosphatase (ecto-PPase) activity of goat spermatozoa during epididymal maturation. The ecto-PPase activity of whole spermatozoa derived from various parts of the epididymis was estimated by the procedure. (b) The specific activity of PPase of isolated sperm plasma membranes (PM) during epididymal maturation. Activity was estimated under standard assay conditions using [^{32}P]-histone as the exogenous substrate. Data are the mean ± SEM for five experiments. Data were analysed by one-way ANOVA. *$p < 0.001$, **$p = 0.009$, ***$p = 0.015$ compared with proximal (prox.) corpus sperm cell or PM. Caput, caput sperm; Dist., distal. (Reproduced from Barua, M., Nath, D., Majumder G.C. 2001. *Reproduction, Fertility and Development* 13:443–450.)

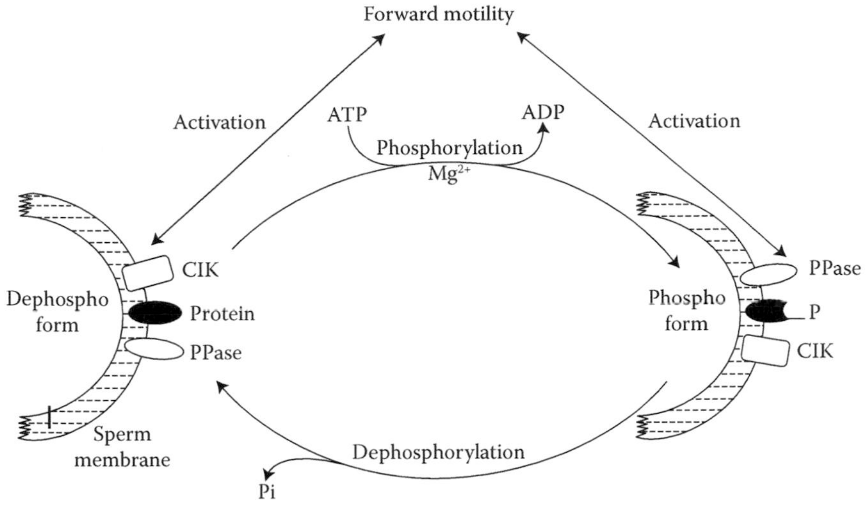

Figure 8.14 Schematic representation of the actions of the coupled-enzyme system (CIK, ecto-cyclic AMP-independent protein kinase; PPase, ecto-phosphoprotein phosphatase) on the sperm outer surface for the phosphorylation and dephosphorylation of ecto- proteins substrate. The conformation and biological activity of the ecto-proteins are likely to undergo alteration as a consequence of their phosphorylation and dephosphorylation. (Reproduced from Majumder, G.C. et al. 2012. *Protein Purification,* Nova Science Publishers, USA, 1–90.)

decrease of PPase activity was also confirmed by the immunofluorescent studies. Results suggest that the ecto-PPase may play an important role in sperm motility by regulating phosphorylation states of the membrane-associated and reproductive fluid phosphoproteins.

It is of interest to note that sperm's vigorous forward motility (i.e., vertical velocity as analysed by the swim-up technique) is associated with higher levels of both ecto-CIK as well as ecto-PPase (Table 8.5). As shown in the schematic diagram in Figure 8.14, the dynamic state of sperm high order of forward movement is associated with rapid phosphorylation and dephosphorylation of specific sperm surface proteins mediated by enhanced levels of the ecto-CIK and ecto-PPase, respectively.

8.6 Sperm Surface Lectins and Their Receptors

Multiple types of lectin have been demonstrated on the cell surface of sperm derived from several species. Human sperm (Goluboff et al. 1995) and sea urchin sperm (Yuuji et al. 1992) have been shown to possess galactose-specific and N-acetyl-D-galactosamine-specific lectins, respectively. Multiple sugar residues on zona pellucida (inducer of acrosome reaction) are recognised by complementary sugar binding lectin like proteins on the mouse sperm surface (Loeser and Tulsiani 1999). Spermadhesin AWN, a sperm-surface-associated lectin on boar spermatozoa, binds *in vivo* to the porcine zona pellucida glycoprotein, suggesting a role in gamete interaction (Rodriguez-Martinez et al. 1998). Some of these lectins have been implicated to play a vital role in fertilisation by mediating sperm–ovum fusion (Tulsiani and Abou-Haila 2001; Rodeheffer and Shur 2004). Although spermatozoa are known to possess several lectins, little is known about their possible role in sperm maturation and regulation of forward motility.

8.6.1 A Novel D-Galactose-Specific Lectin on Sperm Surface

The goat epididymal spermatozoa during epididymal transit specifically at the distal-corpus stage undergo head-to-head autoagglutination when incubated *in vitro* in a chemically defined medium (Figure 8.15) (Banerjee et al. 1992). The sperm autoagglutination process is dependent on Ca^{2+}. Effect of a variety of free sugars have been analysed on this autoagglutinated cells and only D-galactose (50 mM)

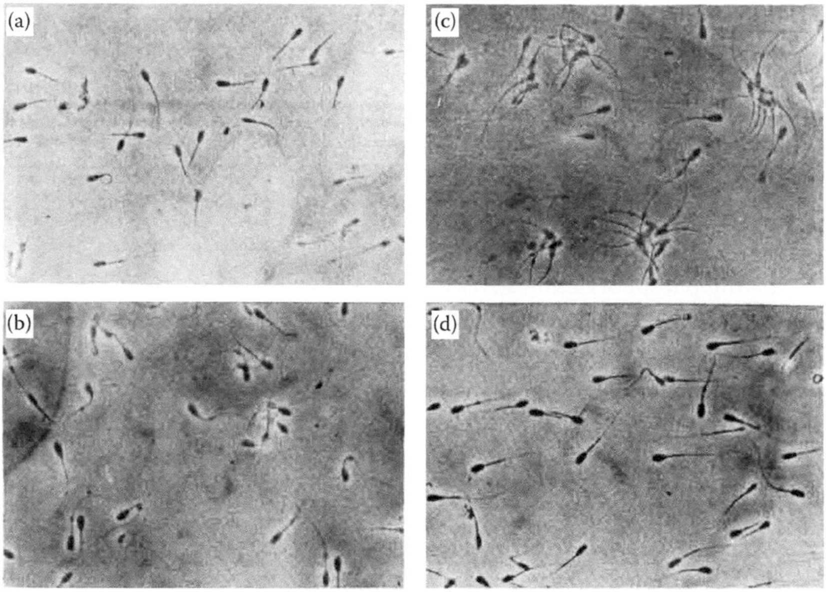

Figure 8.15 Autoagglutination potentiality of goat epididymal maturing spermatozoa. Spermatozoa were isolated from (a) caput, (b) proximal corpus, (c) distal corpus and (d) and cauda ends of epididymis. These cells were then incubated in Medium A to detect autoagglutination under standard assay condition. Microphotographs show spermatozoa with magnification of 800× when observed under a phase-contrast microscope. (Reproduced from Banerjee, S., Dey, N., Majumder, G.C. 1992. *Indian Journal of Experimental Biology* 30:1056–1061.)

served as a potent inhibitor of this sperm–sperm agglutination event (Banerjee et al. 2006). The data provided evidence for the occurrence of a novel calcium-dependent D-galactose-specific lectin on sperm outer surface that binds with its specific receptor of the neighboring cells leading to the observed sperm autoagglutination phenomenon. We have partially purified this D-galactose binding lectin from cauda (mature) sperm plasma membrane by Sepharose-6B affinity chromatography (Banerjee et al. 2006). This lectin agglutinated 100% of the maturing (caput) sperm at 100 μg/mL concentration. The isolated lectin showed high efficacy for inhibiting the distal-corpus sperm autoagglutination event. The sperm surface D-galactose specific lectin has recently been purified to apparent homogeneity in our laboratory from mature cauda sperm plasma membrane (unpublished results). The lectin is a calcium-dependent glycoprotein. It serves as a unique modulator of sperm motility.

8.6.2 D-Galactose-Specific Lectin Receptor on Sperm Surface

Earlier investigators utilising various commercially available lectins (mostly of plant origin) have provided evidence for the presence of lectin receptors on the surface of sperm derived from several mammalian species (Sarkar et al. 1991; for review, see Majumder et al. 1999; Tulsiani and Abou-Halia 2001). Some of the commonly used lectins for these studies are Concanavalin A (ConA), wheat germ agglutinin (WGA) and *Ricinus communis* agglutinin (RCA) that are specific for α-D-mannose, D-N-acetyl glucosamine and β-D-galactose, respectively. Lectin receptors of sperm surface are remarkably altered as a consequence of epididymal sperm maturation, and the extent of receptor modulations varies from species to species as well as the nature of the lectins.

Our results (Banerjee et al. 2006) conclusively demonstrated that epididymal maturation process is associated with a remarkable modulation of a D-galactose-specific lectin and its receptor (Figure 8.16). Immotile caput spermatozoa have undetectable levels of lectin and towards the terminal maturation phase, it increases dramatically whereas the reverse is true in case of the lectin receptor. The immature caput sperm undergo head-to-head agglutination when treated with the partially purified lectin showing that the receptor is primarily localised on the outer surface of the sperm head. Failure of pre- and post-distal corpus sperm to show any appreciable autoagglutination property is due to lack of lectin and its receptors, respectively on the outer surface of sperm head. The results support the view that there is existence of a synchronous modulation of lectin and their receptors on the external surface of homologous cells (e.g. spermatozoa). This postulated cellular regulation constitutes a novel mechanism for the control of the cellular functions.

Receptor of the sperm surface D-galactose-specific lectin has been purified partially from immature sperm membrane by affinity chromatography on a column of D-galactose-specific lectin Sepharose-4B (Banerjee et al. 2006). Using partially purified sperm surface galactose-specific lectin and its receptor, it has been

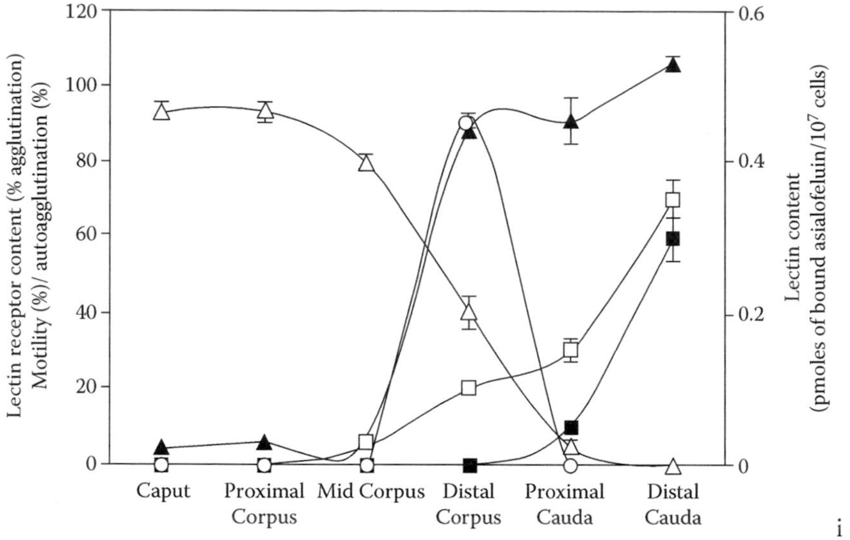

Figure 8.16 Correlation of sperm levels of lectin (-▲-) and its receptor (-△-) with autoagglutination phenomena (-○-) and induction of total motility (-□-) and forward motility (-■-) in the maturing goat epididymal spermatozoa. The data indicate the mean ± SEM of three experiments, $p < 0.01$. (Reproduced from Banerjee, S. et al. 2006. *Experimental Cell Research* 312:2299–2308.)

conclusively demonstrated that epididymal maturation process is associated with a remarkable modulation of lectin and its receptor. The possible modulation of sperm outer-surface lectin and its receptor during the epididymal transit is that lectin receptor is anchored into the lipid bilayer of the sperm plasma membrane. Maturation-dependent acquired additional receptors also get anchored into the lipid bilayer. Lectin that appears on the sperm surface during epididymal maturation seems to be a major biochemical parameter responsible for the induction of flagellar motility during the maturational event.

8.6.3 Novel Copper-Dependent Sialic Acid-Specific Lectin on Sperm Surface

Cu^{2+} is an essential micronutrient for all eukaryotic organisms, as it has a marked positive role in physiological and reproductive processes (Cheah and Yang 2011; Kuo et al. 2001). Recent investigation from our laboratory shows that caprine sperm forward motility is significantly enhanced by Cu^{2+} in a dose-dependent manner; maximal activation (approx 20%) being noted at 5 μM level of the metal (Roy et al. 2014a). Further increase in Cu^{2+} concentration causes decrease in sperm motility, showing that Cu^{2+} exerts a biphasic regulation on sperm motility. A peculiar phenomenon of sperm head-to-head agglutination was noted when the copper concentration was enhanced above 100 μM (Roy et al. 2014b). A systematic study reveals that as compared to other trace elements, copper shows high specificity for the agglutination of the mature sperm cells (Figure 8.17a). To understand the possible mechanism of this peculiar phenomenon, effect of various sugars has been investigated. Of all the sugars tested only free sialic acid showed high specificity to serve as a potent inhibitor of this sperm–sperm association (results not shown). Copper-dependent agglutinated cells can also be dissociated with fetuin that possess terminal sialic acid (Figure 8.17b). However, desialylated fetuin (asialofetuin) loses this antiagglutinin activity nearly completely. The data demonstrate that sialic acid residue of fetuin is essential for its antiagglutinin potency. The results provided evidence for the first time for the occurrence of a novel copper-dependent sialic acid-specific lectin on external sperm surface that binds with its specific receptor of the neighbouring cells thereby causing sperm–sperm agglutination event (Roy et al. 2014b). This is altogether a new class of lectin as no Cu^{2+}-dependent lectin has ever been reported. It may be a new variety of selectin which is sialic acid-specific but is not activated by Ca^{2+}. This novel lectin, like the well-known cell surface adhesive proteins (Marshall et al. 2003) may serve as the secondary force to strengthen the homologous cell–cell contacts, triggered by lectin–receptor interactions. This novel Cu^{2+}-lectin, one of the possible physiological functions of which is to participate in homologous cell–cell adhesion in the tissues, may as well be present in other mammalian cells also. Immature sperm cells (isolated from caput and corpus regions), do not undergo agglutination in presence of copper (Figure 8.18) (unpublished data). It is thus clear that sperm cells acquire this unique copper-dependent lectin/receptor during epididymal maturation process. Appearance of this unique lectin/receptor specifically on the mature goat sperm surface indicates its probable role in sperm motility regulation.

8.7 Sperm Motility Stimulating Proteins

8.7.1 Forward Motility Stimulating Factor (FMSF)

We have reported for the first time purification of a motility-promoting protein to apparent homogeneity from a biological fluid (buffalo serum) and some of its physical and biochemical characteristics (Mandal et al. 2006). FMSF-I (subsequently referred as, FMSF) is the major motility-promoting protein of buffalo blood serum. FMSF is a 66 kDa heat-stable glycoprotein. It is a Mg^{2+} dependent monomeric protein and its protein and sugar parts both are essential for its motility-promoting potential. FMSF has high degree of protein specificity. Radioiodinated FMSF has high affinity for binding to the sperm surface receptors (Figure 8.19). Evidence for the localisation of FMSF on sperm outer surface, has also been obtained by investigation with FMSF antibody that agglutinates the sperm cells. FMSF through its occurrence on the sperm surface may provide continuous stimulation to sustain

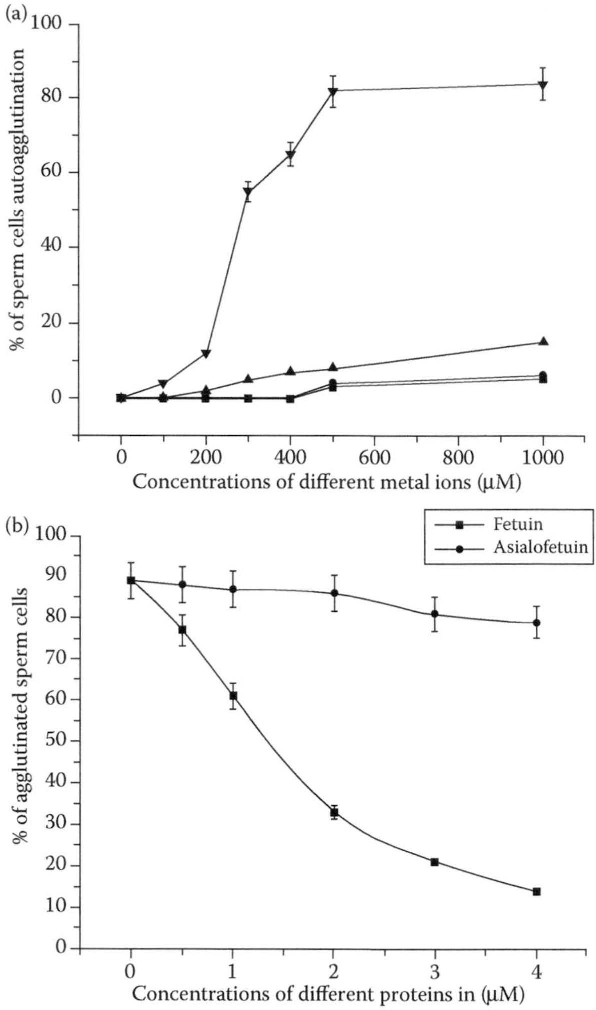

Figure 8.17 (a) Dose courses of Mg^{2+} (■), Mn^{2+} (●), Zn^{2+} (▲) and Cu^{2+} (▼) on caprine mature cauda-sperm cells. Data represents autoagglutination after incubation for 30 min under standard assay condition. The data were mean ± SEM of three different experiments, $p < 0.001$. (b) Dose course of fetuin and asialofetuin (Sigma-Aldrich) on sperm autoagglutination under standard assay condition. Proteins were dissolved in PBS at pH 7.4. Different μM concentrations of fetuin and asialofetuin were included in the assay medium from 200 μM stock solution. The data were mean ± SEM of three different experiments, $p < 0.001$. (Reproduced from Roy, D. et al. 2014. *Glycoconjugate Journal* 31:281–288.)

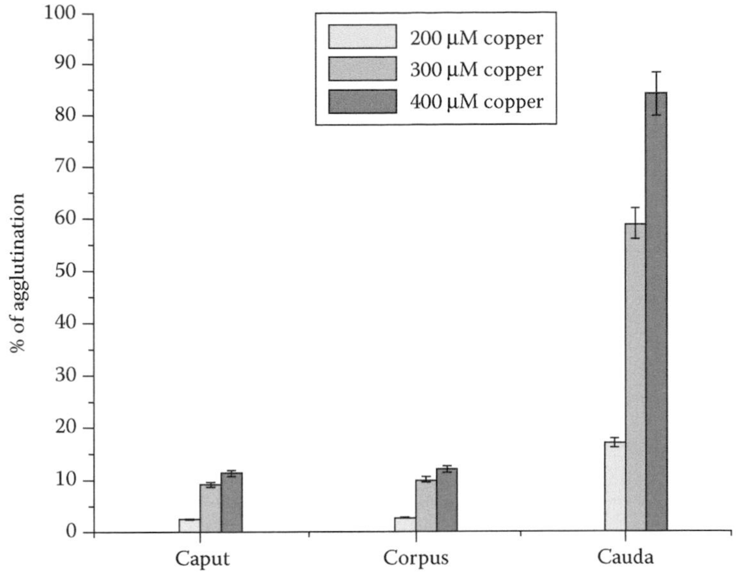

Figure 8.18 Dose course of copper-dependent agglutination of maturing caprine sperm cells isolated from caput, corpus and caudal regions of epididymis. For the agglutination study, isolated spermatozoa were incubated for 10 min under standard assay conditions described earlier. The data are means ± SEM of three different experiments, $p < 0.001$.

sperm motility during the long journey through the female reproductive tract. Its motility promoting efficacy is species non-specific. Liver is the richest source of FMSF but it is also present in testis and epididymal plasma. Recently motility-promoting protein (designated as MSP) has also been purified from caprine blood serum and characterised (Saha et al. 2013). The caprine serum protein activates sperm horizontal forward motility as well as vertical velocity as analysed using the motility analysers CASA and SPERMA, respectively. Antibody of the caprine serum factor markedly agglutinates cauda sperm (Figure 8.20) thereby indicating the occurrence of the motility stimulating protein on the sperm external surface. The serum motility-stimulating factor is clearly different from forward motility protein (FMP) partially purified from seminal plasma by Acott and Hoskins (1978). It is also different from sperm membrane-bound motility activating proteins such as 34 kDa hyaluronic acid binding protein (Ghosh et al. 2002), 36 kDa ecto-phosphoprotein phosphatase (Barua et al. 2001) and 100 kDa ecto-phosphoprotein (Maiti et al. 2004).

Recently, one of our studies elucidated the receptor interaction profile of the buffalo-serum derived FMSF (Dey et al. 2014a). A functionally active covalent conjugate of FMSF with horseradish-peroxidase (HRP) was prepared to check its

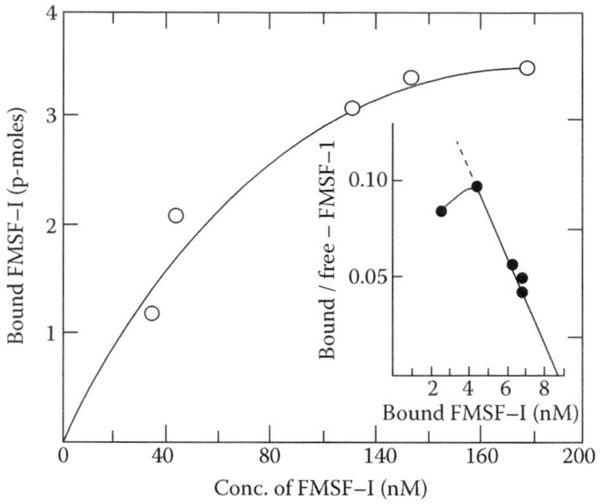

Figure 8.19 Effect of various concentrations of ^{125}I-labelled FMSF-I on their binding to intact spermatozoa (2×10^7 cells/assay) under the standard assay conditions. The insert shows the Scatchard analysis of the data. (Reproduced from Mandal, M. et al. 2006. *Journal of Cellular Physiology* 209:353–362.)

binding efficiency on goat spermatozoa of different maturation stages. Receptor assay was carried out using this FMSF-HRP conjugate in saturating state to bind with spermatozoa isolated from different epididymal segments (Figure 8.21). Activity and binding profile of the conjugate suggested that both these parameters come into play quite late in the maturation stage. Spermatozoa from caput up to mid-corpus regions neither displayed any substantial binding with FMSF nor exhibited significant induction in forward motility. Study of cell surface-bound

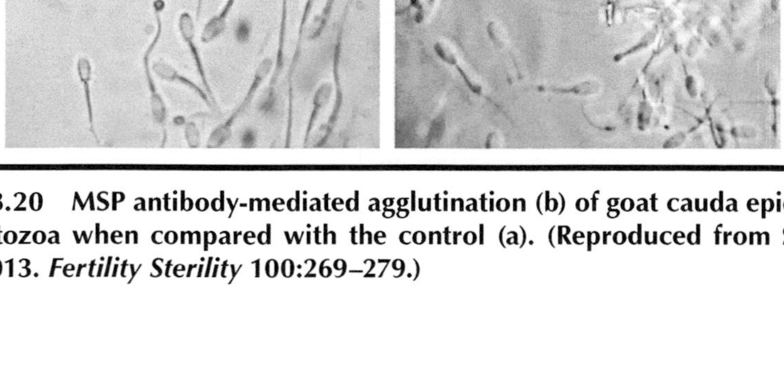

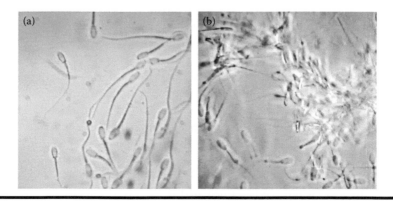

Figure 8.20 MSP antibody-mediated agglutination (b) of goat cauda epididymal spermatozoa when compared with the control (a). (Reproduced from Saha, S. et al. 2013. *Fertility Sterility* 100:269–279.)

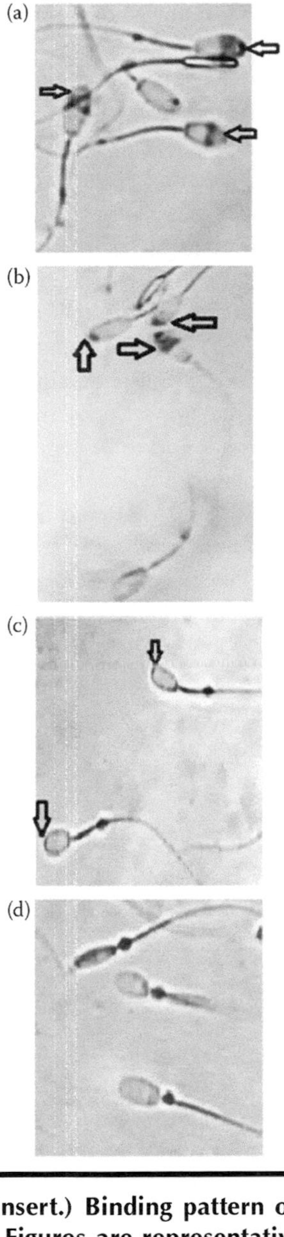

Figure 8.21 (See colour insert.) Binding pattern of the FMSF-HRP conjugate on maturing spermatozoa. Figures are representative for only binding pattern, not frequency. (a) Caudal cells, (b) distal-corpus cells, (c) mid-corpus cells, (d) caput cells. Arrow indicates towards spermatozoa that developed any sort of coloration on the sperm anterior head. (Reproduced from Dey, S. et al. 2014. *Biochemistry and Cell Biology* 92:43–52.)

FMSF on maturing spermatozoa in physiological milieu demonstrated its presence on anterior sperm-head. Thus, immunofluorescence and cyto-chemical examination revealed that the receptor of this factor appears predominantly on mature sperm anterior-head surface; immature spermatozoa are almost devoid of these receptors (Figure 8.22). This study demonstrates for the first time, maturation-dependent expression of FMSF-receptor. Appearance of this receptor on the sperm cell surface thus seems to be the rate-limiting event for FMSF-mediated progressive motility during epididymal maturation.

Cell signalling mechanism of FMSF has been recently investigated (Dey et al. 2014b). FMSF through receptor/G-protein activation promotes the transmembrane adenylyl cyclase (tmAC) activity in a dose-dependent manner to enhance intracellular cAMP and forward motility. Motility boosting effects of FMSF are almost lost in the presence of dideoxyadenosine, the selective tmAC inhibitor. Observed motility under the influence of FMSF was found to be regulated by protein kinase A (PKA) and tyrosine kinase. FMSF initiates a novel signalling cascade by stimulating tmAC activity which augments intracellular cAMP and downstream crosstalk of phosphokinases leading to enhanced forward motility in mature spermatozoa (Figure 8.23). Figure 8.24 shows a schematic diagram of the possible mechanism of cell signalling by the motility promoting protein. FMSF binds to the specific receptors on the external cell surface leading to activation of adenylate cyclase and consequent enhancement of intra-sperm cyclic AMP level that triggers the flagellar movement through a series of cascade molecules including protein kinase A and tyrosine kinase.

Recently, FMSF was immunodetected in epididymal plasma/uterine fluids of the bovine, bubaline and caprine systems (Dey et al. 2014a). This glycoprotein as well activates microscopic motility and vertical velocity of spermatozoa derived from these ruminant species. The results are compatible with the view that FMSF serves as a ubiquitous physiologic entity across the mammalian reproductive-fluids for amplifying sperm forward progression essential for natural fertilisation.

8.7.2 Motility Initiating Protein

Forward motility is induced in previously inactive immature bovine epididymal spermatozoa when incubated *in vitro* in the presence of theophylline, an inhibitor of cyclic adenosine 3′,5′-phosphodiesterase, and seminal plasma or epididymal plasma (EP) (Hoskins et al. 1975). A forward motility protein (FMP) present in seminal plasma has been partially purified (Acott and Hoskins 1978). Using goat caput-epididymal immature spermatozoa as the model we have analysed the biochemical parameters influencing the *in vitro* initiation of forward motility (Jaiswal and Majumder 1998). Our subsequent study demonstrates that the *in vitro*-induced forward motility in caput spermatozoa is transient that is continuous stimulation by theophylline, bicarbonate and EP is essential for sustaining the induced forward motility. Removal of these activators results in total loss of induced motility (Das et al. 2012). Motility initiating protein (designated as MIP) has been purified for the first time to apparent

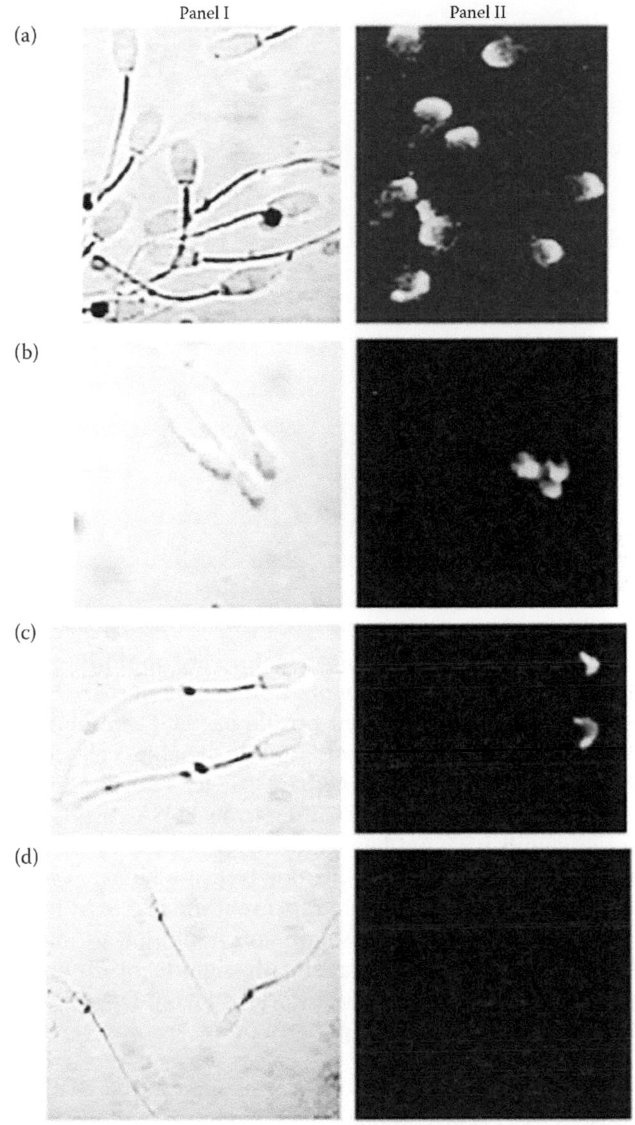

Figure 8.22 **Localisation of the bound (native) FMSF on maturing sperm cells. Figures on Panel II are fluorescent images; cells were treated with anti-FMSF antibody followed by FITC-conjugated secondary antibody. It is representing only binding patterns, not frequency. Panel I is the photographs of the same set of cells under bright-field microscope. (a) Caudal sperm cells, (b) distal-corpus sperm cells, (c) mid-corpus sperm cells, (d) caput sperm cells. (Reproduced from Dey, S. et al. 2014. *Biochemistry and Cell Biology* 92:43–52.)**

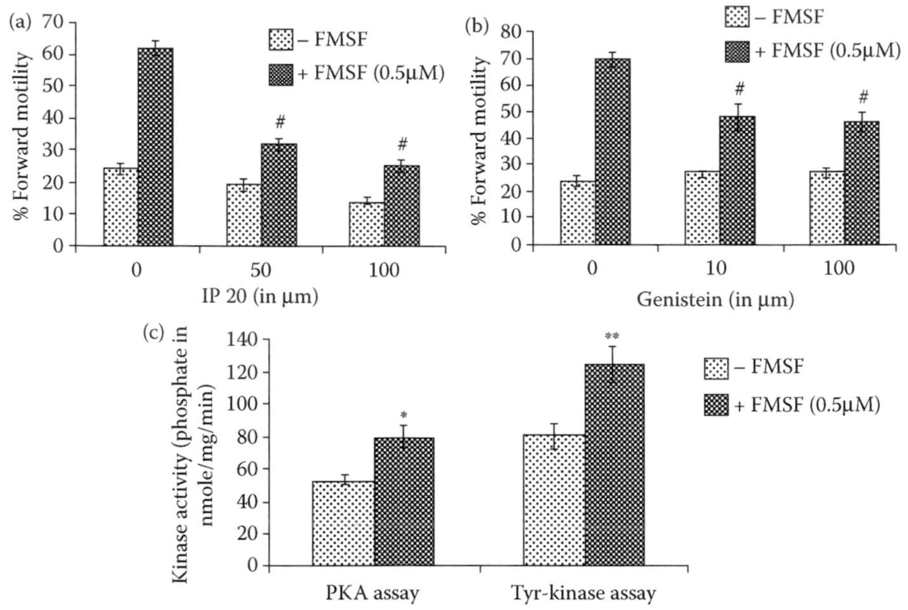

Figure 8.23 (a) Effect of PKA inhibitor in forward motility of the FMSF activated sperm cells. Duration of IP20 incubation was 30 min. (b) Effect of tyrosine-kinase inhibitor on forward movement profile of FMSF stimulated spermatozoa. Duration of Genistein incubation was 30 min. Viable cell count throughout the different assay conditions remained ≥95%. (c) Phospho-kinase stimulating potential of FMSF. Only absorbance values from ELISA results have been plotted, actual activity units have not been shown; anti-phospho-serine/threonine for protein kinase A and tyrosine antibodies for tyrosine kinase were used in ELISA of the assayed treatments. All the data represent mean ± SEM for $n = 3$ samples. Hash mark (#) indicates statistically significant inhibition vs. only FMSF-treated ($p < 0.01$). Asterisks (*, **) denote statistically significant difference vs. control ($p < 0.05$ and $p < 0.01$), respectively. (Reproduced from Dey, S. et al. 2014. *PLOS One* 9, e110669, doi 10.1371/journal.pone.0110669.)

homogeneity from mammalian EP (using caprine model) and characterised (Jaiswal et al. 2010). It is a 125 kDa glycoprotein made up of two subunits (70 and 55 kDa). MIP is an acidic protein with an isoelectric point of 4.75. MIP is heat stable and it is maximally active at pH 8. It is a glycoprotein that contains mannose, galactose and N-acetyl glucosamine approximately in the ratios of 6:1:6. Sugar side chain of the glycoprotein is essential for its biological activity. Epididymal plasma is its richest source. It is also capable of enhancing forward motility of mature cauda-sperm. Studies with MIP antibody, shows that the motility initiator is localised on sperm outer surface. It binds with sperm surface receptor to enhance intra-sperm level of cyclic AMP (Table 8.8). Like FMSF, MIP may also function by elevating intra-sperm cyclic AMP level by the mechanism as shown in Figure 8.24.

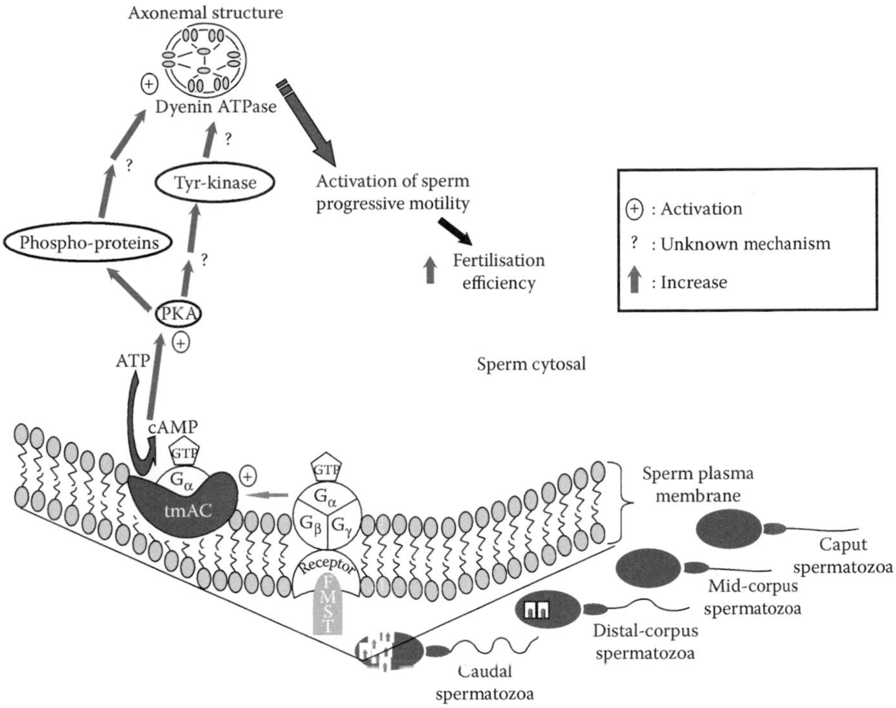

Figure 8.24 Schematic diagram showing the mechanism of action of the FMSF.

Table 8.8 Effect of Purified MIP on Cyclic AMP Content of Goat Caput Sperm

Treatment	*Cyclic AMP Concentration (pmol/10^9 cells) Mean ± SEM*
Control	144 ± 9
+MIP (6 nM)	203 ± 16*

Source: Reproduced from Jaiswal, B.S. et al. 2010. *Journal of Cellular Physiology* 222:254–263.

Note: The control assay system contained caput sperm incubated in RPS medium containing 30 mM theophylline. Cyclic AMP content of sperm incubated in presence or absence of MIP has been measured. The data shown are mean ± SEM of five experiments. *$p < 0.05$ when compared with control.

8.8 Membrane-Bound Sperm Motility Inhibitor

A novel motility-inhibiting factor (MIF) has been purified to apparent homogeneity from caprine sperm plasma membrane and partially characterised (Dungdung and Majumder 2003). MIF is a protein and its molecular weight is 98 kDa. SDS-polyacrylamide gel electrophoresis of MIF gave a single band indicating that the factor is a monomer. MIF is a thermo-stable factor and inhibited the spermatozoa motility in a dose dependent manner (Figure 8.25). MIF, at a concentration as low as 7 μg/mL (70 nM), caused nearly 40%–50% inhibition of sperm flagellar motility. Sperm forward motility was completely inhibited at the concentration 14 μg/mL of MIF. It is a glycoprotein as it binds with high affinity to Sepharose-6B and the affinity matrix-bound factor can be eluted with D-galactose. Data show that the motility inhibiting activity is lost completely when treated with β-galactosidase (Table 8.9) indicating that its sugar side chain is essential for activity. MIF has the capability to inhibit sperm motility of all the species tested. Addition of MIF antibody caused significant enhancement of forward motility of the mature spermatozoa. This antibody may thus be useful for solving some of the problems of human infertility due to low sperm motility. The motility inhibiting protein may also be useful as a vaginal contraceptive. A novel 160 kDa sperm motility inhibiting factor has also been purified to apparent homogeneity from caprine epididymal plasma and characterised (Das et al. 2010). It is a heat labile protein and maximal active at the pH 6.9–7.5. The EP factor inhibits sperm motility by lowering intra-sperm cyclic AMP level presumably by interaction with its specific cell surface receptor.

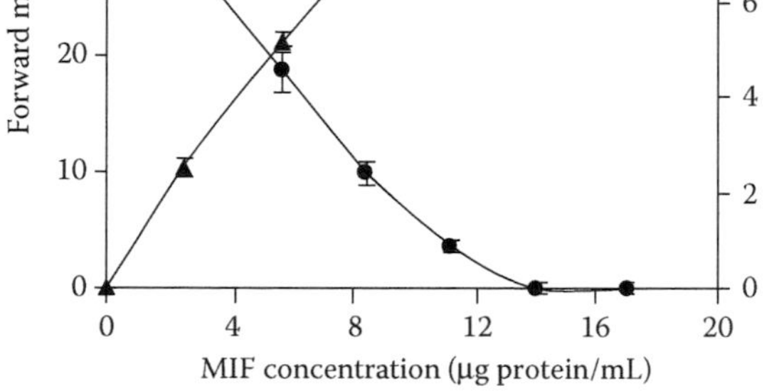

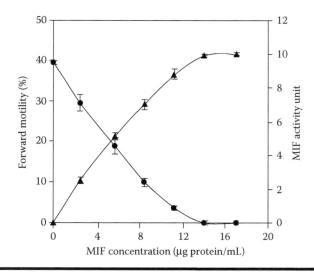

Figure 8.25 Dose course of action of purified MIF under the standard assay conditions. (-●-):% forward motility, (-▲-): MIF activity unit. (Reproduced from Dungdung, S.R., Majumder, G.C. 2003. *Cellular and Molecular Biology* 49:413–420.)

Table 8.9 Effect of Various Glycosidases on MIF Activity

Enzyme Treatment	*MIF Activity (Units) Mean ± SEM*
Nil (Control)	6.2 ± 0.03
+α-mannosidase (27 units/mL)	5.8 ± 0.07
+α-L-fucosidase (1 unit/mL)	6.0 ± 0.03
+α-glucosidase (7 units/mL)	5.5 ± 0.09
+β-galactosidase (5 units/mL)	0
+Neuraminidase (5 units/mL)	5.4 ± 0.07
+β-N-acetyl glucosaminidase (1.6 units/mL)	5.9 ± 0.06

Source: Reproduced from Dungdung, S.R., Majumder, G.C. 2003. *Cellular and Molecular Biology* 49:413–420.

8.9 Ecto-ATPase

Studies from our laboratory provided evidence for the occurrence of a specific ecto-ATPase (Majumder and Biswas 1979) in rat epididymal spermatozoa. The ecto-ATPase activity, an integral plasma-membrane protein, causes hydrolysis of externally added [γ^{32}P] ATP. The ATPase activity of the whole spermatozoa was not due to leakage of the intracellular enzymic activity, contamination of the broken cells or to any possible cell damage during incubation and isolation of spermatozoa. Ecto-ATPase of intact spermatozoa has a high degree of substrate specificity for the hydrolysis of ATP and dATP. It is an Mg-dependent enzyme (Majumder 1981). The activity of the spermatozoal enzyme decreases markedly (approx. threefold) during maturation in epididymis. Orthovanadate is a potent inhibitor of the enzyme of the mature spermatozoa derived from cauda epididymis. Vanadate-sensitivity of the spermatozoal ecto-ATPase markedly increases during the epididymal maturation. However, the sperm ecto-ATPase has not yet been purified and its precise role in sperm flagellar motility is not known.

8.10 Other Studies

Piceatannol, a specific inhibitor of tyrosine kinase inhibits the motility of intact hamster spermatozoa (Uma Devi et al. 2000). Subsequently the tyrosine kinase has been purified from sperm plasma membrane. Piceatannol inhibits as well the activity of the isolated kinase thereby implicating the role of the kinase in hamster sperm motility. The membrane-bound tyrosine kinase causes phosphorylation of multiple tyrosine phosphoproteins that may play an important role in the regulation of sperm capacitation (Kumar et al. 2006; Shivaji et al. 2007; Kota et al. 2009).

Epididymosomes are small membranous vesicles secreted in an apocrine manner in the intraluminal compartment of the epididymis and play a major role in the acquisition of new proteins by the maturing spermatozoa (Sullivan et al. 2007). Some epididymosome-associated proteins including glycosylphosphatidylinositol (GPI)-anchored proteins behave as integral membrane proteins. Epididymal maturation of sperm is associated with the activation of a cAMP-induced tyrosine phosphorylation cascade, which is ultimately associated with the expression of capacitation-dependent sperm functions, such as hyperactivated movement and acrosomal exocytosis (Aitken et al. 2007). As spermatozoa progress through the epididymis they first acquire the capacity to phosphorylate tyrosine on targets on the principal piece, followed by the midpiece. However, little is known regarding the role of these proteins in the regulation of flagellar movement.

Using caprine sperm as the model, it has been shown that radiolabelled p-chloromercuriphenylsulphonic acid (PCMPS) binds with high affinity with the thiol-containing proteins of sperm surface (Roy and Majumder 1986). PCMPS strongly inhibits motility of freshly extracted goat cauda-epididymal sperm preparations (Majumder and Chaudhuri 1984). PCMPS-mediated inhibition of motility cannot be reversed with the addition of excess of various-SH compounds. The data suggest that PCMPS binds with high affinity to the specific -SH group(s) of sperm ecto-sulfhydryl molecules to cause their inactivation irreversibly and thereby destroys sperm motility.

Multiple glycoproteins have been identified to be present in human seminal plasma (Ding et al. 2007; Tomar et al. 2011); later on, these were further characterised by modern proteomic tools. Among them, lactoferrin and prostate specific antigen (PSA) are important cell surface molecules that have been found to be down-regulated in patients with low sperm motility (i.e. in azoospermic patients); another notable glycoprotein is progesterone-associated endometrial protein that has been found to have correlation with azoospermic patients.

8.11 Conclusion

It is well documented that mammalian cell surface molecules with special reference to proteins play a vital role in the regulation of cell functions by modulating cell–cell interactions, effector–receptor interactions, transmembrane signalling etc. However, limited studies have been carried out on the identification of sperm surface molecules and defining their specific roles in sperm physiology, especially epididymal maturation and flagellar motility.

It is essential to follow several guidelines to establish localisation of a protein on outer cell surface i.e. to determine the 'ecto' nature of a protein. Some of the important guidelines have been outlined here. The concerned mammalian cell should be viable and the cell membrane should be intact. There should be little leakage/secretion of the biomolecule in the surrounding extracellular fluid.

The occurrence of the protein on the outer cell surface should be tested using well-documented surface probes such as p-chloromercuriphenylsulfonic acid (PCMPS), diazonium salt of sulphanilic acid (DSS), etc. Finally, it is necessary to confirm the ecto-nature of the protein using the antibody of the purified cell surface protein. The existing scientific literature in the area of ecto-proteins in sperm and other cells is rather confusing as in many studies the aforementioned guidelines have not been followed.

There are a plethora of reports suggesting the presence of undefined biologically active proteins on sperm surface that may have an important role in modulating sperm biology with special reference to epididymal induction of forward progression and its regulation. Some of the broad classes of these cell surface biomolecules that are worth mentioning here are lectins and their receptors, antigens, thiol-containing proteins, ATPase and receptors of several well-characterised exogenous biologically active proteins. As reviewed earlier, evidence has been provided for the occurrence of a Cu^{2+}-dependent sialic acid-specific lectin in a mammalian cell using caprine sperm as the cell model. This is a new class of outer cell surface protein that may participate in the propagation and amplification of the biological functions of copper in sperm and other mammalian cells by modulating cell surface lectin–sugar interactions. Further studies are now necessary to purify and characterise this novel lectin and to define its specific role in cellular regulation.

A sperm surface coupled-enzyme system consisting of a protein kinase (CIK), PPase and their protein substrate (MPS) is the most extensively studied ecto-molecular system. These proteins are the classical ecto-molecules as they conform to the aforementioned guidelines of the ecto-molecules. These proteins in a coordinated manner have the potential to play a vital role in sperm motility regulation. Further studies are now needed to identify multiple physiological substrate proteins of the ecto-CIK and PPase and to delineate their precise role in the induction and regulation of flagellar movement (*in vivo* and *in vitro*). Studies are also necessary to elucidate their sites of biosynthesis, hormonal regulation and specific role in various sperm lifecycle events including capacitation, acrosome reaction and fertilisation.

Acknowledgements

The authors are grateful to Professor Siddhartha Roy, director of Indian Institute of Chemical Biology, Kolkata, for his support and interest in our research work. Emeritus Fellowship to G. C. Majumder by A. I. C. T. E, New Delhi and the support of Centre for Rural and Cryogenic Technologies, Jadavpur University, Kolkata are also acknowledged. Research Associateship to Sudipta Saha and Kaushik Das by CSIR, Senior Research Fellowship to Sujoy Das by ICMR, Junior Research Fellowship to Abhi Das by DST, Arpita Bhoumik by Labonya Prova Bose Trust, Senior Research Fellowship to Souvik Dey by ICMR, and Research Fellowship to Debarun Roy by UGC are duly acknowledged.

References

Acott, T.S., Hoskins, D.D. 1978. Bovine sperm forward motility protein. Partial purification and characterization. *Journal of Biological Chemistry* 253:6744–6750.

Aitken, R.J., Nixon, B., Lin, M., Koppers, A.J., Lee, Y.H., Baker, M.A. 2007. Proteomic changes in mammalian spermatozoa during epididymal maturation. *Asian Journal of Andrology* 9:554–64.

Atherton, R.W., Khatoon, S., Schoff, P.K., Haley, B.E. 1985. A study of rat epididymal sperm adenosine 3,5, monophosphate-dependent protein kinase: Maturation differences and cellular location. *Biology of Reproduction* 32:155–171.

Aveldaño, M.I., Rotstein, N.P., Vermouth, N.T. 1992. Lipid remodelling during epididymal maturation of rat spermatozoa. Enrichment in plasmenylcholines containing long-chain polyenoic fatty acids of the n-9 series. *Biochemical Journal* 283:235–241.

Banerjee, S., Dey, N., Majumder, G.C. 1992. Maturation-dependent goat epididymal sperm autoagglutination and its inhibition by a glycoprotein factor. *Indian Journal of Experimental Biology* 30:1056–1061.

Banerjee, S., Dungdung, S.R., Das, K., Majumder, G.C. 2006. Synchronous modulation of cell-surface lectin and its receptor in a homologous cell population: A novel mechanism of cellular regulation. *Experimental Cell Research* 312:2299–2308.

Barua, M., Majumder, G.C. 1987. Enzymatic characteristic of ecto-phosphoprotein phosphatase in goat epididymal intact spermatozoa. *Biochemistry and Cell Biology* 65:602–609.

Barua, M., Majumder, G.C. 1990. Dephosphorylation of cell surface phosphoproteins of goat spermatozoa. *Journal of Biosciences* 15:217–221.

Barua, M., Bhattacharyya, U., Majumder, G.C. 1985. Occurrence of an ecto-phosphoprotein phosphatase in goat epididymal spermatozoa. *Biochemistry International* 10:733–741.

Barua, M., Ghosh, A.K., Majumder, G.C. 1999. Partial purification and characterization of phosphoprotein phosphatase from sperm plasma membrane. *Reproduction, Fertility and Development* 11:379–386.

Barua, M., Haldar, S., Majumder, G.C. 1990. Occurrence of a coupled enzyme system on the intact sperm outer surface that phosphorylates and dephosphorylates eto-proteins. *Biochemistry International* 20:1089–1096.

Barua, M., Nath, D., Majumder, G.C. 2001. Alteration of goat sperm ecto- phosphoprotein phosphatase activity and its distribution on the sperm surface during epididymal maturation. *Reproduction, Fertility and Development* 13:443–450.

Bhoumik, A., Saha, S., Majumder, G.C., Dungdung, S.R. 2014. Optimum calcium concentration: A crucial factor in regulating sperm motility in vitro. *Cell Biochemistry and Biophysics* 70:1177–1183.

Bongalhardo, D.C., Somnapan-Kakuda, N., Buhr, M.M. 2002. Isolation and unique composition of purified head plasma membrane from rooster sperm. *Poultry Science* 81:1877–1883.

Brokaw, C.J. 1987. Regulation of sperm flagellar motility by calcium and cAMP dependent phosphorylation. *Journal of Cellular Biochemistry* 35:175–185.

Chakrabarty, J., Banerjee, D., Pal, D., De, J., Ghosh, A., Majumder, G.C. 2007. Shedding off specific lipid constituents from sperm cell membrane during cryopreservation. *Cryobiology* 54: 27–35.

Cheah, Y., Yang, W. 2011. Functions of essential nutrition for high quality spermatogenesis. *Advances in Bioscience and Biotechnology* 2:182–197.

Christova, Y., James, P., Mackie, A., Cooper, T.G., Jones, R. 2004. Molecular diffusion in sperm plasma membranes during epididymal maturation. *Molecular and Cellular Endocrinology* 216: 41–46.

Connor, W.E., Lin, D.S., Neuringer, M. 1997. Biochemical markers for puberty in the monkey testis: Desmosterol and docosahexaenoic acid. *Journal of Clinical Endocrinology and Metabolism* 82:1911–1916.

Cooper, T.G. 1986. Maturation of spermatozoa in the epididymis. In *The Epididymis, Sperm Maturation and Fertilisation*, Heidelberg, Springer, 1–8.

Das, K., Das, S., Bhoumik, A., Jaiswal, B.S., Majumder, G.C., Dungdung, S.R. 2012. In vitro initiated sperm forward motility in caput spermatozoa: Weak and transient *Andrologia* 44:807–12.

Das, S., Saha, S., Majumder, G.C., Dungdung, S.R. 2010. Purification and characterization of a sperm motility inhibiting factor from caprine epididymal plasma. *PLoS ONE* 5(8): e12039.

De Lamirande, E., O'Flaherty, C. 2008. Sperm activation: Role of reactive oxygen species and kinases. *Biochimica et Biophysica Acta* 1784:106–15.

Devi, L.G., Shivaji, S. 1994. Computerized analysis of the motility parameters of hamster spermatozoa during maturation. *Molecular Reproduction and Development* 38:94–106.

Devi, K.U., Ahmad, M.B., Shivaji, S. 1997. A maturation-related differential phosphorylation of plasma membrane proteins of the epididymal spermatozoa of the hamster by endogenous protein kinase. *Molecular Reproduction and Development* 47:341–350.

Dey, C.S., Majumder, G.C. 1987a. Ecto cyclic AMP-receptor in goat epididymal intact spermatozoa and its change in activity during forward motility. *Journal of Cellular Biochemistry* 35:259–269.

Dey, C.S., Majumder, G.C. 1987b. Phosphatidyl inositol inhibition of a sperm cyclic AMP-independent protein kinase. *Biochemical Biophysical Research Communication* 146: 422–429.

Dey, C.S., Majumder, G.C. 1990a. Maturation-specific type II cyclic AMP-dependent protein kinase in goat sperm plasma membrane. *Biochemistry International* 21:659–665.

Dey, C.S., Majumder, G.C. 1990b. Type I and II cAMP-dependent ecto-protein kinases in goat epididymal spermatozoa and their enriched activities in forward motile spermatozoa. *Biochemistry and Cell Biology* 68: 459–470.

Dey, S., Roy, D., Majumder, G.C. and Bhattacharyya, D. 2014a. Receptor expression is essential for forward motility in the course of sperm cell maturation. *Biochemistry and Cell Biology* 92:43–52.

Dey, S., Roy, D., Majumder, G.C. and Bhattacharyya, D. 2014b. Extracellular regulation of sperm transmembrane adenylyl cyclase by a forward motility stimulating protein. *PLOS One* 9, e110669. doi 10.1371/journal.pone.0110669.

Ding, Z., Qu, F., Guo, W., Ying, X., Wu, M., Zhang, Y. 2007. Identification of sperm forward motility-related proteins in human seminal plasma. *Molecular Reproduction and Development* 74:1124–1131.

Dungdung, S.R., Majumder, G.C. 2003. Isolation and identification of a novel motility-inhibiting factor from goat cauda sperm plasma membrane. *Cellular and Molecular Biology* 49:413–420.

Evans, R.W., Setchell, B.P. 1979. Lipid changes in boar spermatozoa during epididymal maturation with some observations on the flow and composition of boar rete testis fluid. *Journal of Reproduction and Fertility* 57:189–196.

Freund, M., Oliveira, N. 1987. Visual versus cinemicrographic evaluation of human sperm motility and morphology. *Archives of Andrology* 19:25–32.

Gagnon, C., de Lamirande, E. 2006. *The Sperm Cell: Production, Maturation, Fertilization and Regeneration,* Cambridge University Press, Cambridge, UK, 108–133.

Ghosh, I., Datta, K. 2003. Sperm surface hyaluronan binding protein (HABP1) interacts with zona pellucida of water buffalo (*Bubalus bubalis*) through its clustered mannose residues. *Molecular Reproduction and Development* 64:235–44.

Ghosh, I., Bharadwaj, A., Datta, K. 2002. Reduction in the level of hyaluronan binding protein 1 (HABP1) is associated with loss of sperm motility. *Journal of Reproductive Immunology* 53:45–54.

Gillis, G., Peterson, R., Russell, L., Hook, L., Freud, M. 1978. Isolation and characterization of membrane vesicles from human and boar spermatozoa: Methods using nitrogen cavitation and ionophore induced vesiculation. *Preparative Biochemistry* 8:363–378.

Glander, H.J. 1984. Effect of the epididymis on male fertility. *Zeitschrift für Urologie und Nephrologie* 77:551–557.

Goluboff, E.T., Mertaz, J.R., Tres, L.L., Kierszenbaum, A.L. 1995. Galactosyl receptor in human testis and sperm is antigenically related to the minor C-type (Ca^{2+}-dependent) lectin variant of human and rat liver. *Molecular Reproduction and Development* 40: 460–466.

Gulaya, N.M., Margitich, V.M., Govseeva, N.M., Klimashevsky, V.M., Gorpynchenko, I.I., Boyko, M.I. 2001. Phospholipid composition of human sperm and seminal plasma in relation to sperm fertility. *Archives of Andrology* 46:169–175.

Haidl, G., Opper, C. 1997. Changes in lipids and membrane anisotropy in human spermatozoa during epididymal maturation. *Human Reproduction* 12:2720–2723.

Haldar, S., Dey, C.S., Majumder, G.C. 1986. An ecto-cyclic AMP-independent protein kinase in goat spermatozoa and its change of activity during forward motility. *Biochemistry International* 13:809–817.

Haldar, S., Majumder, G.C. 1986. Phosphorylation of external cell surface proteins by an endogenous ecto-protein kinase of goat epididymal intact spermatozoa. *Biochimica et Biophysica Acta* 887:291–303.

Hammerstedt, R.H., Parks, J.E. 1987. Changes in sperm surfaces associated with epididymal transit. *Journal of Reproduction and Fertility* 34:133–149.

Hinkovska, V.T., Dimitrov, G.P., Koumanov, K.S. 1986. Phospholipid composition and phospholipid asymmetry of ram spermatozoa plasma membranes. *International Journal of Biochemistry and Cell Biology* 18:1115–1121.

Holt, W.V., North, R.D. 1985. Determination of lipid composition and thermal phase transition temperature in an enriched plasma membrane fraction from ram spermatozoa. *Journal of Reproduction and Fertility* 23:285–294.

Hoskins, D.D., Brandt, H., Acott, T.S. 1978. Initiation of sperm motility in the mammalian epididymis. *Federation Proceedings* 37:2534–2542.

Hoskins, D.D., Hall, M.L., Musterman, D. 1975. Induction of motility in immature bovine spermatozoa by cyclic AMP phosphodiesterase inhibitors and seminal plasma. *Biology of Reproduction* 13:168–176.

Iguer-Ouada, M., Verstegen, J.P. 2001. Validation of the sperm quality analyzer (SQA) for dog sperm analysis. *Theriogenology* 55:1143–1158.

Ivanov, N., Profirov, W.J. 1981. Isolation of plasma membranes from ram spermatozoa by a two-phase polymer system. *Journal of Reproduction and Fertility* 63:25–29.

Jaiswal, B.S., Das, K., Saha, S., Dungdung, S.R., Majumder, G.C. 2010. Purification and characterization of a motility initiating protein from caprine epididymal plasma. *Journal of Cellular Physiology* 222:254–263.

Jaiswal, B.S., Majumder, G.C. 1998. Biochemical parameters regulating forward motility initiation in vitro in goat immature epididymal spermatozoa. *Reproduction, Fertility and Development* 10:299–307.

James, P.S., Wolfe, C.A., Ladha, S., Jones, R. 1999. Lipid diffusion in the plasma membrane of ram and boar spermatozoa during maturation in the epididymis measured by fluorescence recovery after photobleaching. *Molecular Reproduction and Development* 52:207–215.

Johnson, M.H. 1975. The macromolecular organization of membranes and its bearing on events leading up to fertilization. *Journal of Reproduction and Fertility* 44:167–84.

Kamidono, S., Hazama, M., Matsumoto, O., Takada, K., Tomioka, O., Ishigami, J. 1983. Study on human spermatozoal motility: Preliminary report on newly developed multiple exposure photography method. *Andrologia* 15:111–119.

Katz, D.F., Drobnis, E.Z., Overstreet, J.W. 1989. Factors regulating mammalian sperm migration through the female reproductive tract and oocyte vestments. *Gamete Research* 22:443–469.

Kota, V., Dhople, V.M., Shivaji, S. 2009. Tyrosine phosphoproteome of hamster spermatozoa: Role of glycerol-3-phosphate dehydrogenase 2 in sperm capacitation. *Proteomics* 9:1809–1826.

Kumar, V., Rangaraj, N., Shivaji, S. 2006. Activity of pyruvate dehydrogenase A (PDHA) in hamster spermatozoa correlates positively with hyperactivation and is associated with sperm capacitation. *Biology of Reproduction* 75:767–777.

Kuo, Y.M., Zhou, B., Cosco, D., Gitschier, J. 2001. The copper transporter CTR1 provides an essential function in mammalian embryonic development. *Proceedings of the National Academy of Sciences* 98:6836–6841.

Lee, H.C., Johnson, C., Epel, D. 1983. Changes in internal pH associated with initiation of motility and acrosome reaction of sea urchin sperm. *Developmental Biology* 95: 31–45.

Loeser, C.R., Tulsiani, D.R.P. 1999. The role of carbohydrates in the induction of the acrosome reaction in mouse spermatozoa. *Biology of Reproduction* 60:94–101.

Lunstra, D.D., Clegg, E.D., Morre, D.J. 1974. Isolation of plasma membrane from porcine spermatozoa. *Preparative Biochemistry* 4:341–352.

Maiti, A., Mishra, K.P., Majumder, G.C. 2004. Identification of goat sperm ecto-cyclic AMP independent protein kinase substrate localized on sperm outer surface. *Journal of Cellular Biochemistry* 92:164–177.

Maiti, A., Mishra, K.P., Majumder, G.C. 2008. Role of the major ecto-phosphoprotein in sperm flagellar motility using a cell electroporation method. *Molecular Reproduction and Development* 75:1185–1195.

Maiti, A., Nath, D., Dungdung, SR., Majumder, G.C. 2009. Sperm ecto-protein kinase and its protein substrate: Novel regulators of membrane fusion during acrosome reaction. *Journal of Cellular Physiology* 220:394–400.

Majumder, G.C. 1978. Occurrence of a cyclic AMP-dependent protein kinase on the outer surface of rat epididymal spermatozoa. *Biochemical and Biophysical Research Communications* 83:829–836.

Majumder, G.C. 1981. Enzymic characteristics of an ecto-cyclic AMP-dependent protein kinase in rat epididymal spermatozoa. *Biochemical Journal* 195:111–117.

Majumder, G.C., Biswas, R. 1979. Evidence for the occurrence of an ecto- (adenosine triphosphatase) in rat epididymal spermatozoa. *Biochemical Journal* 183:737–743.

Majumder, G.C., Chakrabarti, C.K. 1984. A simple spectrophotometric method of assay of forward motility of goat spermatozoa. *Journal of Reproduction and Fertility* 70:235–241.

Majumder, G.C., Chaudhuri, D.P. 1984. Occurrence of -SH containing molecules on the goat sperm external surface that are essential for flagellar motility. *Andrologia* 16:219–223.

Majumder, G.C., Das, K., Saha, S., Nath, D., Maiti, A., Das, S., Dey, N. et al. 2012. Purification and characterization of novel sperm motility-related proteins. In *Protein Purification*, eds. M. Benitez, V. Aguiree, Nova Science Publishers, USA, 1–90.

Majumder, G.C., Dey, C.S., Haldar, S., Barua, M. 1990. Biochemical parameters of initiation and regulation of sperm motility. *Archives of Andrology* 24:287–303.

Majumder, G.C., Jaiswal, B.S., Nath, D., Banerjee, S., Barua, M., Sarkar, M., Rana, A.P.S. et al. 1999. Biochemistry of sperm motility initiation during epididymal maturation. In *Comparative Endocrinology and Reproduction*, eds. K.P. Joy, A. Krishna, C. Haldar, Narosa Publishing House, New Delhi, India, 242–259.

Majumder, G.C., Haldar, S., Dey, C.S., Barua, M., Roy, N. 1988. Occurrence of several ecto-proteins on goat spermatozoal surface that may regulate flagellar motility. *Indian Journal of Biochemistry and Biophysics* 25:215–218.

Makan, N.R. 1979. Phosphoprotein phosphatase activity at the outer surface of intact normal and transformed 3T3 fibroblasts. *Biochimica et Biophysica Acta* 585:360–73.

Mandal, M., Saha, S., Ghosh, A.K., Majumder, G.C. 2006. Identification and characterization of a sperm motility promoting glycoprotein from buffalo blood serum. *Journal of Cellular Physiology* 209:353–362.

Marshall, B.T., Long, M., Piper, J.W., Yago, T., McEver, R.P., Zhu, C. 2003. Direct observation of catch bonds involving cell-adhesion molecules. *Nature* 423:190–193.

Martin, G., Sabido, O., Durand, P., Levy, R. 2005. Phosphatidylserine externalization in human sperm induced by calcium ionophore A23187: Relationship with apoptosis, membrane scrambling and the acrosome reaction. *Human Reproduction* 20:3459–3468.

Mastro, A.M., Rosengurt, E. 1976. Endogenous protein kinase in outer plasma membrane of cultured 3T3 cells. *Journal of Cellular Biochemistry* 251:7899–7906.

Mitra, S., Majumder, G.C. 1991. Alteration of the ecto-protein phosphorylation profile of intact goat spermatozoa during epididymal maturation. *Biochemistry International* 23:611–618.

Mitra, S., Nath, D., Majumder, G.C. 1994. Purification and characterization of a protein kinase from goat sperm plasma membrane. *Biochemistry and Cell Biology* 72:218–226.

Müller, K., Pomorski, T., Müller, P., Zachowski, A., Herrmann, A. 1994. Protein-dependent translocation of aminophospholipids and asymmetric transbilayer distribution of phospholipids in the plasma membrane of ram sperm cells. *Biochemistry* 33:9968–9974.

Nath, D., Bhoumik, A., Das, S., Bhattacharyya, D., Dungdung, S.R., Majumder, G.C. 2012. A novel membrane protein-specific serine/threonine kinase: Tissue distribution and role in sperm maturation. *ISRN Urology*, doi:10.5402/2012/789105.

Nath, D., Maiti, A., Majumder, G.C. 2008. Cell surface phosphorylation by a novel ecto-protein kinase: A key regulator of cellular functions in spermatozoa. *Biochimica et Biophysica Acta* 1778:153–165.

Neumann, E., Schaefer-Ridder, M., Wang, Y., Hofschneider, P.H. 1982. Gene transfer into mouse lyoma cells by electroporation in high electric fields. *European Molecular Biology Organization Journal* 1:841–845.

Nissen, H.P., Kreysel, H.W. 1983. Polyun saturated fatty acid sin relation to sperm motility. *Andrologia* 15:264–269.

Ollero, M., Powers, R.D., Alvarez, J.G. 2000. Variation of docosahexaenoic acid content in subsets of human spermatozoa at different stages of maturation: Implications for sperm lipoperoxidative damage. *Molecular Reproduction and Development* 55:326–334.

Orgebin-Crist, M.C., Tichenor, P.L. 1972. A technique for studying sperm maturation in vitro. *Nature* 239:227–228.

Pariset, C.C.C., Roussel, S.J., Weinman, Damaille, J.D. 1983. Calmodulin intracellular concentration and cAMP- dependent protein kinase activity in human sperm samples in relation to sperm morphology and motility. *Gamete Research* 8:171–182.

Parks, J.E., Lynch, D.V. 1992. Lipid composition and thermotropic phase behavior of boar, bull, stallion, and rooster sperm membranes. *Cryobiology* 29:255–266.

Paul, D., Majumder, G.C., Saha, S., Mukherjee, A., Banerjee, S. 2004. A unique computer based spectrophotometric system with multiple height exposures of The cuvette to determine vertical velocity of spermatozoa for clinical and biological applications: Application for Indian Patent done through CSIR. (Patent File No. 1605DEL2004 Dated: 26.08.2004 and CSIR Reference No. 261NF2004.)

Paul, D., Majumder, G.C., Saha, S., Mukherjee, A., Banerjee, S. 2005. Computer based spectrophotometric system to determine vertical velocity of clinical and biological samples: International patent application file no.: 0261NF2004/WO-PCT/IB05/02541–26/08/2005.

Perez-Sanchez, F., Tablado, L., Yeung, C.H., Cooper, T.G., Soler, C. 1996. Changes in the motility patterns of spermatozoa from the rabbit epididymis as assessed by computer-aided sperm motion analysis. *Molecular Reproduction and Development* 45:364–371.

Prasad, M.R.N., Singh, S.P., Rajalakshmi, M. 1970. Fertility control in male rats by continuous release of microquantitics of cyproterone acetate from subcutaneous silastic capsules. *Contraception* 2:165–178.

Rana, A.P.S., Majumder, G.C. 1987. Factors influencing the yield and purity of goat sperm plasma membranes isolated by means of an aqueous two-phase polymer system. *Preparative Biochemistry* 17:261–281.

Rana, A.P.S., Majumder, G.C. 1990. Changes in the fluidity of the goat sperm plasma membrane in transit from caput to cauda epididymis. *Biochemistry International* 21:797–803.

Rana, A.P.S., Majumder, G.C., Misra, S., Ghosh, A. 1991. Lipid changes of goat sperm plasma membrane during epididymal maturation. *Biochimica et Biophysica Acta* 1061:185–196.

Rana, A.P.S., Misra, S., Majumder, G.C., Ghosh, A. 1993. Phospholipid asymmetry of goat sperm plasma membrane during epididymal maturation. *Biochimica et Biophysica Acta* 1210:1–7.

Redegeld, F.A., Caldwell, CC., Sitkovsky, M.V. 1999. Ecto-protein kinases: Ecto-domain phosphorylation as a novel target for pharmacological manipulation? *Trends in Pharmacological Sciences* 20:453–459.

Rodeheffer, C., Shur, B.D. 2004. Characterization of a novel ZP3-independent sperm-binding ligand that facilitates sperm adhesion to the egg coat. *Development* 131:503–512.

Rodriguez-Martinez, H., Iborra, A., Martinez, P., Calvete, J.J. 1998. Immunoelectronmicroscopic imaging of spermadhesin AWN epitopes on boar spermatozoa bound in vivo to the zona pellucida. *Reproduction, Fertility and Development* 10:491–497.

Rothman, R.E., Lenard, J. 1977. Membrane asymmetry. *Science* 195:743–753.

Roy, D., Dey, S., Majumder, G.C., Bhattacharyya, D. 2014a. Copper: A biphasic regulator of caprine sperm forward progression. *System Biology in Reproductive Medicine* 60:52–57.

Roy, D., Dey, S., Majumder, G.C., Bhattacharyya, D. 2014b. Occurrence of novel Cu^{2+}–dependent sialic acid-specific lectin, on the outer surface of mature caprine spermatozoa. *Glycoconjugate Journal* 31:281–288.

Roy, N., Majumder, G.C. 1986. A simple quantitative method for the estimation of free ecto-sulfhydryl groups of spermatozoa. *Experimental Cell Research* 164:415–425.

Saha, S., Das, S., Bhoumik, A., Ghosh, P., Majumder, G.C., Dungdung, S.R. 2013. Identification of a novel sperm motility-stimulating protein from caprine serum: its characterization and functional significance. *Fertility Sterility* 100:269–279.

Saha, S., Paul, D., Mukherjee, A., Banerjee, S., Majumder, G.C. 2007. A computerized spectrophotometric instrumental system to determine the 'vertical velocity' of sperm cells: A novel concept. *Cytometry Part A* 71A:308–316.

Sarkar, M., Majumder, G.C., Chatterjee, T. 1991. Goat sperm membrane: Lectin-binding sites of sperm surface and lectin affinity chromatography of the mature sperm membrane antigens. *Biochimica et Biophysica Acta* 1070:198–204.

Schieferstein, G., Hook-Vervier, B., Schwarz, M. 1998. Sperm motility index. *Archives of Andrology* 40:43–48.

Schlaeger, E., Kohler, G. 1976. External cyclic AMP-dependent protein kinase activity in rat C-6 glioma cells. *Nature* 260:705–707.

Schoff, P.K, Forrester, I.T., Haley, B.E., Atherton, R.W. 1982. A study of cAMP binding proteins on intact and disrupted sperm cells using 8-azidoadenosine 3′,5′-cyclic monophosphate. *Journal of Cellular Biochemistry* 19:1–15.

Shivaji, S., Kumar, V., Mitra, K., Jha, K.N. 2007. Mammalian sperm capacitation: Role of phosphotyrosine protein. *Society of Reproduction and Fertility Supplement* 63:295–312.

Sokoloski, J.E., Blasco, L., Storey, B.T., Wolf, D.P. 1977. Turbidimetric analysis of human sperm motility. *Fertility and Sterility* 28:1337–1341.

Stubbs, C.H., Smith, P.D. 1984. The modification of mammalian membrane polyunsaturated fatty acid composition in relation to membrane fluidity and function. *Biochimica et Biophysica Acta* 779:89–137.

Sullivan, R., Frenette, G., Girouard, J. 2007. Epididymosomes are involved in the acquisition of new sperm proteins during epididymal transit. *Asian Journal of Andrology* 9:483–491.

Tash, J.S., Means, A.R. 1983. Cyclic adenosine 3′,5 monophosphate, calcium and protein phosphorylation in flagellar motility. *Biology of Reproduction* 28:75–104.

Tomar, A.K., Sooch, B.S., Raj, I., Singh, S., Singh, T.P., Yadav, S. 2011. Isolation and identification of Concanavalin A binding glycoproteins from human seminal plasma: A step towards identification of male infertility marker proteins. *Disease Markers* 31:379–386.

Tsong, T.Y. 1991. Electroporation of cell membranes. *Biophysical Journal* 60:297–306.

Tulsiani, D.R., Abou-Haila, A. 2001. Mammalian sperm molecules that are potentially important in interaction with female genital tract and egg vestments. *Zygote* 9:51–69.

Uma Devi, K., Jha, K., Patil, S.B., Padma, P., Shivaji, S. 2000. Inhibition of motility of hamster spermatozoa by protein tyrosine kinase inhibitors. *Andrologia* 32:95–106.

Vestweber, D., Blanks, J.E. 1999. Mechanisms that regulate the function of the selectins and their ligands. *Physiological Reviews* 79:181–213.

Yamada, K.M., Olden, K., Halm, L.H.E. 1980. Cell surface protein and cell interactions. In *The Cell Surface: Mediator of Developmental Processes*, eds. S. Subtenly, N.K. Wessells, Academic Press, New York, 43–77.

Yanagimachi, R. 1994. Mammalian fertilization. In *The Physiology of Reproduction*, Vol. 2, eds. E. Knobil, J.D. Neill, Raven Press, New York, 189–317.

Yoshida, M., Kawano, N., Yoshida, K. 2008. Control of sperm motility and fertility: Diverse factors and common mechanisms. *Cellular and Molecular Life Sciences* 65:3446 – 3457.

Yuuji, S., Harumi, S., Takashi, S. 1992. Purification of a sperm lectin extracted from spermatozoa of the sea urchin *Hemicentrotus pulcherrimus. Development, Growth and Differentiation* 34:285–291.

Chapter 9

Male Accessory Sex Glands: Structure and Function

Margot Flint, Debra A. McAlister,
Ashok Agarwal and Stefan S. du Plessis

Contents

9.1 Introduction ... 246
9.2 Cowper's Glands ... 247
9.2.1 Location ... 247
9.2.2 Structure ... 247
9.2.3 Secretions and Function ... 248
9.2.4 Pathology ... 248
9.2.4.1 Congenital Lesions ... 249
9.2.4.2 Acquired Lesions ... 249
9.3 Prostate Gland ... 250
9.3.1 Location ... 250
9.3.2 Structure ... 251
9.3.3 Secretions and Function ... 252
9.3.4 Pathologies ... 252
9.3.4.1 Prostatitis ... 253
9.3.4.2 Prostatic Hypertrophy ... 253
9.3.4.3 Prostate Cancer ... 254
9.4 Seminal Vesicles ... 254
9.4.1 Location ... 255

9.4.2 Structure 255
9.4.3 Secretion and Function 255
9.4.4 Pathologies 256
9.5 Conclusions 257
References 257

9.1 Introduction

The accessory sex glands can be regarded as part of the male's reproductive system as they play an integral role in the fertility process. The major male accessory sex glands (Figure 9.1), whose secretions provide the bulk of the semen in varying volumes are the seminal vesicles, prostate glands and Cowper's glands (Dunker and Aumuller 2002; Owen and Katz 2005). Also contributing very small volumes to semen are the ampullary, Littre and Tyson's glands; however, these glands are barely studied and hence are poorly understood (Mortimer 1993). Products of these glands which are secreted at the beginning of the ejaculatory phase serve to nourish and activate the spermatozoa, clear the urethral tract prior to ejaculation, and act as a vehicle of transport for the sperm in the female tract and to plug the female tract after placement of sperm to help ensure fertilisation. The male accessory sex glands

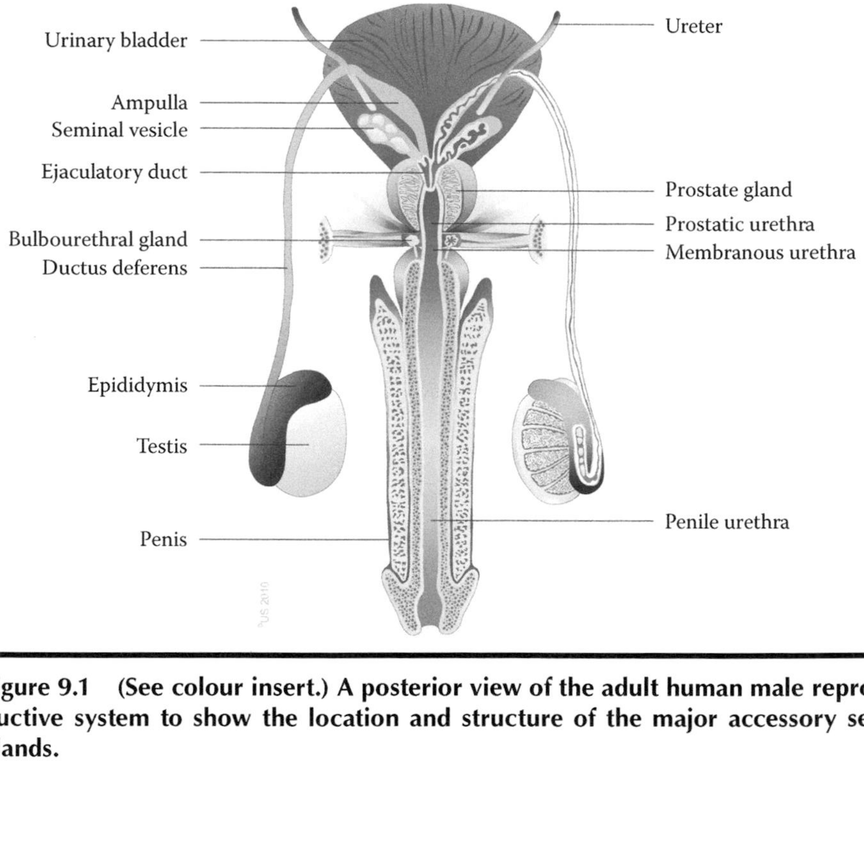

Figure 9.1 (See colour insert.) A posterior view of the adult human male reproductive system to show the location and structure of the major accessory sex glands.

work together in playing an essential role in the reproductive process (Chughtai et al. 2005). The major accessory sex glands will subsequently be discussed in more detail.

9.2 Cowper's Glands

The Cowper's glands, also known as the bulbourethral glands, were discovered in the seventeenth century by an English surgeon, William Cowper (Chughtai et al. 2005). These sex glands, present in the reproductive system of human males, are homologous to Bartholin's glands in females (Woodruff and Friedrich 1985). Although the fluid secreted from the Cowper's glands contributes only a small volume of fluid to the total semen volume in a single ejaculate, it serves an important function prior to ejaculation.

9.2.1 Location

The Cowper's glands exist in pairs and are found in the majority of male mammals. The two Cowper's glands lie side by side and are located beneath the prostate gland in the urogenital diaphragm, posterior and lateral to the membranous urethra. They are positioned between the bulbospongiosus muscles and embedded within the transverse perineal muscle, forming part of the pelvic floor (Dunker and Aumuller 2002). The two glands join the urethra by means of a main duct, to which smaller accessory ducts are attached.

9.2.2 Structure

The pea-sized Cowper's glands are about 1 cm in diameter and yellow in colour (Saboorian et al. 1997; Chughtai et al. 2005). This small pair of male accessory glands is arranged in a network of clearly demarcated lobules of compact tubuloalveolar glands (Saboorian et al. 1997; Awakura et al. 2000; Dunker and Aumuller 2002; Chughtai et al. 2005). Tubuloalveolar glands are characterised by possessing both tubular and alveolar secretory units. These units are composed of mucus-like epithelial cells with basal nuclei (King 1993).

The cluster of lobules are separated by and enclosed in a connective tissue capsule of columnar epithelium and secured within fascicles of thick striated skeletal muscle (Hellgren et al. 1982; Saboorian et al. 1997; Dunker and Aumuller 2002; Chughtai et al. 2005). Small accessory ducts extend from each lobule of the gland which joins and enters the bulbourethra separately or as a united duct (Awakura et al. 2000; Bevers et al. 2000). The accessory ducts may therefore drain into the main duct or enter the urethra directly (Chughtai et al. 2005). The main Cowper's duct which is approximately 3 cm in length enters the ventral surface of the bulbourethra near the midline by penetrating the spongiosum, the mass of spongy tissue surrounding the urethra (Hafez 1977).

9.2.3 Secretions and Function

According to Tikva, the male sexual response consists of four phases: excitement, plateau, orgasm and resolution. During sexual excitement and plateau, the Cowper's glands secrete glycoproteins into the bulbous urethra (Tikva 2003). The secretion is a clear, viscous, alkaline mucus-like fluid and is commonly known as *pre-ejaculate,* as it is secreted prior to ejaculation (Riva et al. 1990; Mortimer 1993; Dunker and Aumuller 2002). The amount of pre-ejaculate emitted varies widely between individual men, averaging about 0.2 mL in most men but can be as much as 5 mL in some men depending largely on the duration of the plateau-phase levels of sexual tension (Tikva 2003). Generally the Cowper's secretions make up less than 1% of the semen composition (Riva et al. 1990).

The fluid secreted functions as a lubricant for the seminal fluid containing sperm which is to follow during ejaculation (Riva et al. 1990). Along with acting as a lubricant, this fluid aids in neutralising the acidity of the urine residue in the urethra which may be unfavourable and harmful to sperm (Mortimer 1993). The fluid also assists in neutralising the acidity of the vagina and provides some lubrication for the tip of the penis during sexual intercourse (Chughtai et al. 2005). Along with neutralisation of an acidic environment, the secretion is also well documented for its contribution to coagulation and the characteristic jelly-like appearance of semen (Riva et al. 1990; Brocker 1998).

Moreover, the Cowper's glands have been shown to contribute to the immune defence of the genitourinary tract by secreting many glycoproteins, such as prostate-specific antigens (PSA) (Migliari et al. 1992; Cina et al. 1997; Pedron et al. 1997). Studies have confirmed the presence of a specific distribution of immunocompetent cells in the Cowper's glands, which seems to offer the first immunological barrier against the entry of antigens (Migliari et al. 1992).

It has been long speculated that the fluid from the Cowper's glands may contain sperm, and that coitus interruptus is not a reliable method of contraception. Although it is possible for this fluid to pick up sperm remaining in the urethral bulb from previous ejaculations, the Cowper's gland secretions have been shown to be absent of sperm. In a study by Tikva, it was shown that pre-ejaculatory fluid secreted at the tip of the urethra during sexual stimulation did not contain sperm and therefore cannot be responsible for pregnancies during coitus interruptus (Tikva 2003).

9.2.4 Pathology

The Cowper's glands can be affected by acquired and congenital lesions (Colodny and Lebowitz 1978). Congenital lesions of these glands mainly consist of syringocele (Masson et al. 1979), which tends to be asymptomatic, and may be easily confused with the more serious conditions during diagnostic tests. Like the prostate, acquired lesions of the Cowper's glands include infection, calcification and neoplasms (Pedron et al. 1997; Chughtai et al. 2005).

9.2.4.1 Congenital Lesions

9.2.4.1.1 Syringocele

Obstruction to the orifice of the Cowper's duct gives rise to dilatation of the duct, a consequence of stasis and pressure changes (Merchant et al. 1997), commonly known as syringocele. Syringocele assumes a cystic appearance (Merchant et al. 1997; Kickuth et al. 2002; Chughtai et al. 2005) and is an uncommon deformity usually diagnosed in male infants and children presenting with urinary tract infection, macrohematuria and voiding symptoms (Awakura et al. 2000; Bevers et al. 2000). Cowper's syringocele may be closed or open, which is a distended cyst-like swelling in the wall of the urethra or an opening enabling urine reflux into the syringocele respectively. Bacterial colonisation in the pool of mucus or urine that stagnates within the syringocele is inevitable. Although most cases are asymptomatic and may go unnoticed, syringocele may present in adults as urinary tract infections with frequency, pain and discharge, symptoms of urethral obstruction, or chronic post-void incontinence (Chughtai et al. 2005).

Treatment of Cowper's syringocele usually consists of a short outpatient procedure such as marsupialisation, which involves the surgical removal of the cyst. This is performed by opening the Cowper's syringocele transurethrally with a Collins knife (Bevers et al. 2000; Kickuth et al. 2002).

9.2.4.2 Acquired Lesions

9.2.4.2.1 Cowperitis

Cowperitis refers to the bacterial infection and inflammation of the Cowper's gland which may be chronic or acute. The infections are generally found to be caused by the same organisms that cause urinary tract infections, such as *Escherichia coli*, *Neisseria gonorrhoea* and *Chlamydia trachomatis* (Birnstingl et al. 1957; Dunker and Aumuller 2002; Chughtai et al. 2005).

Acute cowperitis presents as fever, malaise, sever pain in the perineum with frequency, urgency, painful defecation and sometimes acute urinary retention. Acute cowperitis is usually treated successfully with a suitable anti-biotic (Chughtai et al. 2005).

Chronic cowperitis is usually associated with an underlying defect such as syringocele which causes the bacterial infection as a result of the bacteria-ladened urine and mucus collecting in the area (Birnstingl et al. 1957). Treatment for a pathology of this nature requires the removal of the defect and subsequent treatment with antibiotics, yet antibiotics have not always been found to be successful (Chughtai et al. 2005).

9.2.4.2.2 Stones

Cowper's gland calcifications are found more commonly in elderly patients. Patients with these Cowper's stones usually experience blockages of secretions and infection. The calcifications usually consist of phosphate salts of calcium, magnesium, potassium, calcium carbonate or calcium oxalate and can be detected through pelvic ultrasound. These stones are rarely infected or cause abscesses. The treatment in the case of large stones usually involves the removal of the Cowper's gland, however in asymptomatic cases, the stones may be observed closely without treatment (White et al. 1983; Chughtai et al. 2005).

9.2.4.2.3 Neoplasms

Adenocarcinoma of the Cowper's gland exhibits irregular-shaped glands lined with anaplastic cells. Symptoms usually include the abnormal narrowing of the urethra with or without a bloody discharge and no increase in PSA. Carcinoma of the Cowper's gland is rare, and may be treated with chemotherapy, radiation therapy or surgical removal. In the few reported cases of this pathology, the most appropriate treatment was the combination of radiation and surgical therapy (Saboorian et al. 1997; Chughtai et al. 2005).

9.3 Prostate Gland

The prostate in a healthy young adult male is a walnut-size gland that weighs up to 20 g and contributes approximately 30% to the total seminal volume (Honda et al. 1988; Lalani et al. 1997). The combinations of compounds that are secreted by the prostate enhance the chances of successful oocyte fertilisation and the gland is of clinical importance due to the link between an increase in age and pathological changes in the prostate (Rui et al. 1986; Sherwood 2007).

9.3.1 Location

The prostate gland is located between the urogenital diaphragm and the neck of the bladder, and connecting the prostate and the bulbourethral glands is the urogenital sinus (Wilson et al. 1997). The prostate completely surrounds both the ejaculatory ducts as well as the urethra and begins its journey at the neck of the bladder and ends by merging with the ejaculatory ducts (Pienta and Esper 1993; Sherwood 2007).

9.3.2 Structure

The prostate is a heterogeneous, multilobulated gland consisting of four morphologically different zones (Figure 9.2): the transition, central and peripheral zone, as well as the anterior fibro-muscular stroma (Rui et al. 1986; Fritjofsson et al. 1988; Lalani et al. 1997). The central zone surrounds the ejaculatory ducts and constitutes approximately 25% of the gland. The peripheral zone is the main contributor of glandular tissue, making up 70% of the prostate volume. The transition zone encircles the proximal prostatic urethra and is a minor supplier of glandular tissue occupying only 5% of the volume of the prostate. Lying anteriorly is the fibro-muscular stroma, which contains no glandular tissue but consists of smooth muscle fibres, dense collagen and fibroblasts (Wilson et al. 1997). The central and peripheral zones are collectively termed the outer or cortical prostate. The inner or periurethral prostate consists of the anterior fibro-muscular stroma and the transition zone (Lalani et al. 1997). Penetrating the posterior surface of the prostate are two ejaculatory ducts (Hafez 1977), and encircling the entire organ is a vascular capsule consisting of an inner layer of smooth muscle and outer fibrous shell (Lalani et al. 1997).

The prostate consists of tubuloalveolar glands which open into the prostatic urethra. Together with the epithelium, the prostatic glands form folds that produce a papillary appearance (Hafez 1977). This irregular shape is to facilitate fluctuations

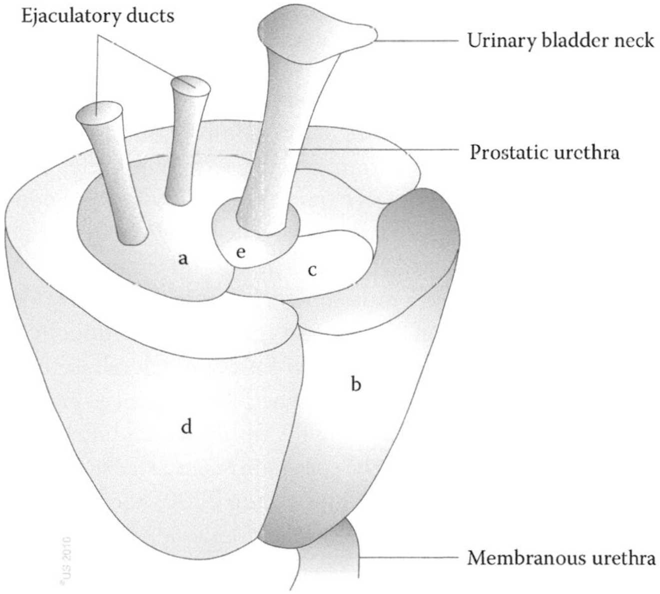

Figure 9.2 (See colour insert.) Schematic representation of the different zones of the prostate gland. a = central zone; b = fibro-muscular zone; c = transitional zone; d = peripheral zone; e = periurethral gland region.

in the gland's size that occurs when the secretions are produced. The epithelial layer of the prostate consists of a double layer of cells: the basement membrane cells and the upper layer of columnar secretory cells. Dividing the two is an intermediate layer of flattened basal cells (Lalani et al. 1997).

9.3.3 Secretions and Function

The prostate secretes a thin, milky, alkaline fluid which is typically low in proteins yet it contains a wide variety of various proteolytic enzymes and electrolytes. The accessory glands supply their secretions in an ordered sequence to the ejaculate. The prostate secretions are released directly after the bulbourethral glands release their secretions (WHO 1992; Nieschlag and Behre 2000). The main compounds secreted by the prostate gland are citric acid and hydrolytic enzymes (Heath and Young 2000). In addition, the following compounds are also secreted: zinc, spermine, cholesterol, magnesium, phospholipids, muramidase, fibrinolysin, fibrinogenase and acid phosphatase (Hafez 1977; Ganong 1981).

Oxytocin, a neurohypophysical hormone produced by the hypothalamus and stored in the posterior pituitary, has also been shown to be synthesised locally in various tissues within the male reproductive system, including the prostate (Robbins and Kumar 1987; Sherwood 2007). Research has shown that the peptide exerts a paracrine effect upon the prostate, stimulating contraction of the gland during ejaculation as well as playing a role in the regulation of the growth of the prostate (Robbins and Kumar 1987).

Prostatic secretory products produce an alkaline fluid with a high pH ranging from about 7.2 to 8.0 which protects the spermatozoa from the acidic cervical mucus and vaginal secretions. This neutralising effect is important as sperm function more advantageously in an alkaline environment and require a buffer from the hostile environment of the female reproductive tract (WHO 1992; Sherwood 2007).

The prostate secretes certain enzymes that interact with the fibrinogen secreted from the seminal vesicles (Lwaleed et al. 2004). This interaction produces the insoluble compound fibrin. Fibrin is a protein which forms the meshwork of a clot and is responsible for the coagulation of semen in the female reproductive tract directly after ejaculation. However, in order to allow for the quick release of motile sperm, the clotted semen needs to be broken down. Fibrinolysin, a proteolytic enzyme produced by the prostate, acts to degrade the fibrin, which liquefies the congealed semen within 5–15 min. Subsequently, spermatozoa are free to now travel through the cervix to reach their target (Heath and Young 2000; Sherwood 2007).

9.3.4 Pathologies

The functional integrity and maintenance of the prostate, as with the other accessory sex glands, is dependent on androgens. The growth of normal prostate epithelium is reliant upon testosterone (Pienta and Esper 1993). However, it is often observed in the category of men over the age of 50 that prostate growth becomes deregulated and the activity of dihydrotestosterone (DHT), a derivative of testicular testosterone, generates a state of hypertrophy (Pienta and Esper 1993). Cases of glandular dysfunction are seldom encountered in men under the age of 40 and this state of malfunctioning results in either benign or malignant overgrowth (Thackare et al. 2006). Inflammation, benign prostatic hypertrophy and carcinoma are the three pathological conditions observed in the prostate (Robbins and Kumar 1987).

9.3.4.1 Prostatitis

Infection of the prostate, prostatitis, is observed in young men and is a focal infection whereby certain areas of the prostate become infected whilst the other zones may not. Prostatitis can be an acute or chronic inflammation of the prostate and is classified in four groups: acute bacterial prostatitis, chronic bacterial prostatitis, achronic bacterial prostatitis and asymptomatic inflammatory prostatitis (Nieschlag and Behre 2000). The condition has a scientifically proven negative impact on the functioning of the reproductive system and is often the common cause for recurrent urinary tract infections (Robbins and Kumar 1987; Jennet 1989). A decrease in sperm motility, semen volume and prostatic secretions is observed in patients presenting with chronic bacterial prostatitis. The predominant organisms responsible for the infection are the bacteria *Ureaplasma urealyticum* and the gram-negative bacteria *Escherichia Coli* (Robbins and Kumar 1987), which reach the prostate either via the blood stream or the urethra. Chronic prostatitis may be responsible for infertility either by its negative influence on spermatozoa motility and viability or through its relation to prostatic secretions (Hafez 1977). The diagnosis of prostatitis is achieved by a digital rectal examination as well as an examination of the urine and glandular secretions to detect the presence of pathogens. The route of treatment for the inflammatory condition is a course of antibiotics (Ullmann 2009).

9.3.4.2 Prostatic Hypertrophy

During a male's reproductive cycle, a common physio-anatomical event when a man reaches reproductive senescence is the enlargement of the prostate (Hafez 1977). This is a result of the non-malignant condition known as benign prostatic hypertrophy which occurs when the mucosal and submucosal glands become enlarged without undergoing cell division (Silverthorn 2009). Benign prostatic hypertrophy most commonly arises in the transitional and central prostatic zones (Pienta and Esper 1993) and difficulty experienced when passing urine is a common symptom (Nieschlag and Behre 2000). This is due to the fact that the prostatic gland completely encircles the urethra, if the gland is swollen; it will impinge on the portion of the urethra that passes though the prostate. Patients present with frequent bouts of urination due to the fact that the wall of the bladder thickens and contraction of the muscular wall of the bladder occurs even with a small amount of urine present in the bladder (Sherwood 2007). The presence of blood in the urine as well as frequent bladder infections can also be used in diagnosing the condition (Jennet 1989). Depending on the severity, there are several ways in which the glandular enlargement can be treated. In the case of partial urinary obstruction, medication is available. However, in severe cases where a patient presents with complete urinary obstruction, surgery is required to relieve the symptoms.

9.3.4.3 Prostate Cancer

Cancer of the prostate has fast become a growing concern in the medical community due to both the increasing incidence of diagnosis, as well as mortality rate in men over the age of 50 years. Although the reason behind the phenomenon remains unclear, the frequency of the condition displays an almost exponential rise in this particular age group and is now the most common form of cancer in men (Pienta and Esper 1993). The disease is most often found in the peripheral zone and manifests itself in two forms: histological and clinical (Pienta and Esper 1993). Researchers have suggested that disturbances in hormonal metabolism may result in the disease progressing from the one state to the other. The initial stages of the disease have been linked to hormonal stimuli, notably the anabolic steroid, testosterone (Pienta and Esper 1993). Increased concentrations of testosterone are present in men who have undergone vasectomies and these men have an increased predisposition for developing the disease (Honda et al. 1988). A rectal examination is used to detect an enlargement of the prostate, as well as an ultrasound examination to obtain a more comprehensive idea of the shape and size of the gland. Prostatic acid phosphatase and PSA are phenotypically expressed by the luminal secretory cells of the prostate into the seminal fluid (Lalani et al. 1997) and the enzyme plays a role in the degradation of the coagulated semen following ejaculation. However, in men with prostatic cancer, the concentrations increase. This makes PSA a sensitive serum marker for detecting the disease and screening for PSA can be used as an alternative diagnostic method (Stenman et al. 1999). Once malignancy is detected, the size of the tumour dictates the line of treatment. For larger tumours, hormone therapy can be used. The aim of hormone treatment is to disable the growth-stimulating influence of the male hormones on the prostate. For smaller tumours, surgery is the first line of treatment. Surgical removal of a tumour involves the removal of the entire prostate gland, the seminal vesicle behind the prostate, a section of the ureter as well as the neighbouring lymph nodes (Ullmann 2009).

9.4 Seminal Vesicles

Seminal fluid is a complex assortment of various substances, which ultimately are responsible for optimising spermatozoa functioning. The seminal vesicles are a pair of accessory sexual glands, which provide a variety of secretions vital to the overall composition of semen. The seminal vesicles are an important part of the male reproductive system as they are the main and final contributors towards the seminal plasma, together supplying up to 85% of the total volume of semen (Heath and Young 2000).

9.4.1 Location

Anatomically, the sac-like seminal vesicles are found in the space between the rectum and the posterior surface of the bladder (Wilson et al. 1997). The glands lie on either side of the last section of the ductus deferens. The ductus deferens is part of the extra-testicular pathway and is two tube-like structures which extend out of the scrotal sac from the epidymis and empties at the base of the bladder into the urethra. This last section of the ductus deferens is where the seminal vesicles deposit their secretions. After this point, the duct continues onto what becomes the ejaculatory duct, which then empties into the urethra (Nieschlag and Behre 2000).

9.4.2 Structure

The seminal vesicles are a pair of two membranous pouches, approximately 7.5 cm in length. Each of the glands is comprised of a tube which is coiled upon itself, creating an irregular glandular diverticulum (Hafez 1977; Heath and Young 2000; Sherwood 2007). Surrounding the glands is a muscular wall organised into an outer longitudinal and an inner circular layer (Heath and Young 2000). Short adrenergic neurons that originate from the pelvic ganglia are found innervating the muscular wall of the seminal vesicles. During the emission phase of ejaculation (Nieschlag and Behre 2000) sympathetic impulses result in sequential contraction of the smooth muscles (Heath and Young 2000). As a result, the secretions of the seminal vesicles are delivered to the urethra in preparation for expulsion from the penis.

9.4.3 Secretion and Function

The seminal vesicles secrete a fluid which is yellowish in colour, viscid and alkaline in nature (Heath and Young 2000). A wide range of compounds are produced by the secretory cells found in the epithelial lining of the glands (Heath and Young 2000). The predominant compounds are fructose, proteins and prostaglandins. Other secretory products include fibrinogen, ascorbic acid, flavins, phosphorylcholine and ergothioneine (Hafez 1977).

The seminal vesicles play no role in the storage of sperm but are however responsible for the final contribution to the seminal plasma. This assists in expelling and diluting the spermatozoa into the urethra and helping them to become mobile. The alkalinity of the secretions functions as a buffer against the acidic environment that spermatozoa will encounter in the female reproductive tract (WHO 1992; Sherwood 2007). However, the primary role of the seminal vesicles is to provide high concentrations of fructose to the seminal plasma. The normal value of fructose is 13 µmol or more per ejaculate and is vital to the functional integrity of spermatozoa as it is the major source of glycolytic energy in order to maintain motility (WHO 1992).

Another function of the seminal vesicles is to secrete fibrinogen, a precursor of the molecule fibrin. Fibrinogen interacts with enzymes produced by the prostate, ultimately resulting in the 'clotting' of semen. This ensures that following coitus and the extraction of the penis, the semen remains in the female reproductive tract (Sherwood 2007).

An important role of the seminal vesicles is to produce prostaglandins. Prostaglandins were first discovered in semen and are found in abundance in human seminal plasma. Originally, the prostate was believed to be responsible for producing acidic lipids known as prostaglandins, hence how the name is derived. However, it is now known that prostaglandins originate from the seminal vesicles. Prostaglandin F2α (PGF2α) is the predominant form of prostaglandins in semen and males presenting with infertility display significantly decreased concentrations of PGF2α (Hafez 1977). Prostaglandins are one of the most biologically active compounds known with numerous pharmacological effects, including the stimulation of smooth muscle. The prostaglandins produced by the seminal vesicles influence both the female and male reproductive systems and are responsible for promoting the transport of sperm. Prostaglandins stimulate the smooth muscle of the male reproductive tract during the process of ejaculation and the resultant contractile action promotes the movement of sperm (Hafez 1977; WHO 1992). In the female reproductive system, seminal prostaglandins cause the uterus and vagina to contract following sexual intercourse which also facilitates the transport of sperm towards the site of fertilisation (Hafez 1976; Sherwood 2007).

9.4.4 Pathologies

Certain abnormalities in semen parameters can be indicative of dysfunctional or overactive seminal vesicles. Sex gland secretions play a vital role in providing spermatozoa with nourishment and a suitable environment, and alterations in the secretory patterns can be detrimental to viability and motility. Conditions such as raised concentrations of zinc and fructose as well as an increase in the volume (>6 mL) of semen are symptomatic of a glandular dysfunction in the seminal vesicles. Diabetic men who often suffer from infertility present semen which can have a fructose concentration that is almost double that of a non-diabetic ejaculate (Hafez 1976, 1977).

The seminal vesicles are androgenic dependent and the primary androgen responsible for maintaining the normal function of the seminal vesicles is a derivative of testosterone: dihydrotestosterone (DHT). The absence of the anabolic steroid can be detrimental to the reproductive system, possibly even causing the condition of aspermia (ejaculation failure). This is a result of the deterioration of the secretory epithelia within the seminal vesicles (Hafez 1977; Nieschlag and Behre 2000).

Patients who present with the condition of hypogonadism can undergo testosterone treatment. This form of therapy can result in the seminal vesicles as well as the prostate enlarging and hence normal functioning of the glands is resumed.

9.5 Conclusions

The male gametes are produced in the testis, but ultimately released during the process of ejaculation in order to enter the female reproductive tract. During this stage it is vital that the male accessory sex glands release their secretions in an orderly manner, allowing for the facilitation of sperm movement, nourishment and protection, which may eventually lead to the successful fertilisation of the oocyte.

References

Awakura Y, Nonomura M, Fukuyama T 2000. Cowper's syringocele causing voiding disturbance in an adult. *Int J Urol* 7:340–2.

Bevers RF, Abbekerk EM, Boon TA 2000. Cowper's syringocele: Symptoms, classification and treatment of an unappreciated problem. *J Urol* 163:782–4.

Birnstingl MA, Griffiths JD, Nicol CS, Redmond A 1957. Two cases of perineal fistula following cowperitis. *Br J Vener Dis* 33:246–8.

Brocker C 1998. *Human Structure and Function: Nursing Applications in Clinical Practice.* Mosby, London.

Chughtai B, Sawas A, O'Malley RL, Naik RR, Ali Khan S, Pentyala S 2005. A neglected gland: A review of Cowper's gland. *Int J Androl* 28:74–7.

Cina SJ, Silberman MA, Kahane H, Epstein JI 1997. Diagnosis of Cowper's glands on prostate needle biopsy. *Am J Surg Pathol* 21:550–5.

Colodny AH, Lebowitz RL 1978. Lesions of Cowper's ducts and glands in infants and children. *Urology* 11:321–5.

Dunker N, Aumuller G 2002. Transforming growth factor-beta 2 heterozygous mutant mice exhibit Cowper's gland hyperplasia and cystic dilations of the gland ducts (Cowper's syringoceles). *J Anat* 201:173–83.

Fritjofsson A, Kvist U, Ronquist G 1988. Anatomy of the prostate. Aspects of the secretory function in relation to lobar structure. *Scand J Urol Nephrol Suppl* 107:5–13.

Ganong WF 1981. *Review of Medical Physiology*, 10 edn. Lange Medical Publications, Los Altos.

Hafez ESE 1976. *Human Semen and Fertility Regulation in Men.* The C.V. Mosby, Company, London.

Hafez ESE 1977. *Techniques of Human Andrology. Human Reproductive Medicine*, vol 1. North-Holland Biomedical Press, Amsterdam.

Heath JW, Young B 2000. *Wheater's Functional Histology*, 4 edn. Churchill Livingstone, London.

Hellgren L, Mylius E, Vincent J 1982. The ultrastructure of the human bulbo-urethral gland. *J Submicrosc Cytol* 14:683–9.

Honda GD, Bernstein L, Ross RK, Greenland S, Gerkins V, Henderson BE 1988. Vasectomy, cigarette smoking, and age at first sexual intercourse as risk factors for prostate cancer in middle-aged men. *Br J Cancer* 57:326–31.

Jennet S 1989. *Human Physiology*. Churchill Livingstone, London.
Kickuth R, Laufer U, Pannek J, Kirchner TH, Herbe E, Kirchner J 2002. Cowper's syringocele: Diagnosis based on MRI findings. *Pediatr Radiol* 32:56–8.
King GJ 1993. *Reproduction in Domesticated Animals*. Amsterdam Elsevier, Amsterdam.
Lalani E-N, Laniado ME, Abel PD 1997. Molecular and cellular biology of prostate cancer. *Cancer Metastasis Rev* 16:29–66.
Lwaleed BA, Greenfield R, Stewart A, Birch B, Cooper AJ 2004. Seminal clotting and fibrinolytic balance: A possible physiological role in the male reproductive system. *Thromb Haemost* 92:752–66.
Masson JC, Suhler A, Garbay B 1979. Cowper's canals and glands. Pathological manifestations and radiologic aspects. *J Urol Nephrol (Paris)* 85:497–511.
Merchant SA, Amonkar PP, Patil JA 1997. Imperforate syringoceles of the bulbourethral duct: Appearance on urethrography, sonography, and CT. *AJR Am J Roentgenol* 169:823–4.
Migliari R, Riva A, Lantini MS, Melis M, Usai E 1992. Diffuse lymphoid tissue associated with the human bulbourethral gland. An immunohistologic characterization. *J Androl* 13:337–41.
Mortimer S 1993. *Practical Laboratory Andrology*. Oxford University Press, New York.
Nieschlag E, Behre H 2000. *Andrology: Male Reproductive Health and Dysfunction*, 2 edn. Springer, Berlin.
Owen DH, Katz DF 2005. A review of the physical and chemical properties of human semen and the formulation of a semen simulant. *J Androl* 26:459–69.
Pedron P, Traxer O, Haab F, Farres MT, Tligui M, Thibault P, Gattegno B 1997. [Cowper's gland: Anatomic, physiological and pathological aspects]. *Prog Urol* 7:563–9.
Pienta KJ, Esper PS 1993. Risk factors for prostate cancer. *Ann Intern Med* 118:793–803.
Riva A, Usai E, Cossu M, Lantini MS, Scarpa R, Testa-Riva F 1990. Ultrastructure of human bulbourethral glands and of their main excretory ducts. *Arch Androl* 24:177–84.
Robbins SL, Kumar V 1987. *Basic Pathology*, 4 edn. W.B Saunders Company, Philadelphia.
Rui H, Thomassen Y, Oldereid NB, Purvis K 1986. Accessory sex gland function in normal young (20–25 years) and middle-aged (50–55 years) men. *J Androl* 7:93–9.
Saboorian MH, Huffman H, Ashfaq R, Ayala AG, Ro JY 1997. Distinguishing Cowper's glands from neoplastic and pseudoneoplastic lesions of prostate: Immunohistochemical and ultrastructural studies. *Am J Surg Pathol* 21:1069–74.
Sherwood L 2007. *Human Physiology from Cells to Systems*, 7 edn. Brooks/Cole, Belmont.
Silverthorn DU 2009. *Human Physiology: An Integrated Approach*, 5 edn. Pearson, San Francisco.
Stenman UH, Leinonen J, Zhang WM, Finne P 1999. Prostate-specific antigen. *Semin Cancer Biol* 9:83–93.
Thackare H, Nicholson HD, Whittington K 2006. Oxytocin – Its role in male reproduction and new potential therapeutic uses. *Hum Reprod Update* 12:437–48.
Tikva P 2003. Short communication: Does preejaculatory penile secretion originating from Cowper's gland contain sperm? *J Assist Reprod Genet* 20:157–9.
Ullmann HF 2009. *Atlas of Anatomy*. Ullmann, H.F, Munich.
White JM, Amis ES, Jr., Cronan JJ 1983. Cowper's gland calcification: A new radiographic finding. *Urol Radiol* 5:107–11.
WHO 1992. *WHO Laboratory Manuel for the Examination of Human Semen and Sperm-Cervical Mucus Interaction*. Cambridge University Press, Cambridge.
Wilson FJ, Kestenbaum MG, Gibney JA, Matta S 1997. *Histology Image Review*. McGraw Hill-Companies, New York.
Woodruff JD, Friedrich EG, Jr. 1985. The vestibule. *Clin Obstet Gynecol* 28:134–41.

Chapter 10

Reproductive Endocrinology and Contraception in the Male

Mannencheril Rajalakshmi, Pramod Chandra Pal and Radhey Shyam Sharma

Contents

10.1 Physiology of Male Reproduction.....261
10.1.1 Histoarchitecture of the Testis and Spermatogenesis.....261
10.1.2 Leydig Cells.....262
10.1.2.1 Testicular Production of Androgens.....263
10.1.2.2 Autocrine and Paracrine Regulation of Leydig Cell Function.....263
10.1.2.3 Role of Androgens in Spermatogenesis.....263
10.1.2.4 Role of Estradiol as Regulatory Factor in the Testis.....264
10.1.3 Sertoli Cells.....264
10.1.3.1 Role of Sertoli Cell in Maintaining Blood–Testis Barrier.....265
10.1.4 Role of Growth Factors in Testis.....266
10.2 Male Contraception.....267
10.2.1 Methods for Prevention of Sperm Transport: Condoms.....267
10.2.2 Vasectomy and Vaso-Occlusive Devices.....268

10.2.3 Recent Advances in Use of Non-Hormonal Agents for Contraception in Males 268
10.2.4 Hormonal Contraception in Males 268
10.2.4.1 Testosterone Buciclate 270
10.2.4.2 Testosterone Undecanoate (TU) 270
10.2.4.3 Alternate Routes of Androgen Delivery 271
10.2.4.4 Progestogen–Androgen Combination Regimens 272
10.2.4.5 Anti-Androgen–Androgen Combination Regimens 272
10.2.4.6 GnRH Analogues-Androgen Combination Regimens 273
10.2.4.7 Side Effects of Use of Male Hormonal Contraceptives 273
10.3 Acceptability of Contraceptives by Men 275
References 276

The expanding population in India is reported to be adversely impacting the national economy in addition to causing environmental pollution and degradation, migration from rural to urban cities with severe stress on the urban infrastructure and immense human suffering due to poverty, malnutrition and disease prevalence. The Indian population has reached more than 1.2 billion and is still rapidly growing. The results of the last National Family Health Survey 2005–2006 (NFHS-3) (International Institute for Population Sciences (IIPS) and Macro International, 2007) has clearly shown that the total fertility rate (TFR) has declined only marginally to 2.7 from 2.9 children per woman at the time of NFHS-2 conducted nearly 7 years back in 1998–1999 and is still above the replacement level. An interesting finding of the NFHS-3 study is that TFR has reached replacement level of 2.1 in urban areas but is high in rural areas. While knowledge of contraception is nearly universal, as per NFHS-3, only 79% of women and 87% of men know about male contraception. Of the two contraceptive methods available for men, the percentage of men undergoing sterilisation (vasectomy) has declined from NFHS-1 (3.5%) to NFHS-2 (1.9%) and the decline has continued in NFHS-3 (1%). But, a marginal improvement in the use of condoms is seen, from NFHS-1 (2.4%), NFHS-2 (3.1%) and NFHS-3 (5.3%). While the unmet need for family planning of 13.2% has not been fulfilled as per NFHS-3, the figures still indicate that if acceptable, safe and effective methods of male contraception are available to meet the unmet demands for contraceptives, a significant decrease in unplanned pregnancies would occur. These facts highlight the need for developing male contraceptive methods that can be used by men, thus relieving the female partner from the disproportionate responsibility of limiting family size. It is in this context that the need for developing new male contraceptives has assumed importance. In spite of worldwide efforts to develop an effective contraceptive option for men, safe, effective, reversible and affordable methods have not become available since the introduction of condoms. In this chapter, recent advances in male contraception are reviewed along with a resume of the male reproductive processes and male reproductive endocrinology which form the basis for the development of male-based contraceptive regimens.

10.1 Physiology of Male Reproduction

The mammalian testis has compartmentalised its two primary functions of production of androgenic steroid hormones and spermatozoa to distinct cellular entities: Leydig cells and spermatogenic cells. Research in the past decade has shown that these two major functions of the testis are regulated by endocrine factors and local control mechanisms/factors with inter-communication between the two compartments. In this chapter, the current information on these regulatory mechanisms is briefly outlined.

10.1.1 Histoarchitecture of the Testis and Spermatogenesis

The adult human testis, approximately 4.5–5.1 cm long and oval in shape, is located within the scrotal sac and covered on all sides, except at its posterior border, by the thick fibrous capsule called tunica albuginea. The testicular parenchyma in human is divided into lobules by very thin septa called trabeculae, arising from the tunica albuginea, and form the connective tissue support for the blood vessels. Each lobule contains 1–4 highly convoluted seminiferous tubules except at the apex, where the tubules are straight and have only Sertoli cells, narrow abruptly and pass into the system of excurrent ducts called tubuli recti or vasa recta. These converge on the rete testis which is an anastamosing network of tubules. From the upper part of the rete testis, about 6–12 efferent ductules join to form a single canal, called the epididymis, which is closely apposed to the posterior part of the testis and continues at its distal end as the vas deferens. The epididymis can be divided into three major parts: the caput, corpus and cauda epididymides. The vas deferens, after crossing the ureter in the abdominal cavity forms a fusiform enlargement, called the ampulla, where it is joined by the duct of the seminal vesicle, on each side, to form the ejaculatory duct. The ejaculatory duct pierces the body of the prostate at the base of the urinary bladder and opens into the prostatic part of the urethra called the colliculus seminalis.

The seminiferous tubules provide the milieu for the production of spermatozoa by a complex series of divisions of the precursor cells termed as spermatogonia. The seminiferous epithelium has two main types of cells: the germ cells and the supporting cells. The supporting cells include the peritubular cells of the basement membrane which form the myoid layer and the Sertoli cells. The germinal elements are composed of (1) a slowly dividing population of primitive stem cells, (2) the rapidly dividing spermatogonia, (3) spermatocytes which undergo meiosis and (4) spermatids which undergo differentiation to form the spermatozoa.

The gonocytes or primitive germ cells are located in a central position at the time of differentiation of gonad into testes and are called as spermatogonia when they migrate to the periphery of the seminiferous tubules. Mitotic activity in the gonocytes is visible from 7 to 9 years of life of the child. Spermatogonia populate the base of the seminiferous tubule in numbers equal to that of Sertoli cells.

Two major types of cells, the germ cells and Sertoli cells, are present in the seminiferous epithelium. Sertoli cells extend from the basement membrane to the lumen of the seminiferous tubules and have a close morphological and physiological relationship with the germ cells. During spermatogenesis, spermatogonia by a series of divisions form spermatozoa. This process has been divided into four major phases: (1) proliferation and division of spermatogonia to form spermatocytes, (2) formation of spermatids by reduction division (meiosis) of spermatocytes, (3) differentiation of the round spermatid arising from the final division of meiosis into the spermatozoon – a process termed spermiogenesis and (4) spermiation, the process of release of spermatozoon from the Sertoli cell. These are inter-dependent processes but each process is also dependent on regulatory molecules from other components of the testis like Sertoli cells, Leydig cells and peritubular cells. For details of the processes involved in spermatogenesis, the reader may refer to a number of published papers (Clermont, 1966a, 1966b, 1969; Dadoune, 2007; Dym and Clermont, 1970; Ehmcke et al., 2005; Ehmcke and Schlatt, 2006; Ogawa et al., 2005; van Alphen et al., 1988a, 1988b; Yoshida et al., 2007). But, the mechanisms regulating spermatogenesis at the molecular and biochemical levels are not well understood (Mok et al., 2012).

10.1.2 Leydig Cells

Leydig cells are present in the intertubular space between seminiferous tubules. The intertubular space has blood vessels surrounded by Leydig cells, loose connective tissue, fibroblasts, macrophages, lymphocytes and lymphatic sinusoids (Fawcett, 1973). In humans, Leydig cells are of four possible types: fetal, infantile, pubertal and adult (Ariyaratne and Mendes-Handagama, 2000). Fetal Leydig cells disappear by about 3–6 months, along with proliferation of infantile Leydig cells until about 1 year. Infantile Leydig cells continue until 8 years of age and are then replaced by pubertal and adult Leydig cells (Nistal et al., 1986).

The adult human Leydig cell is polygonal, 15–20 μm in diameter and surrounded by a plasma membrane which is thrown frequently into folds or microvilli. The nucleus is large, round or irregularly oval with a thin peripheral rim of heterochromatin broken at the sites of the pores in the nuclear membrane. One or two prominent nucleoli are seen. The cytoplasm of the Leydig cell is characterised by the presence of large amounts of smooth endoplasmic reticulum (Christensen, 1975; de Kretser, 1967), lipid droplets, microtubules, filaments and lysosomes. A strong correlation exists between androgen secretion and the amount of smooth endoplasmic reticulum and Golgi membranes in the Leydig cell (Ewing et al., 1979). The rough endoplasmic reticulum is scattered and seen in patches. Mitochondria vary in size and number and are usually lamellar. Golgi is well developed and consists of 4–6 flattened sacs closely pressed together with small vesicles at their periphery.

10.1.2.1 Testicular Production of Androgens

Leydig cells are the source of almost all of androgens produced by the testis. Cholesterol for steroid biosynthesis is sourced from plasma and also by local production. Cholesterol is transported from outer to inner mitochondrial membrane with the help of StAR protein (Clark et al., 1994; Seedorf et al., 2000), undergoes side chain cleavage and is first converted to pregnenolone (C_{27} side chain cleavage) which is the precursor of all steroid hormones. In the human testis, pregnenolone is hydroxylated to dehydroepiandrosterone by cytochrome P450 17α hydroxylase and to androstenedione; these are converted to testosterone by 17β hydroxysteroid dehydrogenase. Androgen production by the Leydig cells is dependent on luteinising hormone (LH) secreted by the pituitary. LH acts on cholesterol transport to the mitochondria. But in addition to the action of LH, Leydig cell functions are also modulated by intra-testicular paracrine and autocrine factors (Huhtaniemi et al., 1998).

10.1.2.2 Autocrine and Paracrine Regulation of Leydig Cell Function

The control of secretion of androgenic hormones by LH from the pituitary is well known. In addition, several paracrine and autocrine factors have been implicated in the regulation of Leydig cell function.

The observations of Aoki and Fawcett, as early as in 1978, that damage to the spermatogenic cells locally causes hyperplasia or hypertrophy of Leydig cells in the vicinity suggested a paracrine control of these cells. Another observation supporting the paracrine role of Leydig cells was the observation that Leydig cells surrounding seminiferous tubules at stage VII of the cycle (which is androgen dependent) are larger in size (Bergh, 1982). In contrast, Fouguet (1987) observed larger smooth endoplasmic reticulum in Leydig cells around stage VII tubule but did not observe an increase in the size of Leydig cells. Numerous studies have reported that Leydig cell function can be influenced by follicle-stimulating hormone (FSH) by indirect action mediated through the Sertoli cells, possibly through estradiol.

10.1.2.3 Role of Androgens in Spermatogenesis

Sertoli cells have androgen receptors (AR) and androgen is considered to influence spermatogenesis via Sertoli cells (Zirkin, 1993) but the role of androgen on Sertoli cells is not clear. Since androgen receptors have been detected in human spermatogonia and spermatocytes, androgen may directly act on the germ cells (Komori et al., 2007). The AR gene is located on the long arm of the X chromosome at Xq 11–12 (Quigley et al., 1995). Defects in the AR gene are reported to result in testicular feminisation, male infertility and abnormalities of sexual development (Komori et al., 2007). Androgens regulate transcription of transition proteins TP1 and TP2. Thus androgen affects spermatogenesis by action on both germ cells and Sertoli cells.

10.1.2.4 Role of Estradiol as Regulatory Factor in the Testis

Estradiol is considered as one of the regulatory factors which mediates between the Sertoli cell and the Leydig cell. It has been suggested that prepubertal Sertoli cells which synthesise estradiol (Dorrington and Armstrong, 1975) negatively regulate Leydig cell function by acting as a growth suppressor (Moore et al., 1992), thus preventing the early onset of androgen biosynthesis. The aromatase activity shifts to the Leydig cells after puberty (Rommerts and van der Molen, 1989) and estradiol would act as an LH-dependent autocrine regulator. An *in vivo* interaction between Sertoli cells and Leydig cells has been demonstrated in rodents (Vihko et al., 1991). Adult Leydig cells express aromatase P450 and actively synthesise estradiol at a rate much greater than that seen in adult Sertoli cell (Carreau et al., 1999, 2003). Evidence indicate that germ cells also synthesise oestrogen and are likely to be the major source of oestrogens in the male reproductive tract (Carreau et al., 2003). P450 aromatase has been detected in the cytoplasmic droplet of sperm and represents approximately 62% of the total testicular aromatase (Carreau et al., 1999; Levallet and Carreau, 1997; Levallet et al., 1998). However, it must be mentioned that the concentration of oestrogens in peripheral blood of males is low except in the horse in which high levels of estrone sulfate are found.

Evidence to support a major role for oestrogen in adult testis are limited (O'Donnell et al., 2001). Of the two estradiol receptors, ERα is not present in the seminiferous epithelium (Zhou et al., 2002). Though ERβ is found in Sertoli cells and germ cells (Scobie et al., 2002; Zhou et al., 2002), ERβ knockout male testis appear normal and the males are fertile (Krege et al., 1998).

It has been suggested that oestrogen may have subtle functions in the testis, at the level of Leydig cells and possibly at seminiferous epithelium (Hess, 2003).

10.1.3 Sertoli Cells

Sertoli cells provide an immunologically protected intra-luminal environment for the germ cells to undergo various stages of development. Sharpe (1994) stated that without the physical and metabolic support of the Sertoli cells, germ cell differentiation, meiosis and transformation into spermatozoa would not occur. The number of germ cells that can be supported through spermatogenesis is fixed for each Sertoli cell (Orth et al., 1988). Only immature Sertoli cells undergo proliferation. The proliferation of Sertoli cells in all species occurs during two periods of life, first during the fetal or neonatal period and second in the peripubertal period. The factors that determine the number of Sertoli cells include genetic factors (fragile X gene), hormonal factors (FSH, thyroid hormones) and growth factors (Sharpe et al., 2003). The maturation of Sertoli cells from fetal to adult life involves loss of ability to undergo multiplication, formation of Sertoli–Sertoli cell junctions and development of functional capabilities (Sharpe et al., 2003).

The Sertoli cell has distinct cellular boundaries but alter their morphology in relation to the stages of the spermatogenic cycle to accommodate the changes in structure and the movement of germ cells from the base to the surface of the seminiferous epithelium (Parvinen, 1982). The margins of the Sertoli cell undergo transformation to remain associated with the different types of germ cells and the changes taking place in them during different stages of spermatogenesis including spermiation. The characteristic structure of the Sertoli cell is well described and is not repeated here.

10.1.3.1 Role of Sertoli Cell in Maintaining Blood–Testis Barrier

At the time of onset of puberty, Sertoli cells change from the immature proliferative stage to a mature non-proliferative stage. The cytological changes include change in nuclear shape to a tripartite form, increased prominence of nucleolus and formation of tight junctions between adjacent Sertoli cells leading to the formation of a blood–testis barrier. Tracer substances that reach seminiferous epithelium remain in the basal region of the epithelium and rarely penetrate beyond the zone occupied by spermatogonia due to the presence of these specialised junctional complexes (Flickinger and Fawcett, 1967) where the lateral cytoplasmic processes of adjacent Sertoli cells arch over the spermatogonia but below the spermatocytes. As a consequence of the blood–testis barrier, the bioavailability of non-hormonal contraceptives like adjudin is poor (Cheng and Mruk, 2010). These Sertoli cell–Sertoli cell junctions consist of symmetrical specialisations of neighbouring Sertoli cells each having subsurface cisternae of endoplasmic reticulum separated from opposing cell membranes by parallel bundles of filaments (Flickinger and Fawcett, 1967). The distance between membranes is usually ~200A^0 resembling gap junctions or at regular intervals along the junctional complex, the membranes are fused to form tight junctions. In laboratory rodents, there are two compartments to the blood–testis barrier: (1) the adventitial compartment, an incomplete barrier constituted by the myoid cells and (2) the Sertoli cell–Sertoli cell junctions forming a more effective intraepithelial compartment. In primates, the adventitial compartment is absent. The Sertoli cell–Sertoli cell junctional complexes divide the seminiferous epithelium into two compartments: (1) a basal compartment containing spermatogonia and preleptotene and leptotene spermatocytes and (2) an adluminal compartment beyond the level of the tight junctions that sequesters the more differentiated germ cells like more advanced spermatocytes and spermatids. Substances from blood directly reach the basal compartment but due to the occluding junctions, substances have to pass through the Sertoli cell cytoplasm to reach the germ cells in the adluminal compartment.

The actual physiological function of the blood–testis barrier is not known. The processes of meiosis and spermiogenesis occur in an immunologically privileged adluminal compartment, ensured by the presence of Sertoli cell–Sertoli cell junctions (Waites and Gladwell, 1982). But, the barrier is flexible accommodating the needs of the germ cells as they migrate from basal to adluminal compartment. The Sertoli cell is the only cell that extends from the basal to the adluminal compartment.

The Sertoli cell also plays a major role in conferring stability to the seminiferous epithelium. The germ cells are prevented from detachment and extrusion into the lumen of seminiferous tubules by regions of ectoplasmic specialisations of the Sertoli cell that face the surface of the germ cells. These ectoplasmic specialisations are seen facing mid-pachytene spermatocytes and round spermatids but during elongation of spermatids, a covering of the ectoplasmic specialisation is seen around the spermatid head (Russel, 1980). The ectoplasmic specialisations have been implicated in (a) adhesion between Sertoli and germ cells, (b) as structural support, (c) in spermiation and (d) as contractile elements.

Sertoli cells secrete a number of products including androgen-binding protein (ABP), transport proteins like transferrin and ceruloplasmin, acidic glycoprotein, inhibin, activin, growth factors (fibroblast growth factor, somatomedin), H-Y protein, CMB-21, Eppin and anti-Müllerian hormone (Bardin et al., 1988; Josso et al., 1993; Sivashanmugam et al., 2003).

10.1.4 Role of Growth Factors in Testis

Growth factors play an important role in germ cell development by providing a suitable environment for the growth, multiplication and differentiation of germ cells. Clermont (1972) has suggested that the interaction between Sertoli cells and germ cells is mainly unidirectional and that Sertoli cells function as nurse cells, supplying growth factors and nutrients to the developing germ cells; according to this, meiosis, spermiogenesis and spermiation are genetically predetermined. Sertoli cells interact with spermatogonial stem cells and control their proliferation and differentiation through the secretion of specific growth factors like glial cell line-derived neutrotrophic factor (GDNF), fibroblast growth factor 2 (van Disssel-Emiliani et al., 1996) and Sertoli cell transcription factors (Chen and Liu, 2014). But, the studies to understand the paracrine regulation of spermatogonial stem cell (SSC) renewal and their fate are mainly done in rodents, and transcription factors with a role in either SSC self-renewal or differentiation in humans have not been identified (Chen and Liu, 2014). One of the growth factors that mediate specific and distinct events during spermatogenesis is nerve growth factor, a protein which mediates communication of germ cells with Sertoli cells (Ayer-Lelievre et al., 1988). Sertoli cells and germ cells produce fibroblast growth factor-like proteins (Mullaney and Skinner, 1992), and fibroblast growth factor (FGF) receptors are localised in the Sertoli cells (LeMargueresse-Battistoni et al., 1995). Fibroblast growth factors (acidic and basic) act as potent mitogens. Basic FGF is present in bovine and human testis (Story et al., 1988). FGF gene expression is maximum during puberty and is under FSH control (Mullaney and Skinner, 1992).

Sertoli cells also secrete other growth factors like insulin growth factor (IGF) and transforming growth factor (TGF α and β) which may have paracrine/autocrine action on Leydig cell functions (Huhtaniemi et al., 1998), but the physiological role of most of these factors are still to be demonstrated.

Three criteria are considered necessary to establish a paracrine/autocrine role for regulatory factors: (1) presence of receptors and biological action on local cells, (2) local secretion regulated by physiological signals and (3) blockade of the factor or its receptors must modify the function of local cells (Roy et al., 1999). Insulin-like growth factor 1 (IGF-1) and transforming growth factor β1 (TGFβ1) fulfilled the three criteria. Both factors play an autocrine but opposite role, on Leydig cell function. IGF-1 stimulated the transcription of genes encoding Leydig cell differentiated function whereas TGFβ acts as a potent inhibitor. Binding sites for IGF-1 have been found in rat and human Sertoli cells, Leydig cells and pachytene spermatocytes (Borland et al., 1984; Vanelli et al., 1988).

The TGF-β family of growth factors including inhibin and activin act as potent differentiation factors and as mitogens (Vale et al., 1994). Inhibin and activin show essentially antagonistic actions. Inhibin acts as a tumour suppressor in the gonad and inhibits spermatogonial proliferation *in vivo* (Matzuk et al., 1992; Van Dissel-Emiliani et al., 1989), but the precise role of this family of growth factors in testicular physiology is still to be defined.

10.2 Male Contraception

10.2.1 Methods for Prevention of Sperm Transport: Condoms

The only currently available reversible contraceptive option for men is the use of condoms. Gabriele Falloppio was the first to describe a condom which he used as a sheath as a barrier method to prevent syphilis. Condoms coated with microbicides like nonoxynol 9 have been tried to improve efficacy but this was found to lead to increase in HIV infection (Kreiss et al., 1992) due to cell membrane-rupturing action of nonoxynol (Roddy et al., 1993). Condoms have relatively poor contraceptive efficacy. This is due to a variety of factors like improper use, inconsistent use, breakage, slippage, allergic reactions to latex condoms etc. Polyurethane condoms developed to mitigate latex allergy have a slightly higher rate of condom breakage and slippage compared to latex condoms but with similar efficacy rates (Potter and de Villemeur, 2003) and difficulty in usage.

In India, the use of condoms has shown a marginal increase during the three National Family Health Surveys conducted. The latest NFHS-3 survey (International Institute for Population Sciences (IIPS) and Macro International, 2007) showed a 5.3% use of condoms, compared to 3.1% and 2.4% in NFHS-2 and NFHS-1, respectively. Condom use has poor long-term compliance due to a variety of reasons. To improve acceptability, polyurethane condoms have been marketed in place of latex condoms but these have a slightly higher rate of breakage and slippage. In a limited open-label acceptability study comparing lubricated natural rubber latex condom, polyurethane condom and a non-latex styrene ethylene butylene styrene (SEBS) condom, a statistically higher proportion of couples preferred the natural rubber latex condom and the non-latex SEBS condom above the polyurethane condom (Frezieres and Walsh, 2000). Using baseline data of the Measurement, Learning and Evaluation (MLE) project for the Urban Health Initiative in Uttar Pradesh, India, data collected from 6431 currently married men in this state were evaluated regarding current use of contraception and contraceptive method choice. In contrast to the NFHS data, 33.2% of currently married men aged 18–54 years used condoms (Mishra et al., 2014).

10.2.2 Vasectomy and Vaso-Occlusive Devices

Vasectomy is an economical and effective method of preventing sperm transport through the vas deferens, which has been in use for many decades. The 'no scalpel technique' introduced in 1990s has minimised blood loss, risk of infection and duration of hospital stay and is considered an improved technique over conventional vasectomy. While reversal of vasectomy (vasovasostomy) is a feasible option for restoration of fertility, post-reversal pregnancy rates vary from 30% to 60% (Schwingl and Guess, 2000).

The intravasal injection of a bioactive drug called RISUG (formulated by combining styrene maleic anhydride with DMSO) renders spermatozoa incapable of fertilisation. It acts by an electrical charge effect and pH lowering effect (Guha, 2001). A single injection is effective for many years and reversal is possible by a non-invasive technique (Guha, 2001). Phase II and limited Phase III clinical trials showed the pregnancy protection efficacy of RISUG without any serious side effects and RISUG is now undergoing Phase III clinical trials in India.

10.2.3 Recent Advances in Use of Non-Hormonal Agents for Contraception in Males

Adjudin, an analog of lonidamine, is a potent anti-spermatogenic agent and depletes germ cells. The compound disrupts adhesion of spermatids to Sertoli cells resulting in infertility. It acts on apical ectoplasmic specialisation (ES) at interphase between Sertoli cells and steps 8–19 spermatids resulting in premature spermiation; defoliation of almost all germ cells except the spermatogonia/spermatogonial stem cells take place (Mok et al., 2011; Wong et al., 2008). Adjudin initially affects the integrin-based adhesion protein complex at the apical ES but it also impacts other signalling and/or adaptor proteins at the site (Mok et al., 2012). But, due to the blood–testis barrier, the bioavailability is poor. The authors have suggested modification of adjudin to induce transient disruption of the blood–testis barrier.

10.2.4 Hormonal Contraception in Males

Male hormonal contraception (MHC) is based on gonadotropin suppression by the use of steroidal or non-steroidal agents like androgens, progestogens, anti-androgens, GnRH analogues, either alone or in combination with an androgen to suppress spermatogenesis but maintain androgen levels sufficiently for normal sexual functions. While initial clinical trials using a limited number of subjects on androgen alone (daily injection of testosterone propionate) showed uniform azoospermia (Reddy and Rao, 1972), the two large multicentre efficacy trials carried out by the World Health Organization (WHO) using injectable testosterone enanthate showed that (1) the cumulative life table to achieve azoospermia was only 64.5% (WHO, 1990), (2) the subjects showed variable response in achievement of azoospermia within 6 months of exposure to weekly injections of TE and (3) differences in contraceptive efficacy in Caucasian men (60%) compared to Chinese men (91%). Subsequent studies also confirmed the greater contraceptive efficacy of MHC in non-Caucasian men (Arsyad, 1999; WHO, 1993). A second major contraceptive efficacy trial was conducted by WHO (1996) in which TE-induced suppression of sperm count to 3 million/mL was taken as the cut-off limit to enter efficacy phase. However, a small percentage of men failed to achieve azoospermia even when gonadotrophin levels were undetectable (McLachlan et al., 2004).

The results of the first WHO study showed a conception rate of 0.8 per 100 person years (WHO, 1990). In the second study (WHO, 1996), the azoospermic men achieved a conception rate of 0.0 and oligozoospermic men a conception rate of 8.1 per 100 person years but the CIs were wide in the oligozoospermic group (0.1–3 million/mL). These data show that these regimens had a better efficacy than condoms (12 per 100 person years; Trussell and Kost, 1987). But, the worrisome aspects were the high discontinuation rate of 29%–42% and the inability of such high doses of TE to suppress spermatogenesis adequately in a percentage of men who showed sperm counts >3 million/mL and thus have retained fertilising potential. The consequences of supra-physiological levels of blood testosterone induced by repeated injections of TE were also a major area of concern. This issue of side effects of sustained supra-physiological levels of blood testosterone on various parameters was evaluated in a non-human primate model and is described elsewhere in this chapter. These studies highlighted the impracticability of using weekly injections of an androgen for MHC and indicated the need for (1) long-acting androgens and (2) combining androgen with other gonadotrophin-suppressing agents to improve the regimen.

While these studies proved the concept that hormonal suppression of spermatogenesis can be a feasible option for the successful development of a male contraceptive, the availability of long-acting steroids for such use was essential for the implementation of such strategies. These felt needs translated into the development and trial of long-acting androgens and progestogens. The long-acting androgens tested included (1) testosterone buciclate (Rajalakshmi and Ramakrishnan, 1989) used either alone or in combination with a long-acting progestational ester, levonorgestrel butanoate (Rajalakshmi et al., 2000) and (2) testosterone undecanoate (Gu et al., 2004; Zhang et al., 1999).

10.2.4.1 Testosterone Buciclate

The pharmacokinetics and pharmacodynamics of testosterone buciclate (TB), synthesised under the WHO Steroid Synthesis Programme, showed its potentiality for use in male contraception since it was one of the longest-acting testosterone esters available (Rajalakshmi and Ramakrishnan, 1989). It was also effective in suppressing spermatogenesis to azoospermia or oligozoospermia when combined with the long-acting levonorgestrel butanoate (Rajalakshmi et al., 2000) while androgen levels were maintained at the normal range.

10.2.4.2 Testosterone Undecanoate (TU)

Injectable testosterone undecanoate (TU) in tea seed oil has favourable pharmacokinetics in men (Zhang et al., 1998). Limited contraceptive efficacy studies in Chinese men using 1000 mg/4w of TU showed azoospermia could be induced in all 12 men but lowering the dose to 500 mg/4w (Zhang et al., 1999) or 1000 mg at 8-week intervals lowered the efficacy and induced azoospermia in only 67% of men (Gu et al., 2004). A subsequent large contraceptive efficacy trial in 308 Chinese men using TU in tea seed oil at a loading dose of 1000 mg/i.m. and a sustaining dose of 500 mg/i.m. also could not induce azoospermia; the contraceptive efficacy rate was 2.3/100 couple years in azoospermic/oligozoospermic men (Gu et al., 2003). Three per cent of men did not attain azoospermia/oligozoospermia. Two hundred ninety-six of these men continued to use TU injections as a sole means of contraception for one year; one pregnancy was reported. In 6 men, sperm concentration rebounded to >3 million/mL and discontinued from the study. The overall contraceptive efficacy was 97%. At this schedule of injection, supra-physiological levels of testosterone were reported in some men and the lowest levels of testosterone were 31% higher than baseline testosterone levels. Other side effects included weight gain, increase in haemoglobin (9%) and decline in HDL-cholesterol (14%). These effects were reversible. A multicentre phase III contraceptive efficacy clinical trial of injectable TU in 1045 Chinese men, in which 500 mg of TU was injected monthly for 30 months (Gu et al., 2009), 4.8% of participants did not achieve azoospermia or severe oligozoospermia (<1 million/mL) within the 6-month suppression phase. Of the 855 subjects who entered the efficacy phase, 733 completed TU injections and follow-up. Nine pregnancies were reported in 1554.1 person years of exposure in the 24 months of efficacy phase. Cumulative contraceptive failure rate was 1.1 per 100 men. The combined method failure rate was 6.1% (4.8% inadequate suppression and 1.3% post-suppression sperm rebound). Spermatozoa returned to normal fertility range except in two subjects. Serious adverse effects were not seen. However, results of a phase II clinical study using norethisterone enanthate and TU, conducted by WHO and CONRAD, the review panel in early 2011 recommended discontinuation of trial due to risks of possible side effects outweighing potential benefits (reported in Mok et al., 2012).

Taking all these facts into consideration, it is unlikely that universal suppression of sperm output in all men can be achieved by use of male hormonal contraceptives since a minority of men may not achieve complete suppression of sperm production.

10.2.4.3 Alternate Routes of Androgen Delivery

Alternate routes of androgen delivery explored by investigators include transdermal (Ilani et al., 2012) application in scrotal and non-scrotal areas, oral and buccal delivery and subdermal implants. Scrotal application of testosterone patches did not uniformly suppress sperm counts to azoospermic levels (Buchter et al., 1999). Even when combined with progestational agents, azoospermia was attained in less than 50% of subjects (Gonzalo et al., 2002; Hair et al., 2001). Further, skin reactions to the patches were reported in 24% of subjects who withdrew from the study (Hair et al., 2001). Azoospermia/oligozoospermia was reported using testosterone gels in combination with depot medroxyprogesterone acetate (Page et al., 2006) and the gel was found to be acceptable.

A contraceptive efficacy trial in Australia using testosterone implants (800 mg/4 monthly) along with 300 mg medroxyprogesterone acetate (MPA) once in 3 months, with entry criteria of <1 million sperm/mL, reported that sperm counts of <1 million/mL was achieved in 53 of 55 men who entered the contraceptive efficacy phase during which pregnancy was not reported (Turner et al., 2003). The drawbacks in use of testosterone implants included need for expertise to insert implants, ~10% extrusion, infection, bruising and haematoma (Handelsman et al., 1990).

Daily oral use of TE (800 mg) combined with the 5α-reductase inhibitor dutasteride for 28 days significantly suppressed gonadotrophins without untoward side effects. Serum DHT was suppressed and serum estradiol increased during treatment in both groups. High-density lipoprotein cholesterol was suppressed during treatment, but liver function tests, hematocrit, creatinine, mood and sexual functions were unaffected (Amory et al., 2008).

7α-methyl-19-nor-testosterone (MENT), perhaps the first steroidal androgen receptor modulator developed by the Population Council, has the advantage that it does not undergo 5α-reduction, can be aromatised, is not bound to SHBG and is rapidly cleared from circulation (review: Rajalakshmi, 2005). About 25 μg of MENT delivered through silastic implants, induced oligozoospermia in bonnet monkeys but increasing the dose supported spermatogenesis (Ramachandra et al., 2002). MENT has a less prostate stimulating effect than other androgens. MENT implants combined with etonogestrel could not induce effective sperm suppression (Walton et al., 2007). A number of non-steroidal small molecules which may act as selective androgen receptor modulators are undergoing clinical evaluation (Higuchi et al., 2007).

10.2.4.4 Progestogen–Androgen Combination Regimens

Clinical trials conducted since 1965 to identify the most effective combination regimen of progestogen–androgen could not come to a consensus on the dose, schedule or duration of administration or on the identity of the components of the regimen that would induce uniform azoospermia in all subjects on the trial (Rajalakshmi, 2005). Progestogens can be administered orally, by injection or as implants. Progestogen in a combination regimen acts as the primary anti-spermatogenic agent while androgen potentiates the anti-spermatogenic action of the progestogen. Further, only relatively smaller amounts of the androgen need to be used, compared to an androgen-only regimen (Rajalakshmi and Sharma, 2001). Progestins used in efficacy trials include MPA, depot medroxyprogesterone acetate (DMPA), norethindrone, norethisterone enanthate, levonorgestrel, desogestrel and etonogestrel in combination with TE/TU/testosterone cypionate/methyl testosterone, 19-nor-testosterone or testosterone decanoate (TD). The clinical data using these regimens have been extensively reviewed (Amory, 2009; Matthiesson and McLachlan, 2006; Page et al., 2008; Rajalakshmi, 2005; Rajalakshmi and Sharma, 2001). The earlier studies stressed the importance on achieving a greater degree of sperm suppression for a longer duration. The emphasis shifted later to reducing the steroid dose and use of alternate modalities of drug delivery. While these studies established the proof of concept that MHC based on a progestogen–androgen combination is feasible, these exploratory trials are based usually on a small number of subjects; their power to detect important differences are limited and their results imprecise, with imprecise or inconsistent definitions of oligozoospermia (Grimes et al., 2007). These studies also underlined the ethnic differences in response to male hormonal contraceptives. The results of a recent multicentre trial using 750–1000 mg of TU every 10–12 weeks combined with two doses of etonogestrel implants for 42–44 weeks in 354 men randomised to 1–7 groups showed that 3% of men (largely Caucasian-European) could not attain sperm count <1 million/mL, even after 42 weeks of treatment (Mommers et al., 2008). About 6% of all treated men ($n = 17$) showed rebound sperm output. This has raised concern about possible unprotected pregnancies in such couples who would be relying only on this regimen as the sole method of contraception. Further, safety data showed a significant doubling of the frequency of adverse effects (increased acne, weight gain, changes in mood/libido) compared to placebo controls. Before a male hormonal contraceptive can be marketed, large-scale Phase III studies with pregnancy prevention as the end point need to be carried out.

10.2.4.5 Anti-Androgen–Androgen Combination Regimens

Cyproterone acetate (CPA), a powerful anti-androgen, underwent clinical trials in the 1970s. It induced marked suppression of spermatogenesis but also inhibited androgen action by competitive binding of testosterone and dihydrotestosterone to androgen receptors. It is also a progestational compound and has direct effects on testis. Hence, use of CPA in MHC requires simultaneous administration of an androgen. CPA given along with TE or testosterone buciclate (TB) suppressed spermatogenesis in non-human primates (Lohiya et al., 1987; Sharma et al., 2000). In clinical trials orally administered CPA with TE/TU caused rapid arrest of spermatogenesis (Merrigiola et al., 1996, 1997). CPA did not induce any adverse effects (Sharma et al., 2000, 2001). CPA, if available for use as a male hormonal contraceptive, would have been a promising compound in combination with a suitable androgen for clinical use as a contraceptive agent.

10.2.4.6 GnRH Analogues-Androgen Combination Regimens

GnRH agonists and antagonists have been tested for male contraception but only antagonists were capable of suppressing sperm concentration for prolonged periods (Matthiesson and McLachlan, 2006; Rajalakshmi, 2005). A number of GnRH antagonists were evaluated in primate models and men; uniform azoospermia could not be induced either in non-human primates or in the majority of clinical trials. Administration of Cetrolix combined with 19-nor-testosterone could induce azoospermia in the limited subjects tested in 12 weeks (Behre et al., 2001). In both non-human primates and men, complete suppression of spermatogenesis to azoospermia/severe oligozoospermia occurred by simultaneous administration of GnRH antagonist and androgen (Bremner et al., 1991; Rajalakshmi et al., 1995).

The currently available GnRH antagonists are expensive. Effective use of these drugs for contraception will depend on their availability at costs that can be affordable in developing countries. Studies in non-human primates and in men also have shown weight reduction in GnRH antagonist exposed animals/subjects which would be a cause for concern for use of this regimen in developing countries with low nutritional status.

10.2.4.7 Side Effects of Use of Male Hormonal Contraceptives

The development of a hormonal male contraceptive is clearly dependent on androgen supplementation. The requirements of an ideal androgen for male fertility regulation have been described earlier (Rajalakshmi and Bajaj, 1999) and are (1) the ability to induce spermatogenic arrest in one cycle or even less, (2) the recovery of spermatogenesis in one cycle or even less with the presence of normal spermatozoa possessing full fertilising potential in the ejaculate, (3) the maintenance of steady-state physiological levels of circulating androgens, (4) the absence of adverse effects on libido and sexual functions, (5) the absence of serious or major adverse side effects on systemic and metabolic functions when given for at least 2 years, (6) ease of administration, (7) long acting requiring infrequent administration thereby increasing practicability and acceptability and (8) affordable so that the introduction of the method into the family planning programmes of developing countries does not require major financial input.

The maintenance of physiological androgen levels while using an MHC is important to avoid androgen-deprivation effects like fatigue, loss of muscle, low libido, mood changes, anemia, adverse effects on lipid profile and loss of bone mineral density. Excess androgen-induced effects also should be avoided including acne, fluid retention, weight gain, raised haematocrit, mood changes, gynaecomastia, prostate hypertrophy/hyperplasia or adverse effects on lipid profile.

A critical evaluation of the safety of androgens, when used for male contraception, is of great importance. Androgens cause major shifts in blood lipoprotein fractions (Bajaj et al., 1989; Plymate and Swerdloff, 1992) and act as a risk factor for ischemic heart disease (Bajaj et al., 1989). Plasma high-density lipoprotein (HDL) cholesterol is decreased in conditions associated with ischemic heart disease and preceded the occurrence of coronary heart disease (Miller and Miller, 1975). Further, a high incidence of cardiovascular disease was linked to high levels of triglycerides in blood and serum levels of very light-density lipoprotein cholesterol (Lipid Research Clinics Coronary Primary Prevention Trial Results, 1984).

A non-human primate study was done in adult male rhesus monkeys kept under controlled dietary conditions to investigate the side effects of repeated long-term use of TE, an androgen used extensively in clinical practice. Adult male rhesus monkeys were injected with 50 mg of TE once in 14 days, for ~32 months while a group of control animals received olive oil injections. In TE-injected animals, serum testosterone levels at 10.00 h increased sharply to supra-physiological range within 24 h of first injection and peaked on day 3 followed by gradual decrease until day 14 (Tyagi et al., 1999a). A second injection of TE increased serum testosterone levels to peak values on day 1 followed by decrease until day 14 and remained at this supra-physiological range by subsequent injections of TE. In TE-injected animals, HDL-cholesterol levels decreased gradually but significantly from the 19th month of injection until the 1st month of recovery. Compared to control olive oil-injected animals, low-density lipoprotein (LDL) cholesterol levels and the LDL/HDL cholesterol ratio increased significantly from the 12th month until the end of the treatment period, followed by a return to baseline values by the end of the recovery period (Tyagi et al., 1999b). TE injections did not change the levels of alkaline phosphatase or bilirubin, but serum transaminase (SGOT and SGPT) levels increased (Tyagi et al., 1999b). The weight of cranial and caudal lobes of prostate in the TE-injected animals increased along with increase in secretory activity, cellular hypertrophy, increase in fibromuscular stroma, marked increase in prostatic acid phosphatase, marginal increase in prostate specific antigen and increase in DNA index in the caudal lobe of the prostate (Udayakumar et al., 1998). This non-human primate study is the only study conducted so far in which an androgen has been administered resulting in supra-physiological androgen levels for long duration whereas clinical data that has emanated are from studies involving small number of healthy men on MHC for a relatively short duration. The primate study involved the use of an equal number of vehicle-treated control animals.

The results of this study provide convincing evidence that long-term androgen use decreased HDL-cholesterol levels and increased LDL/HDL cholesterol ratio. Similar data has been reported in men on hormone therapy (Anderson et al., 1995; Handelsman et al., 1992; Wu et al., 1996). In the WHO efficacy trial, HDL cholesterol levels decreased by 13% and weight increased by 4 kg (WHO, 1996). Physiological doses of testosterone, combined with a progestin adversely affected HDL-cholesterol (Bebb et al., 1996). The route of administration of the steroid also influences metabolic effects induced by steroids. Low HDL-cholesterol levels impair the clearance of cholesterol from the arterial walls, accelerate ischemic heart disease and have been associated with pathogenesis of coronary atherosclerosis (Miller and Miller, 1975). MPA is reported to increase cardiovascular hyperactivity probably acting via an increase in thromboxane prostanoid receptor expression in the vasculature with a proposed predisposition to cardiovascular hyperactivity mediated myocardial ischemia (Mishra et al., 2005).

The efficacy trials using various male hormonal contraceptives have been of relatively short duration whereas the pathogenesis of coronary artery disease is of long duration. Hence, it is imperative that men on MHC are monitored long term.

In contrast to the large number of studies related to androgen use on lipid metabolism, information available on the adverse effects of androgen administration on liver are few. An increase in SGOT and SGPT levels was reported only in ethnic Chinese men who were administered 200 mg TE in a prospective multinational contraceptive efficacy study conducted by WHO (Wu et al., 1996).

10.3 Acceptability of Contraceptives by Men

The last National Family Health Survey conducted in India recorded that 98% of women and 99% of men of age 15–49 are aware of one or more methods of contraception. While 94% of women and men know about female sterilisation, only 79% women and 87% men know about male sterilisation and 93% of men know about condoms, compared to 74% women (International Institute for Population Sciences (IIPS) and Macro International, 2007).

There have been misconceptions about the acceptability of male-based contraceptive methods. A four-continent survey showed that the majority of men indicated acceptability of use of a male contraceptive, if available (Heinemann et al., 2005a, 2005b) and this is further supported by other surveys (Glasier et al., 2000; Martin et al., 2000) to a certain extent. Some of these surveys also showed that oral forms of male contraceptive is more favoured (Heinemann et al., 2005b). Similar information from developing countries also is needed.

References

Amory JK. 2009. Progress and prospects in male hormonal contraception. *Current Opinion in Endocrinology, Diabetes, Obesity*, 15: 255–260.

Amory JK, Kalhorn TF and Page ST. 2008. Pharmacokinetics and pharmacodynamics of oral testosterone enanthate plus dutasteride for 4 weeks in normal men: Implications for male hormonal contraception. *Journal of Andrology*, 29: 260–271.

Anderson RA, Wallace EM, Wu FCW. 1995. Effect of testosterone enanthate on serum lipoproteins in man. *Contraception*, 52: 115–119.

Aoki A and Fawcett DW. 1978. Is there a local feedback from the seminiferous tubules affecting activity of the Leydig cells? *Biology of Reproduction*, 19: 144–158.

Ariyaratne HB and Mendis-Handagama CS. 2000. Changes in the testis interstitium of Sprague Dawley rats from birth to sexual maturity. *Biology of Reproduction*, 62: 680–690.

Arsyad KM. 1999. Progestogen-testosterone combinations: Results of Indonesian clinical trials. In: *Male Contraception: Present and Future*, Eds. Rajalakshmi M and Griffin PD. New Age International Publishers, New Delhi.

Ayer-Lelievre C, Olson L, Ebendal T, Hallbrook F and Persson H. 1988. Nerve growth factor mRNA and protein in the testis and epididymis of mouse and rat. *Proceedings of the National Academy of Sciences, USA*, 85: 2628–2632.

Bajaj JS, Rajalakshmi M and Madan R. 1989. Metabolic effects of contraceptive steroids. In: *Safety Requirements for Contraceptive Steroids*, Ed. Michal F. Cambridge University Press, New York, 360–384.

Bardin CW, Cheng CY, Musto NA and Gunsalus GL. 1988. The Sertoli cell. In: *The Physiology of Reproduction*, 2nd edn., Eds. Knobil E and Neill J. Raven Press, New York, 933–974.

Bebb RA, Anawalt BD, Christensen RB, Paulsen CA, Bremner WJ and Matsumoto AM. 1996. Combined administration of levonorgestrel and testosterone induces more rapid and effective suppression of spermatogenesis than testosterone alone: A promising male contraceptive approach. *Journal of Clinical Endocrinology and Metabolism*, 81: 757–762.

Behre HM, Kliesch S, Lemcke B, von Eckardstein S and Nieschlag E. 2001. Suppression of spermatogenesis to azoospermia by combined administration of GnRH antagonist and 19-nortestosterone cannot be maintained by this non-aromatizable androgen alone. *Human Reproduction*, 16: 2570–2577.

Bergh, A. 1982. Local differences in Leydig cell morphology in the adult rat testis: Evidence for a local control of Leydig cells by adjacent seminiferous tubules. *International Journal of Andrology*, 5: 325–330.

Borland K, Mita M, Oppenheimer CL, Blinderman LA, Massague J, Hall PF and Czech MP. 1984. The actions of insulin-like growth factors I and II on cultured Sertoli cells. *Endocrinology*, 114: 240–246.

Bremner WJ, Bagatell CJ and Steiner RA. 1991. Gonadotropin releasing hormone antagonist plus testosterone: A potential male contraceptive. *Journal of Clinical Endocrinology and Metabolism*, 73: 465–469.

Buchter D, von Eckardstein S, von Eckardstein A, Kamischke A, Simoni M, Behre HM and Nieschlag E. 1999. Clinical trial of transdermal testosterone and oral levonorgestrel for male contraception. *Journal of Clinical Endocrinology and Metabolism*, 84: 1244–1249.

Carreau S, Genissel C, Bilinska B and Levallet J. 1999. Sources of estrogen in the testis and reproductive tract of the male. *International Journal of Andrology*, 22: 211–223.

Carreau S, Lambard S, Delalande C, Denis-Galeraud I, Bilinska B and Bourguiba S. 2003. Aromatase expression and role of estrogens in male gonad: A review. *Reproductive Biology and Endocrinology*, 1: 35–42.

Chen SR and Liu YX. 2014. Regulation of spermatogonial stem cell renewal and spermatocyte meiosis. *Reproduction*, December 12, pii: REP-14–0481.

Cheng CY and Mruk DD. 2010. New frontiers in non-hormonal male contraception. *Contraception*, 82: 476–482.

Christensen AK. 1975. Leydig cells. In: *Handbook of Physiology*, section 7, *Endocrinology*, vol. 5, *Male Reproductive System*, Eds. Hamilton DW and Greep RO. Williams and Wilkens, Baltimore, 21–55.

Clark BJ, Wells J, King SR and Stocco DM. 1994. The purification, cloning, and expression of a novel luteinising hormone-induced mitochondrial protein in MA-10 mouse Leydig tumour cells. Characterization of steroidogenic acute regulatory protein (StAR). *Journal of Biological Chemistry*, 269: 28314–28322.

Clermont Y. 1966a. Renewal of spermatogonia in man. *American Journal of Anatomy*, 118: 509–524.

Clermont Y. 1966b. Spermatogenesis in man: A study of the spermatogonial population. *Fertility and Sterility*, 17: 705–721.

Clermont Y. 1969. Two classes of spermatogonial stem cells in the monkey (Cercopithecus aethiops). *American Journal of Anatomy*, 126: 57–71.

Clermont Y. 1972. Kinetics of spermatogenesis in mammals: Seminiferous epithelium cycle and spermatogonial renewal. *Physiological Reviews*, 52: 198–235.

Dadoune JP. 2007. New insights into male gametogenesis: What about the spermatogonial stem cell niche? *Folia Histochem et Cytobiol*, 45: 141–147.

de Kretser DM. 1967. The fine structure of the testicular interstitial cells in men of normal androgenic status. *Z Zellforschung*, 80: 594–609.

Dorrington JF and Armstrong DT. 1975. Follicle-stimulating hormone stimulates estradiol-17B synthesis in cultured Sertoli cells. *Proceedings of the National Academy of Sciences USA*, 72: 2677–2681.

Dym M and Clermont Y. 1970. Role of spermatogonia in the repair of the seminiferous epithelium following X-irradiation of the rat testis. *American Journal of Anatomy*, 128: 265–282.

Ehmcke J and Schlatt S. 2006. A revised model for spermatogonial expansion in man: Lessons from non-human primates. *Reproduction*, 132: 673–680.

Ehmcke J, Simorangkir DR and Schlatt S. 2005. Identification of the starting point for spermatogenesis and characterization of the testicular stem cell in adult male rhesus monkeys. *Human Reproduction*, 20: 1185–1193.

Ewing LL, Zirkin BR, Cochran RC, Kromann N, Peters C and Ruiz-Bravo N. 1979. Testosterone secretion by rat, rabbit, guinea pig, dog and hamster testes perfused in vitro: Correlation with Leydig cell mass. *Endocrinology*, 105: 1135–1142.

Fawcett DW. 1973. Observations on the organization of the interstitial tissue of the testis and on the occluding cell junctions in the seminiferous epithelium. *Advances in BioSciences*, 16: 83–99.

Flickinger C and Fawcett DW. 1967. Junctional specializations of the Sertoli cells in the seminiferous epithelium. *The Anatomical Record*, 158: 207–222.

Fouguet JP. 1987. Ultrastructural analysis of a local regulation of Leydig cells in the adult monkey *(Macaca fascicularis*) and rat. *Journal of Reproduction and Fertility*, 79: 49–56.

Frezieres RG and Walsh TL. 2000. Acceptability evaluation of a natural rubber latex, a polyurethane, and a new non-latex condom. *Contraception*, 61: 369–377.

Glasier AF, Anakwe R, Everington D, martin CW, van der Spuy Z, Cheng L, Ho PC and Anderson RA. 2000. Would women trust their partners to use a male pill? *Human Reproduction*, 15: 646–649.

Gonzalo IT, Swerdloff RS, Nelson AL, Clevenger B, Garcia R, Berman N and Wang C. 2002. Levonorgestrel implants (Norplant II) for male contraception clinical trials: Combination with transdermal and injectable testosterone. *Journal of Clinical Endocrinology and Metabolism*, 87: 3562–3572.

Grimes DA, Lopez LM, Gallo MF, Halpern V, Nanda K and Schulz KF. 2007. Steroid hormones for contraception in men. *Cochrane Database System Review*, 18: CD004316.

Gu Y, Liang X, Wu W, Liu M, Song S, Cheng L, Bo L et al. 2009. Multicenter contraceptive efficacy trial of injectable testosterone undecanoate in Chinese men. *Journal of Clinical Endocrinology and Metabolism*, 94: 1910–1915.

Gu YQ, Tong JS, Ma DZ, Wang XH, Yuan D, Tang WH and Bremner WJ. 2004. Male hormonal contraception: Effects of injections of testosterone undecanoate and depot medroxyprogesterone acetate at eight-week intervals in Chinese men. *Journal of Clinical Endocrinology and Metabolism*, 89: 2254–2262.

Gu YQ, Wang XH, Xu D, Peng L, Cheng LF, Huang MK, Huang ZJ and Zhang GY. 2003. A multicenter contraceptive efficacy study of injectable testosterone undecanoate in healthy Chinese men. *Journal of Clinical Endocrinology and Metabolism*, 88: 562–568.

Guha SK. 2001. Towards an injectable contraceptive for the male. In: *Current Status in Fertility Regulation: Indigenous and Modern Approaches*, Eds. Chowdhury SR, Gupta CM and Kamboj VP. Central Drug Research Institute, Lucknow, 225–232.

Hair WM, Kitteridge K, O'Conner DB and Wu FC. 2001. A novel male contraceptive pill-patch combination: Oral desogestrel and transdermal testosterone in the suppression of spermatogenesis in normal men. *Journal of Clinical Endocrinology and Metabolism*, 86: 5201–5209.

Handelsman DJ, Conway AJ and Boylan LM. 1990. Pharmacokinetics and pharmacodynamics of testosterone pellets in man. *Journal of Clinical Endocrinology and Metabolism*, 71: 216–222.

Handelsman DJ, Conway AJ and Boylan LM. 1992. Suppression of human spermatogenesis by testosterone implants. *Journal of Clinical Endocrinology and Metabolism*, 75; 1326–1332.

Heinemann K, Saad F, Wiesemes M and Heinemann LA. 2005a. Expectations toward a novel male fertility control method and potential user types: Results of a multinational survey. *Journal of Andrology*, 26: 155–162.

Heinemann K, Saad F, Wiesemes M, White S and Heinemann LA. 2005b. Attitudes towards male fertility control: Results of a multinational survey on four continents. *Human Reproduction*, 20: 549–556.

Hess RA. 2003. Estrogen in the adult male reproductive tract: A review. *Reproductive Biology and Endocrinology*, 1: 1–52.

Higuchi Arienti KL, López FJ, Mani NS, Mais DE, Caferro TR, Long YO, Jones TK et al. 2007. Novel series of potent, nonsteroidal, selective androgen receptor modulators based on 7H-[1,4]oxazino[3,2-g]quinolin-7-ones.*Journal of Medical Chemistry*, 50: 2486–2496.

Huhtaniemi I, El-Hefnawy, Zhang FP, Markkula M and Toppari J. 1998. Paracrine and autocrine control of Leydig cell function. In: *Male Contraception: Present and Future*, Eds. Rajalakshmi M and Griffin PD. New Age international (P) Ltd., New Delhi.

Ilani N, Roth MY, Amory JK, Swerdloff RS, Dart C, Page ST, Bremner WJ et al. 2012. A new combination of testosterone and Norethisterone transdermal gels for male hormonal contraception. *Journal of Clinical Endocrinology and Metabolism*, 97: 3476–86.

International Institute for Population Sciences (IIPS) and Macro International. 2007. National Family Health Survey (NFHS-3), 2005–2006: India: Volume I. Mumbai: IIPS.

Josso N, Cate RL, Vigier JY, di Clemente N, Wilson C, Imbeaud S, Pepinsky RB, Guerrier D, Boussin L et al. 1993. Anti-Müllerian hormone, the Jost factor. *Recent Progress in Hormone Research*, 48: 1–59.

Komori S, Kasumi H, Sakata K and Koyama K. 2007. The role of androgens in spermatogenesis. In: *Gamete Biology, Emerging Frontiers in Fertility and Contraceptive Development*, Eds. Gupta SK, Koyama K and Murray JF. Nottingham University press, UK, 25–30.

Krege JH, Hodgin JB, Couse JF, Enmark E, Warner M, Mahler JF, Sar M, Korach KS, Gustafsson JA and Smithies O. 1998. Generation and reproductive phenotypes of mice lacking estrogen receptor beta. *Proceedings of the National Academy of Sciences USA*, 95: 15677–82.

Kreiss J, Ngugi E, Holmes K, Ndinya-Achola J, Waiyaki P, Roberts PL, Ruminjo I, Sajabal R, Kimata J and Fleming TR. 1992. Efficacy of nonoxynol 9 contraceptive sponge use in preventing heterosexual acquisition of HIV in Nairobi prostitutes. *JAMA*, 268: 477.

LeMargueresse-Battistoni B, Wolff J, Morera A-M and Benahmed M. 1995. Fibroblast growth factor receptor type 1 expression during rat testicular development and its regulation in cultured Sertoli cells. *Endocrinology*, 135: 2404–2411.

Levallet J, Bilinska B, Mittre H, Genissel C, Fresnel J and Carreau S. 1998. Expression and immunolocalisation of functional cytochrome P450 aromatase in mature rat testicular cells. *Biology of Reproduction*, 58: 919–926.

Levallet J and Carreau S. 1997. In vitro gene expression of aromatase in rat testicular cells. *C R Academy of Sciences III*, 320: 123–129.

Lipid Research Clinics Coronary Primary Prevention Trial Results. II. 1984. The relationship of reduction in incidence of coronary heart disease to cholesterol lowering. *Journal of American Medical Association*, 25: 365–374.

Lohiya NK, Sharma OP, Sharma RC and Sharma RS. 1987. Reversible sterility by cyproterone acetate plus testosterone enanthate in langur monkey with maintenance of libido. *Biomedica Biochimica Acta*, 46, 259–266.

Matthiesson KL and McLachlan RI. 2006. Male hormonal contraception: Concept proven, product in sight? *Human Reproduction Update*, 12: 463–482.

Martin CW, Anderson RA, Cheng L, Ho PC, van der Spuy Z, Smith KB, Glasier AF, Everington D and Baird DT. 2000. Potential impact of hormonal male contraception: Cross cultural implications for development of novel preparations. *Human Reproduction*, 15: 637–645.

Matzuk MM, Finegold MJ, Su JG, Hsueh AJ and Bradley A. 1992. Alpha inhibin is a tumour suppressor gene with gonadal specificity in mice. *Nature*, 360: 313–319.

McLachlan RI, Robertson DM, Pruysers E, Ugoni A, Matsumoto AM, Anawalt BD, Bremner WJ and Merrigiola C. 2004. Relationship between serum gonadotropins and spermatogenic suppression in men undergoing steroidal contraceptive treatment. *Journal of Clinical Endocrinology and Metabolism*, 89: 142–149.

Merrigiola MC, Bremner WJ, Constantino A, Pavani A, Capelli M and Flamigni CA. 1997. An oral regimen of cyproterone acetate and testosterone undecanoate for spermatogenic suppression in men. *Fertility and Sterility*, 68: 844–850.

Merrigiola MC, Bremner WJ, Paulsen CA, Valdiserri A, Incorvaia L, Mota R, Pavani A, Capelli M and Flamigni CA. 1996. A combined regimen of cyproterone acetate and testosterone enanthate as a potentially highly effective male contraceptive. *Journal of Clinical Endocrinology and Metabolism*, 81: 3018–3023.

Miller GJ and Miller NE. 1975. Plasma high density lipoprotein concentration and development of ischemic heart disease. *Lancet*, 1: 16–19.

Mishra A, Nanda P, Speizer IS, Calhoun LM, Zimmerman A and Bhardwaj R. 2014. Men's attitudes on gender equality and their contraceptive use in Uttar Pradesh India. *Reproductive Health*, 11: 41. doi: 10.1186/1742–4755-11-41.

Mishra RG, Hermsmeyer RK, Miyagawa K, Sarrel P, Uchida B, Stanczyk FZ, Burry KA, Illingworth DR and Nordt FJ. 2005. Medroxyprogesterone and dihydrotestosterone induce coronary hyperactivity in intact male rhesus monkeys. *Journal of Clinical Endocrinology and Metabolism*, 90: 3706–3714.

Mok KW, Lie PPY, Mruk DD, Mannu J, Mathur PP, Silvestrini B and Cheng CY. 2012. The apical ectoplasmic specialization-blood-testis barrier functional axis is a novel target for male contraception. *Advances in Experimental Medicine and Biology*, 763: 334–355.

Mok KW, Mruk DD, Lee WM, Cheng CY. 2011. Spermatogonial stem cells alone are not sufficient to re-initiate spermatogenesis in the rat testis following adjudin-induced infertility. *International Journal of Andrology*, 35: 86–101.

Mommers E, Kersemaekers WM, Elliesen J, Kepers M, Apter D, Behre HM, Beynon J et al. 2008. Male hormonal contraception: A double-blind, placebo-controlled study. *Journal of Clinical Endocrinology and Metabolism*, 93: 2572–2580.

Moore A, Findlay JK and Morris ID. 1992. In vitro DNA synthesis in Leydig and other interstitial cells of the rat testis. *Journal of Endocrinology*, 134: 247–255.

Mullaney BP and Skinner MK. 1992. Basic fibroblast growth factor (FGF2) gene expression and protein production during pubertal development of the seminiferous tubule: Follicle-stimulating hormone induced Sertoli cell FGF2 expression. *Endocrinology*, 131: 2928–2934.

National Family Health Survey 2005–2006 (NFHS-3). International Institute for Population Sciences, Mumbai.

Nistal M, Paniagua R, Regadera J, Santamaria L and Amat P. 1986. A quantitative morphological study of human Leydig cells from birth to adulthood. *Cell and Tissue Research*, 246: 229–236.

O'Donnell L, Robertson KM, Jones ME and Simpson ER. 2001. Estrogen and spermatogenesis. *Endocrine Reviews*, 2001, 22: 289–318.

Ogawa T, Ohmura M and Ohbo K. 2005. The niche for spermatogonial stem cells in the mammalian testis. *International Journal of Hematology*, 82: 381–388.

Orth JM, Gunsalus GM and Lamperti AA. 1988. Evidence from Sertoli cell-depleted rats indicates that spermatid numbers in adults depends on numbers of Sertoli cells produced during perinatal development. *Endocrinology*, 122: 787–794.

Page ST, Amory JK and Bremner WJ. 2008. Advances in male contraception. *Endocrine Reviews*, 29: 465–493.

Page ST, Amory JK, Anawalt BD, Irwig MS, Brockenbrough AT, Matsumoto AM and Bremner WJ. 2006. Testosterone gel combined with depomedroxyprogesterone acetate (DMPA) is an effective male hormonal contraceptive regimen not enhanced by the addition of the GnRH antagonist acycline. *Journal of Clinical Endocrinology and Metabolism*, 91: 4374–4380.

Parvinen M. 1982. Regulation of the seminiferous epithelium. *Endocrine Reviews*, 3: 404–417.

Plymate SR and Swerdloff RS. 1992. Androgens, lipids, and cardiovascular risk. *Annals of Internal Medicine*, 117: 871–872.

Potter WD and de Villemeur 2003. Clinical breakage, slippage and acceptability of a new commercial polyurethane condom: A randomized, controlled study. *Contraception*, 68: 39–45.

Quigley CA, De Bellis A, Marschke KB, El-Awady MK, Wilson EM and French FS 1995. Androgen receptor defects: Historical, clinical and molecular perspectives. *Endocrine Reviews*, 16: 271–321.

Rajalakshmi M. 2005. Male contraception: Expanding reproductive choice. *Indian Journal of Experimental Biology*, 43: 1032–1041.

Rajalakshmi M and Bajaj JS. 1999. Preclinical evaluation of androgens for the development of male contraceptive. In: *Male Contraception: Present and Future*, Eds. Rajalakshmi M and Griffin PD. New Age International (P) Ltd, New Delhi, 177–187.

Rajalakshmi M, Pal PC, Jeyaraj DA, Griffin PD and Waites GM. 2000. Evaluation of the ability of levonorgestrel butanoate alone or in combination with testosterone buciclate to suppress spermatogenesis in bonnet monkeys (*Macaca radiata*). *International Journal of Andrology*, 23: 95–105.

Rajalakshmi M and Ramakrishnan PR. 1989. Pharmacokinetics and pharmacodynamics of a new long acting androgen ester: Maintenance of physiological androgen levels for 4 months after a single injection. *Contraception*, 40: 399–412.

Rajalakshmi M and Sharma RS. 2001. Methods for the regulation of male Fertility. In: *Current Status in Fertility Regulation: Indigenous and Modern Approaches*, Eds. Chowdhury SR, Gupta CM and Kamboj VP. Army Printing Press, Lucknow, 179–210.

Rajalakshmi M, Suresh Kumar PK, Kinger S, Pal PC, Pruthi JS and Bajaj JS. 1995. Suppression of testicular and epididymal functions in a non-human primate (bonnet monkey) by combined administration of a gonadotropin-releasing hormone antagonist and testosterone buciclate. *Contraception*, 52: 381–388.

Ramachandra SG, Ramesh V, Krishnamurthy HN, Kumar N, Sundaram K, Hardy MP and Rao AJ. 2002. Effect of chronic administration of 7α-methyl-19-nor-testosterone on serum testosterone, number of spermatozoa and fertility in adult male bonnet monkeys (*Macaca radiata*). *Reproduction*, 124: 301–309.

Reddy PRK and Rao JM. 1972. Reversible antifertility action of testosterone propionate in human males. *Contraception*, 5: 295–301.

Roddy RE, Cordero M, Cordero C and Fortney JA. 1993. A dosing study of nonoxynol 9 and genital irritation. *International Journal of STD/AIDS*, 4: 165.

Rommerts FFG and van der Molen HJ. 1989. Testicular steroidogenesis. In: *The Testis*, 2nd edn., Eds. Burger H and de Kretser D. Raven Press Ltd., New York, 303–328.

Roy CL, Lejeune H, Chuzel F, Saez JM and Langlois D. 1999. Autocrine regulation of Leydig cell differentiated functions by insulin-like growth factor 1 and transforming growth factor beta. *Journal of Steroid Biochemistry and Molecular Biology*, 69: 379–384.

Russel LD. 1980. Sertoli-germ cell interactions: A review. *Gamete Research*, 3: 179–202.

Schwingl PJ and Guess HA. 2000. Safety and effectiveness of vasectomy. *Fertility and Sterility*, 73: 923–936.

Scobie GA, Macpherson S, Millar MR, Groome NP, Romana PG and Saunders PT. 2002. Human estrogen receptors: Differential expression of ER alpha and beta and the identification of ER beta variants. *Steroids*, 67: 985–992.

Seedorf U, Ellinghaus P and Roch Nofer J. 2000. Sterol carrier protein-2. *Biochim Biophysics Acta*, 1486: 45–54.

Sharma RS, Rajalakshmi M and Jeyaraj DA. 2001. Current status of fertility control methods in India. *Journal of Biosciences*, 26(suppl 4): 391–405.

Sharma RS, Rajalakshmi M, Pal PC, Roy S, Behal VK, Sharma DN, Chaturvedi PK and Pruthi JS 2000. Evaluation of efficacy, safety, and reversibility of combination regimen of cyproterone acetate and testosterone buciclate in bonnet monkey. *Contraception*, 62:195–201.

Sharpe RM. 1994. Regulation of spermatogenesis. In: *The Physiology of Reproduction*, 2nd edn., Eds. Knobil E and Neill JD. Raven Press, New York, 1363–1436.

Sharpe RM, McKinnell, C, Kivlin C and Fisher JS. 2003. Proliferation and functional maturation of Sertoli cells, and their relevance to disorders of testis function in adulthood. *Reproduction*, 125: 769–784.

Sivashanmugam P, Hall SH, Hamil KG, French FS, O'Rand MG and Richardson RT. 2003. Characterization of mouse Eppin and a gene cluster of similar protease inhibitors on mouse chromosome 2. *Gene*, 17: 125–134.

Story MT, Sasse J, Kakuska D, Jacobs SC and Lawson RK. 1988. A growth factor in bovine and human testes structurally related to basic fibroblast growth factor. *Journal of Urology*, 140: 422–427.

Trussell J and Kost K. 1987. Contraceptive failure in the United States: A critical review of the literature. *Studies in Family Planning*, 18: 237–283.

Turner L, Conway AJ, Jimenez M, Liu PY, Forbes E, McLachlan RI and Handelsman DJ. 2003. Contraceptive efficacy of a depot progestin and androgen combination in men. *Journal of Clinical Endocrinology and Metabolism*, 88: 4659–4667.

Tyagi A, Rajalakshmi M, Bajaj JS and Mohan Kumar V. 1999a. Effects of long-term treatment with testosterone enanthate in rhesus monkeys. I. Pharmacokinetics of testosterone, testicular volume and liver metabolism of testosterone. *International Journal of Andrology*, 22: 139–147.

Tyagi A, Rajalakshmi M, Jeyaraj DA, Sharma RS and Bajaj JS. 1999b. Effects of long-term use of testosterone enanthate. II. Effects on lipids, high and low density lipoprotein cholesterol and liver function parameters. *International Journal of Andrology*, 22: 347–355.

Udayakumar TS, Tyagi A, Rajalakshmi M, Das SN, Hashim S and Bajaj JS. 1998. Changes in prostate structure and function by long-term administration of an androgen, testosterone enanthate. *The Anatomical Record*, 252: 637–645.

Vale W, Bilezikjian LM and Rivier C. 1994. Reproductive and other roles of inhibins and activins. In: *The Physiology of Reproduction*, Eds. Knobil E and Neill JD. Raven Press, New York, 1861–1878.

van Alphen MM, van de Kant HJ and de Rooij DG. 1988a. Depletion of the spermatogonia from the seminiferous epithelium of the rhesus monkey after X irradiation. *Radiation Research*, 113: 473–486.

van Alphen MM, van de Kant HJ and de Rooij DG. 1988b. Repopulation of the seminiferous epithelium of the rhesus monkey after X irradiation. *Radiation Research*, 113: 487–500.

van Dissel-Emiliani FMF, Grootenhuis AJ, de Jong FH and de Rooij DG. 1989. Inhibin reduces spermatogonial numbers in testes of adult mice and Chinese hamsters. *Endocrinology*, 125: 1898–1903.

van Dissel-Emiliani FMF, de Boer-Brouwer M and de Rooij DG. 1996. Effect of fibroblast growth factor 2 on Sertoli cells and gonocytes in coculture during the prenatal period. *Endocrinology*, 137: 647–654.

Vanelli BG, Barni T, Orlando C, Natali A, Serio M and Balboni GC. 1988. Insulin-like growth factor (IGF-1) and IGF-1 receptor in human testes: An immunohistochemical study. *Fertility and Sterility*, 49: 666–669.

Vihko KK, LaPol PS, Nishimori K and Hsueh AJW. 1991. Stimulatory effect of recombinant follicle-stimulating hormone on Leydig cell function and spermatogenesis in immature hypophysectomized rats. *Endocrinology*, 129: 1926–1932.

Waites GMH and Gladwell RT. 1982. Physiological significance of fluid secretion in the testis and blood testis barrier. *Physiological Reviews*, 62: 624–671.

Walton MJ, Kumar N, Baird DT, Ludlow H and Anderson RA. 2007. 7 (alpha)-methyl-19-nortestosterone (MENT) vs testosterone in combination with etonogestrel implants for spermatogenic suppression in normal men. *Journal of Andrology*, 28:679–688.

Wong EWP, Mruk DB and Lee WM. 2008. Par 3/Par 6 polarity complex coordinates apical ectoplasmic specialization and blood testis barrier restructuring during spermatogenesis. *Proceedings of the National Academy of Science* USA, 105: 9657–9662.

World Health Organization Task Force on Methods for the Regulation of Male Fertility. 1990. Contraceptive efficacy of testosterone-induced azoospermia in normal men. *Lancet*, 336: 955–959.

World Health Organization Task Force on Methods for the Regulation of Male Fertility. 1993. Comparison of two androgens plus depot medroxyprogesterone acetate for suppression to azoospermia in Indonesian men. *Fertility and Sterility*, 60: 1062–1068.

World Health Organization Task Force on Methods for the Regulation of Male Fertility. 1996. Contraceptive efficacy of testosterone-induced azoospermia and oligozoospermia in normal men. *Fertility and Sterility*, 65: 821–829.

Wu FCW, Farley TMM, Peregoudov A and Waites GMH. 1996. Effects of testosterone enanthate in normal men: Experience from a multicentre contraceptive efficacy study. World Health Organization Task Force on methods for the regulation of male fertility. *Fertility and Sterility*, 65: 626–636.

Yoshida S, Nabeshima Y and Nakagawa T. 2007. Stem cell heterogeneity: Actual and potential stem cell compartments in mouse spermatogenesis. *Annals of New York Academy of Sciences*, 1120: 47–58.

Zhang GY, Gu YQ, Wang ZH, Cui YG and Bremner WJ. 1998. A pharmacokinetic study of injectable testosterone undecanoate in hypogonadal men. *Journal of Andrology*, 19: 761–768.

Zhang GY, Gu YQ, Wang ZH, Cui YG and Bremner WJ. 1999. A clinical trial of injectable testosterone undecanoate as a potential male contraceptive in normal Chinese men. *Journal of Clinical Endocrinology and Metabolism*, 84: 3642–3647.

Zhou Q, Nie R, Prins GS, Saunders PT, Katzenellenbogen BS and Hess RA. 2002. Localization of androgen and estrogen receptors in adult male mouse reproductive tract. *Journal of Andrology*, 23: 870–881.

Zirkin BR. 1993. Regulation of spermatogenesis in the adult mammal. Gonadotropins and androgens. In: *Cell and Molecular Biology of the Testis*, Eds. Desjardins C and Ewing LL. Oxford University Press, Oxford, UK, 166–188.

Chapter 11

Molecular Mechanism of Androgen Action

Mahendra Kumar Thakur

Contents

11.1 Introduction.....285
11.2 Mechanism of Androgen Action.....287
11.2.1 Genomic.....287
11.2.2 Non-Genomic.....289
11.3 Androgen Receptor (AR).....289
11.4 Coregulators.....290
11.5 Diseases Related to AR Coregulator.....292
11.5.1 Androgen Insensitivity Syndrome (AIS).....292
11.5.2 Rubinstein–Taybi Syndrome (RTS).....292
11.5.3 CAG Repeat Disorder.....293
11.5.4 Dementia.....293
11.5.5 Depression.....294
11.6 Conclusions.....294
Acknowledgements.....294
References.....295

11.1 Introduction

Androgens are crucial for the development and maintenance of male sexual characteristics and induction of male-specific behaviours. Their action is mediated through an androgen receptor (AR) which is a member of the steroid receptor superfamily. AR exhibits a conserved modular structure consisting of the amino-terminal transactivation domain (TAD), central DNA-binding domain (DBD) and carboxy-terminal ligand-binding domain (LBD). These domains play an important role in receptor function and androgen signalling via interactions with distinct DNA sequences and a host of proteins termed coregulators. A large number of coregulators have been identified during the last decade. Defects in AR or its coregulators lead to disruptions of transcriptional activity causing a wide spectrum of syndromes, prostate cancer, cardiovascular disease, cognitive impairments and neurodegenerative disorders. This chapter reviews the current progress in understanding the molecular mechanism of androgen action with emphasis on coregulators and their role in physiology and pathology, particularly with a focus on the brain.

Androgens, testosterone (T) and its metabolites such as dihydrotestosterone (DHT), play a key role in developmental, behavioural and physiological processes. They exert divergent effects on their target tissues like the reproductive organs, muscle, skin, bone and brain. In the central nervous system (CNS), androgens are involved in the organisation or programming of brain circuits during critical periods of development and they activate a variety of reproductive and non-reproductive behaviour patterns dealing with sexuality, aggression and emotion. Together with oestrogens, androgens are responsible for many gender-related differences including behavioural dimorphisms. Androgens modify several functions by altering the expression of specific genes that are relevant for cell-to-cell communication, neuronal structure and differentiation, somatic, axonal and dendritic growth, synaptogenesis, neurogenesis, neuroplasticity and neuroprotection (MacLusky et al. 2006; Pike et al. 2008). DHT also shows neuroprotection against β-amyloid, a molecular culprit of Alzheimer's disease (AD), in PC12 cells stably transfected with AR but not in PC12 cells stably transfected with empty vector (Pike et al. 2008). Androgens are responsible for male sexual differentiation and development as well as the maintenance and support of sexual tissues in the adult (Matsuda et al. 2008). They affect the ability of the brain to process, store and retrieve sensory information (Genazzani et al. 2007). They help in improvement of cognition, mood, learning and memory (Fink et al. 1999; Janowsky 2006), and are important for the development and progression of age-associated pathologies like prostate cancer, cardiovascular disease, cognitive impairment and psychiatric and neurodegenerative disorders (Chmelar et al. 2007; Edinger and Frye 2007). Recently, Gámez-Del-Estal et al. (2014) reported that testosterone impaired the behavioural pattern of *C. elegans* and this impairment was attenuated by RNAi against AR gene. Interestingly, the testosterone effect was observed during four generations in the absence of hormone, but it was abolished following treatment with sodium butyrate, a histone deacetylase inhibitor. These results suggest the epigenetic effect of testosterone in the behaviour function.

Such complexity of androgen action is associated with (1) regional distribution of AR in the CNS and co-localisation of targets of steroid hormone action in the neuronal populations, (2) cross-talk between molecular pathways of steroid hormone signalling and (3) chemical nature and conversion of androgens to several biologically active metabolites in the CNS. In neurons as well as glia, testosterone is converted by aromatase to estradiol and by 5α-reductase to DHT. Further metabolism of DHT results in the formation of 5α-androstan-3α, 17β-diol (3α A-diol) as well as its 3β-isomer (3β A-diol), both of which are capable of eliciting oestrogen receptor-dependent responses (Pak et al. 2005). Other weak circulating androgens, such as dehydroepiandrosterone (DHEA) and its principal circulating metabolite, DHEA sulfate, exert a wide range of effects on neuronal function (Sullivan and Moenter 2003). Plasma levels of DHEA are significantly increased in the patients of schizophrenia (Michele et al. 2005). The study of Mo et al. (2009) identified neurogenomic effects of DHEA treatment on a subset of genes which are directly implicated in the regulation of appetite, energy utilisation, alertness, apoptosis and cell survival.

In both men and women, androgens comprise a substantial component of the total circulating pool of gonadal steroids in young adults, but then decline markedly with age. Circulating levels of DHEA, in particular, undergo an abrupt decline over the course of middle age, decreasing more than 70% between the third and sixth decades of life (Labrie et al. 2003). As compared to normal controls, AD patients show a reduction in the circulating concentrations of testosterone and DHEA sulfate. A study from the National Institute of Aging (USA) reported that older men with low levels of free circulating testosterone are at higher risk of developing AD than men with higher serum levels of this hormone (Moffat et al. 2004). Recently, Jayaraman and colleagues (2014) demonstrated neuroprotective effect of selective AR modulator RAD140 in cultured hippocampal neurons and kainite-lesioned male rats. Further investigations may provide tempting hints that maintenance of androgen action may be as important as that of oestrogens, in terms of minimising the risks of neurodegenerative disorders.

11.2 Mechanism of Androgen Action

The primary mechanism of androgen action is direct regulation of gene transcription (Thakur 1995; Li and Al-Azzawi 2009; Brinkmann 2011). This is mediated through both genomic as well as non-genomic mechanisms (Figure 11.1).

11.2.1 Genomic

In the genomic mechanism, androgens mediate their effects through AR, also known as NR3C4 (nuclear receptor subfamily 3, group C, member 4), which regulates a complex network of genes that coordinate nearly all the activities of homeostasis, growth and reproduction. AR is present in all target tissues with predominance in brain regions such as the hypothalamus, hippocampus, amygdala and prefrontal cortex, which are essential for learning and memory (Patchev et al. 2004).

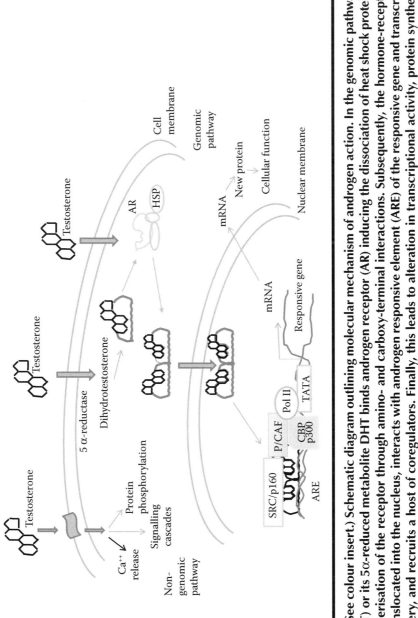

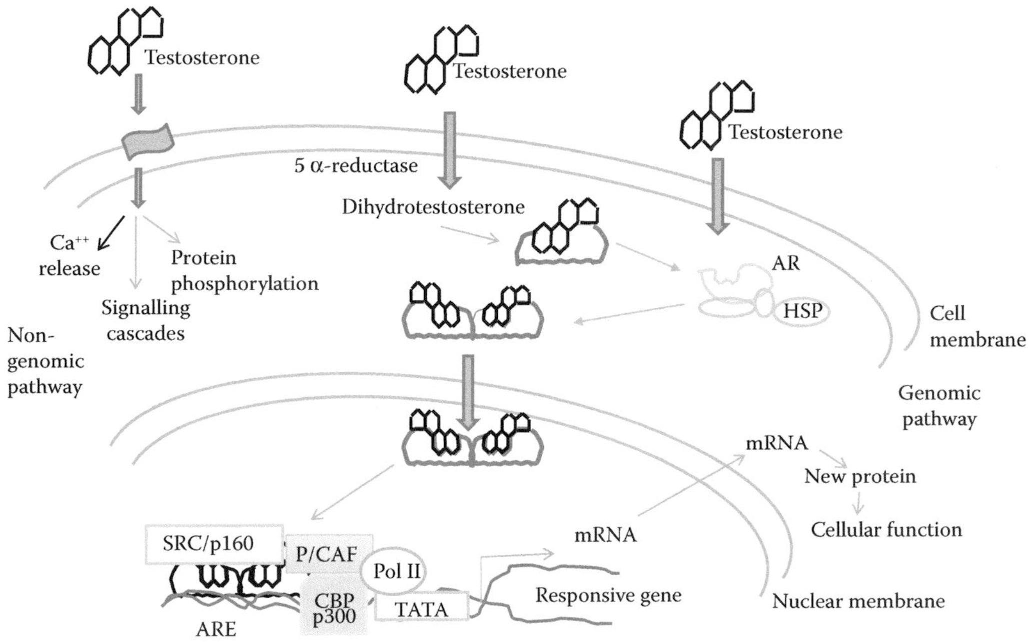

Figure 11.1 (See colour insert.) Schematic diagram outlining molecular mechanism of androgen action. In the genomic pathway, testosterone (T) or its 5α-reduced metabolite DHT binds androgen receptor (AR) inducing the dissociation of heat shock proteins (HSP) and dimerisation of the receptor through amino- and carboxy-terminal interactions. Subsequently, the hormone-receptor complex is translocated into the nucleus, interacts with androgen responsive element (ARE) of the responsive gene and transcriptional machinery, and recruits a host of coregulators. Finally, this leads to alteration in transcriptional activity, protein synthesis and cellular functions. In non-genomic pathway, androgens act at the level of cell membrane and interact with uncharacterised receptor proteins to influence the release of calcium, phosphorylation of proteins and other signalling cascades.

The binding of androgens to AR initiates a series of events leading to the regulation of target genes in a temporal and tissue-specific manner. The receptor undergoes a conformational change in its LBD, and is dissociated from heat shock proteins, such as hsp90, hsp70 and hsp56, transported from the cytosol into the cell nucleus and dimerised. The AR dimer binds to a specific sequence of DNA known as androgen responsive elements (ARE), recruits a host of other proteins known as coregulators to form an active pre-initiation complex and interacts with basal transcription machinery. This results in up- or down-regulation of specific gene transcription and consequently altered synthesis of mRNA which in turn produces specific proteins and controls cellular response.

11.2.2 Non-Genomic

Besides the classical mode, AR can modulate the transcription of specific genes by the direct protein–protein interaction with other factors in a ligand-dependent manner, for which DNA binding by AR is not necessary. In this mechanism, AR binds to androgens and rapidly induces second messenger signal transduction cascades, including increase in free intracellular calcium, and activation of protein kinase A (PKA), protein kinase C (PKC) and mitogen-activated protein kinase (MAPK), leading to diverse cellular effects such as smooth muscle relaxation, neuromuscular and junctional signal transmission and neuronal plasticity (Heinlein and Chang 2002a; Nguyen et al. 2005). Second messenger induction by non-genomic steroid action is insensitive to inhibitors of transcription and translation, suggesting a mechanism independent of transcriptional activity. Commonly, these effects occur within seconds to minutes, considered to be too rapid to involve changes in transcription and protein synthesis. Most non-genomic effects involve a membrane receptor (Pascal and Wang 2014); however, the evidence for a membrane-bound AR is not uniformly accepted. Shihan et al. (2014) reported that such signalling of testosterone is mediated by a G-protein-coupled receptor (GPCR) interaction with Gna11. The non-genomic changes include ion transport and phosphorylation of transcription factors, which can indirectly lead to alterations in gene transcription.

11.3 Androgen Receptor (AR)

The AR gene is located on chromosome Xq11-12.12,13 and spans ~90 kb containing eight exons that code for a ~2757 bp open reading frame within a 10.6 kb mRNA. The first exon codes for a hypervariable amino-terminal TAD, exons 2 and 3 code for a highly conserved central DBD while exons 4–8 code for a conserved carboxy-terminal LBD. These functional domains, like other nuclear receptors, assign a modular structure to AR. They have distinct but diverse functions in AR-mediated signalling (Claessens et al. 2008). During the last decade, the three-dimensional structure of AR LBD has been determined (Matias et al. 2000). It folds to give rise to a hydrophobic pocket in which the ligands are bound; the pocket is made of twelve α-helices and four short β-strands arranged in two anti-parallel β sheets. The ligands, agonists or antagonists, modify the carboxy-terminal helix H12 leading to two opposite orientations; in the agonist-bound conformation, helix H12 serves as a 'lid' to close the ligand-binding hydrophobic pocket containing the steroid, whereas in the antagonist-bound conformation, helix H12 is positioned in a different orientation thus keeping open the entrance to the hydrophobic pocket. Coregulators can also bind to this pocket in LBD, but its predominant binding site is in TAD. The structure of TAD changes upon binding to proteins or to DNA (Brodie and McEwan 2005). This raises the possibility that AR-TAD serves as a flexible platform for the recruitment and assembly of coregulators and members of the transcriptional machinery, and may serve as the primary mediator of the cell and gene-specific effects of androgens (Chmelar et al. 2007). Two autonomous transactivation functions, a constitutively active activation function (AF)-1 originating in amino-terminal and a ligand-dependent AF-2 arising in LBD, are responsible for the transcriptional activity of nuclear receptors. A FXXLF motif in AR amino-terminal interacts with different regions of AF-2/LBD to stabilise the hormone-receptor complex, and mediate AR-specific interactions with FXXLF motif-containing coregulators (He et al. 2001). Recently, Jehle et al. (2014) identified a novel receptor-binding coregulator motif which controls AR action.

AR is regulated by androgen itself, oestrogen, DHEA and other hormones (Lu et al. 2003; Kumar and Thakur 2004; Thakur and Kumar 2007). Upon synthesis, AR undergoes several different covalent post-translational modifications including phosphorylation, sumoylation and ubiquitination which are necessary for receptor function. The majority of phosphorylation sites are located in TAD and, therefore, phosphorylation is linked to the activation and stabilisation of AR. The findings of Yang et al. (2007) suggest that androgens can enhance the phosphorylation state of AR either by negatively regulating the ability of LBD to bind phosphatases or by inducing an AR conformation that is resistant to phosphatase action. Phosphorylation also provides cross-talk links to the numerous cytoplasmic kinase signalling cascades of a cell, such as the epidermal growth factor receptor-2/Her2, MAPK and phosphatidylinositol 3-OH kinase (PI3K)/AKT/protein kinase B (PKB)/phosphatase and tensin homologue pathways.

11.4 Coregulators

The transcriptional activity of AR is affected by coregulators that influence a number of functional properties of AR, including ligand selectivity and DNA binding capacity. An important new mechanism whereby coregulators can change androgen sensitivity was identified in men with recurrent prostate cancer after castration, in which up-regulation of two AR coactivators potently increases cellular androgen sensitivity (Gregory et al. 2001). Cloning and characterisation of several AR coregulators have allowed for cellular and molecular analysis of many different aspects of androgen physiology and pathophysiology.

Coregulators use multiple mechanisms to influence nuclear receptor transcription. Based on their functional characteristics, they are categorised into three groups (Matsumoto et al. 2008; Thakur and Paramanik 2009): (1) directly regulate transcriptional activity through physical interaction with general transcription factors and RNA polymerase II; (2) modify histone tails covalently and (3) remodel chromatin structure which involves ATP-dependent dynamic process. Some of the characterised coregulators are members of the p160 family, ARA70, ARA55, ARA54, ARA267-α, Smad-3 and AIB1. ARA55 and ARA70 allow the activation of AR by 17β-estradiol. Furthermore, both ARA55 and Smad-3 have been suggested to function as bridges for cross-talk between transforming growth factor-β signalling pathway and androgen action.

In general, these coregulators bridge the AR dimer with some of the general transcription factors involved in the pre-initiation complex on the TATA box, and/or with other inducible transcription factors acting on a specific promoter context, with the goal to stabilise the transcription factor machinery and allow the initiation of transcription by RNA polymerase II (Brinkmann 2001). Among the coactivators that mediate AR transcriptional activities, several are general coactivators involved in the control of many genes; others seem to be specific for AR. All the coactivators act in a cell-specific manner and form complexes of multiple proteins. Some proteins of these complexes have intrinsic enzyme activities like histone acetyl transferase (HAT), while corepressors have histone deacetylase (HDAC) activities. HAT and HDAC alter the nucleosome structure, leading to chromatin remodelling; in fact, acetylation of histone tails relaxes chromatin packaging, facilitating gene transcription (Heinlein and Chang 2002b), whereas deacetylation of histones leads to condensation of nucleosomal structures and repression of transcription. Therefore, the AR/coactivators complex, on one side, recruits the general transcription factors to the TATA box, and on the other hand, exerts HAT activity de-structuring the nucleosomes to allow RNA polymerase II to transcribe the genes (Spencer et al. 1997). The two mechanisms act jointly and activate transcription of target genes.

The coactivators involved in AR functions are SRC-1 (steroid receptor coactivator 1 or NCoA-1), the first protein identified as a nuclear receptor coactivator (Onate et al. 1995), SRC-2 (or TIF2/GRIP) and SRC-3 (or p/CIP/RAC3/ACTR/AIB-1/TRAM) which also serve as coactivators for many other nuclear receptors; CREB-binding protein (CBP), which modulates the activity of CREB and a wide variety of other transcription factors; and p160 family of coactivators, which also interact with most, if not all, nuclear receptors. Some of these coactivators perform HAT activity and may also recruit P/CAF, another HAT, enhancing histone acetylation at target genes. Several other more selective proteins are able to enhance transcriptional activity of AR (ARA, androgen receptor associated proteins; SNURF, small nuclear RING finger protein; Tip60, a coactivator for human immunodeficiency virus TAT protein, etc.). The complex of coactivators is utilised in a cell and promoter specific context, creating a highly dynamic process, based on the transient recruitment of a single coactivator by AR, and therefore of the general transcription factors on the TATA box. RNA polymerase II recognises this protein assembly (transcriptosome) on the promoters, starting fast cyclic activation of the AR transcription complex controlling transcription. The antagonist-bound receptor complex binds to DNA, but does not recruit coactivators, blocking the function of RNA polymerase II. Rather the antagonist-bound AR dimer binds to ARE, recruits corepressors such as N-CoR and SMRT and, indirectly, HDAC such as RPD3, leading to transrepression, presumably due to chromatin compaction. Further crystallographic analysis of one of the major protein–protein interaction surfaces on the AR has raised expectations that it will be possible to develop small-molecule antagonists that would block cofactor interactions (Chang and McDonnell 2005).

11.5 Diseases Related to AR Coregulator

Aberrant AR coregulator activity due to mutation or altered expression levels may be a contributing factor in the deregulation of androgen signalling and consequently the progression of diseases related to AR activity.

11.5.1 Androgen Insensitivity Syndrome (AIS)

Of the several steroid hormone resistance syndromes, AIS is thought to be the most common. It occurs in genetically 46XY males who have testis but suffer from various degrees of virilisation failure because of androgen insensitivity. In this syndrome, androgens are synthesised but the receptor fails to respond. This disease can be classified depending on the degree of virilisation failure into four types: complete AIS, partial AIS, Reifenstein syndrome and male infertility. Patients with complete AIS show male pseudohermaphroditism, that is, they have female external genitalia and female type breast development. Patients with Reifenstein syndrome have hypospadias, a micropenis and gynecomastia. Most cases of AIS are the result of AR gene mutations. Single missense mutation in LBD can weaken or eliminate either ligand binding or DNA binding. More than 300 AIS mutations have been identified. The degree of AR dysfunction due to AR gene abnormality generally correlates with the clinical severity (feminisation). The complete AIS syndrome may also result without AR gene mutation. In such case, the pathogenesis occurs due to absence of AF-1-binding 90 kDa coactivator protein which is crucial for the transactivation activity of AR.

11.5.2 Rubinstein–Taybi Syndrome (RTS)

Rubinstein–Taybi syndrome (RTS) presents autosomal dominant inheritance and is clinically characterised by short stature, craniofacial malformation, heart malformation and mental retardation. It is caused by defects in the CBP gene which codes for the CREB-binding protein. CBP was originally identified as a coactivator that stimulates the phosphorylation-dependent transcriptional activation of CREB and also acts as a nuclear integrator when various classes of transcription factors cross-talk with each other.

11.5.3 CAG Repeat Disorder

The first exon of AR gene contains several regions of repetitive DNA sequences, of which the most noteworthy is a CAG triplet repeat that codes for a polyglutamine stretch. The normal repeat length of CAG ranges from 14 to 35 in males and can affect AR activity and risk to prostate cancer. The excessive extension of CAG repeats (40–62) decreases the transactivation potential of AR (yielding relative androgen insensitivity) and can result in an inherited neuromuscular degenerative disease called Kennedy's disease or spinal and bulbar muscular atrophy (SBMA) (Kennedy et al. 1968). Typical features of SBMA are the death of motor neurons in the spinal cord and in the bulbar region of the brainstem, leading to muscle atrophy. The disease affects only men and is characterised by an age of onset between 30 and 50 years. Clinical symptoms include cramps, fasciculations (twitching), weakness and waste of the limbs and face muscles; the weakness mainly affects the shoulders and legs. Studies of CAG repeat length in patients with SBMA have established a correlation between the size of the expansion and the severity of the disease; in general, CAG repeat length correlates inversely with the age of onset of muscle weakness and directly with the progression rate, with the difficulty of climbing stairs (Poletti 2004). The mechanism by which the expanded polyglutamine domain causes motor neuron damage is not understood (Greenland et al. 2004). Recently, Manucka et al. (2010) examined whether amygdala reactivity to threat-related facial expressions (fear, anger) differs as a function of variation in the AR CAG length and endogenous (salivary) testosterone in serum samples of 41 healthy men, and observed that testosterone correlated inversely with the participant age and positively with the number of CAG repeats (mean age: 45.6 years; range: 34–54 years; CAG repeats: 19–29).

11.5.4 Dementia

Several investigators have examined the effect of androgen substitution and withdrawal upon various study populations (Tan et al. 2004; Thakur 2009). However, most studies are associational in nature and the intervention studies involve short duration of testosterone exposure in small samples of subjects. Age-associated decrease in testosterone levels increases the risk for cognitive decline and possibly for dementia and AD (Moffat et al. 2004; Rosario et al. 2004; Raber 2008). Conversely, the maintenance of higher testosterone levels either endogenously or through exogenous supplementation may prove beneficial for cognitive and brain function in elderly men. There is strong evidence from basic sciences and epidemiological studies that both oestrogens and androgens play a protective role in neurodegeneration and cognitive impairment. Patients have reported improvements in both declarative and procedural memory after being treated with hormones. Moreover, testosterone reduces anxiety in male mice through the conversion of testosterone to neurosteroids that interact with gamma-aminobutyric acid (GABA) receptors (Aikey et al. 2002). Androgen administration to aged male mice increases anti-anxiety behaviour and enhances cognitive performance (Frye et al. 2008). Further in-depth research is required to investigate whether there is a window of time in which hormone therapy could be beneficial for middle-aged men, and whether androgens, in particular testosterone, can be prescribed as the future treatment for mild cognitive impairment and related cognitive illnesses, including dementia.

11.5.5 Depression

Epidemiological studies have consistently demonstrated a sex difference in the prevalence of depression, with women experiencing major depression at twice the rate of men (Kessler et al. 1994). Women are particularly vulnerable to mood disturbances at key periods in their reproductive life cycle, that is, the premenstruum, puerperium and perimenopause (Pariser 1993), suggesting oestrogen as an important pharmacologic agent in the treatment of postnatal and perimenopausal depression (Epperson et al. 1999). Many hormones and neurotransmitters are involved in the aetiology and the course of depression including serotonin, dopamine, noradrenaline, vasopressin and cortisol. Testosterone is known to interact with them and its lower level has been linked with higher risk for depression. Pre-clinical data suggest that testosterone administration has antidepressant potential, but the clinical studies are not consistent (Ebinger et al. 2009).

11.6 Conclusions

Androgen action is mediated through both genomic as well as non-genomic mechanisms. In the genomic mechanism, coregulators play a key role in determining tissue-specific actions and a variety of biological functions. It is expected that the extensive research on coregulators will elucidate the mechanisms associated with the action of nuclear receptors as well as the pathogenesis of hormone-dependent diseases. Thus the understanding of molecular mechanism of androgen action will be helpful to advance our knowledge of transcriptional regulation and hormonal influences on disease and to facilitate the design of drugs with greater therapeutic values.

Acknowledgements

The research work cited in this chapter from the author's laboratory has been financially supported by the Department of Science and Technology, Department of Biotechnology, Indian Council of Medical Research, and University Grants Commission – Centre of Advanced Study research program to the Department of Zoology, Banaras Hindu University.

References

Aikey, J. L., Nyby, J. G., Anmuth, D. M. and James, P. J. 2002. Testosterone rapidly reduces anxiety in male house mice (*Mus musculus*). *Hormones and Behavior* 42:448–460.

Brinkmann, A. O. 2001. Molecular basis of androgen insensitivity. *Molecular and Cellular Endocrinology* 179:105–109.

Brinkmann, A. O. 2011. Molecular mechanisms of androgen action—A historical perspective. *Methods in Molecular Biology* 776:3–24.

Brodie, J. and McEwan, I. J. 2005. Intra-domain communication between the N-terminal and DNA-binding domains of the androgen receptor: Modulation of androgen response element DNA binding. *Journal of Molecular Endocrinology* 34:603–615.

Chang, C. and McDonnell, D. P. 2005. Androgen receptor–cofactor interactions as targets for new drug discovery. *Trends in Pharmacological Sciences* 26:225–228.

Chmelar, R., Buchanan, G., Need, E. F., Tilley, W. and Greenberg, N. M. 2007. Androgen receptor coregulators and their involvement in the development and progression of prostate cancer. *International Journal of Cancer* 120:719–733.

Claessens, F., Denayer, S., Tilborgh, N. V., Kerkhofs, S., Helsen, C. and Haelens, A. 2008. Diverse roles of androgen receptor (AR) domains in AR-mediated signaling. *Nuclear Receptor Signaling* 6:e008.

Ebinger, M., Sievers, C., Ivan, D., Schneider, H. J. and Stalla, G. K. 2009. Is there a neuroendocrinological rationale for testosterone as a therapeutic option in depression? *Journal of Psychopharmacology* 23:841–853.

Edinger, K. L. and Frye, C. A. 2007. Androgens' effects to enhance learning may be mediated in part through actions at estrogen receptor β in the hippocampus. *Neurobiology of Learning and Memory* 87:78–85.

Epperson, N. C., Wisner, K. L. and Yamamoto, B. 1999. Gonadal steroids in the treatment of mood disorders. *Psychosomatic Medicine* 61:676–697.

Fink, G., Sumner, B., Rosie, R., Wilson, H. and McQueen, J. 1999. Androgen actions on central serotonin neurotransmission: Relevance for mood, mental state and memory. *Behavioural Brain Research* 105:53–68.

Frye, C. A., Edinger, K. and Sumida, K. 2008. Androgen administration to aged male mice increases anti-anxiety behavior and enhances cognitive performance. *Neuropsychopharmacology* 33:1049–1061.

Gámez-Del-Estal, M. M., Contreras, I., Prieto-Pérez, R. and Ruiz-Rubio, M. 2014. Epigenetic effect of testosterone in the behavior of C. elegans. A clue to explain androgen-dependent autistic traits? *Frontiers in Cellular Neuroscience* 8:69. doi: 10.3389/fncel.2014.00069.

Genazzani, A. R., Pluchino, N., Freschi, L., Ninni, F. and Luisi, M. 2007. Androgens and the brain. *Maturitas* 57:27–30.

Greenland, K. J., Beilin, J., Castro, J., Varghese, P. N. and Zajac, J. D. 2004. Polymorphic CAG repeat length in the androgen receptor gene and association with neurodegeneration in a heterozygous female carrier of Kennedy's disease. *Journal of Neurology* 251:35–41.

Gregory, C. W., He, B., Johnson, R. T., Ford, O. H., Mohler, J. L., French, F. S. and Wilson, E. M. 2001. A mechanism for androgen receptor mediated prostate cancer recurrence after androgen deprivation therapy. *Cancer Research* 61:4315–4319.

He, B., Bowen, N. T., Minges, J. T. and Wilson, E. M. 2001. Androgen induced NH_2- and COOH-terminal interaction inhibits p160 coactivator recruitment by activation function 2. *Journal of Biological Chemistry* 276:2293–2301.

Heinlein, C. A. and Chang, C. 2002a. The roles of androgen receptors and androgen-binding proteins in nongenomic androgen actions. *Molecular Endocrinology* 16:2181–2187.

Heinlein, C. A. and Chang, C. 2002b. Androgen receptor (AR) coregulators: An overview. *Endocrine Review* 23:175–200.

Janowsky, J. S. 2006. The role of androgens in cognition and brain aging in men. *Neuroscience* 138:1015–1020.

Jayaraman, A., Christensen, A., Moser, V. A., Vest, R. S., Miller, C. P., Hattersley, G. and Pike, C. J. 2014. Selective androgen receptor modulator RAD140 is neuroprotective in cultured neurons and kainate-lesioned male rats. *Endocrinology* 155:1398–1406.

Jehle, K., Cato, L., Neeb, A., Muhle-Goll, C., Jung, N., Smith, E. W., Buzon, V. et al. 2014. Coregulator control of androgen receptor action by a novel nuclear receptor-binding motif. *Journal of Biological Chemistry* 289:8839–8851.

Kennedy, W. R., Alter, M. and Sung, J. H. 1968. Progressive proximal spinal and bulbar muscular atrophy of late onset. A sex-linked recessive trait. *Neurology* 18:671–680.

Kessler, R. C., McGonagle, K. A., Zhao, S., Nelson, C. B., Hughes, M., Eshleman, S., Wittchen, H. U. and Kendler, K. S. 1994. Lifetime and 12-month prevalence of DSM-III-R psychiatric disorders in the United States. *Archives of General Psychiatry* 51:8–19.

Kumar, R. C. and Thakur, M. K. 2004. Androgen receptor mRNA expression is inversely regulated by testosterone and estradiol in adult mouse brain. *Neurobiology of Aging* 25:925–933.

Labrie, F., Luu-The, V., Labrie, C., BéLanger, A., Simard, J., Lin, S.-X. and Pelletier, G. 2003. Endocrine and intracrine sources of androgens in women: Inhibition of breast cancer and other roles of androgens and their precursor dehydroepiandrosterone. *Endocrine Review* 24:152–182.

Li, J. and Al-Azzawi, F. 2009. Mechanism of androgen receptor action. *Maturitas* 63:142–148.

Lu, S. F., Mo, Q., Hu, S., Garippa, C. and Simon, N. G. 2003. Dehydroepiandrosterone upregulates neural androgen receptor level and transcriptional activity. *Journal of Neurobiology* 57:163–171.

MacLusky, N. J., Hajszan, T., Prange-Kiel, J. and Leranth, C. 2006. Androgen modulation of hippocampal synaptic plasticity. *Neuroscience* 138:957–965.

Manucka, S. B., Marsland, A. L., Flory, J. D., Gorka, A., Ferrell, R. E. and Hariri, A. R. 2010. Salivary testosterone and a trinucleotide (CAG) length polymorphism in the androgen receptor gene predict amygdala reactivity in men. *Psychoneuroendocrinology* 35:94–104.

Matias, P. M., Donner, P., Coelho, R., Thomaz, M., Peixoto, C., Macedo, S., Otto, N. et al. 2000. Structural evidence for ligand specificity in the binding domain of the human androgen receptor: Implications for pathogenic gene mutations. *Journal of Biological Chemistry* 275:26,164–26,171.

Matsuda, K., Sakamoto, H. and Kawata, M. 2008. Androgen action in the brain and spinal cord for the regulation of male sexual behaviors. *Current Opinion in Pharmacology* 8:747–751.

Matsumoto, T., Shiina, H., Kawano, H., Sato, T. and Kato, S. 2008. Androgen receptor functions in male and female physiology. *Journal of Steroid Biochemistry and Molecular Biology* 109:236–241.

Michele, F., Caltagirone, C., Bonaviri, G., Romeo, E. and Spalletta, G. 2005. Plasma dehydroepiandrosterone levels are strongly increased in schizophrenia. *Journal of Psychiatric Research* 39:267–273.

Mo, Q., Lua, S., Garippaa, C., Brownsteinb, M. J. and Simon, N. G. 2009. Genome-wide analysis of DHEA- and DHT-induced gene expression in mouse hypothalamus and hippocampus. *Journal of Steroid Biochemistry and Molecular Biology* 114:135–143.

Moffat, S. D., Zonderman, A. B., Metter, E. J., Kawas, C., Blackman, M. R., Harman, S. M. and Resnick, S. M. 2004. Free testosterone and risk for Alzheimer disease in older men. *Neurology* 62:188–193.

Nguyen, T. V., Yao, M. and Pike, C. J. 2005. Androgens activate mitogen-activated protein kinase signaling: Role in neuroprotection. *Journal of Neurochemistry* 94:1639–1651.

Onate, S. A., Tsai, S. Y., Tsai, M. J. and O'Malley, B. W. 1995. Sequence and characterization of a coactivator for the steroid hormone receptor superfamily. *Science* 270:1354–1357.

Pak, T. R., Chung, W. C., Lund, T. D., Hinds, L. R., Clay, C. M. and Handa, R. J. 2005. The androgen metabolite, 5alpha-androstane-3beta, 17beta-diol, is a potent modulator of estrogen receptor-beta1-mediated gene transcription in neuronal cells. *Endocrinology* 146:147–155.

Pariser, S. F. 1993. Women and mood disorders: Menarche to menopause. *Annals of Clinical Psychiatry* 5:249–253.

Pascal, L. E. and Wang, Z. 2014. Unzipping androgen action through ZIP9: A novel membrane androgen receptor. *Endocrinology* 155:4120–4123.

Patchev, V. K., Schroeder, J., Goetz, F., Rohde, W. and Patchev, A. V. 2004. Neurotropic action of androgens: Principles, mechanisms and novel targets. *Experimental Gerontology* 39:1651–1660.

Pike, C. J., Nguyen, T. V., Ramsden, M., Yao, M., Murphy, M. P. and Rosario, E. R. 2008. Androgen cell signaling pathways involved in neuroprotective actions. *Hormone and Behavior* 53:693–705.

Poletti, A. 2004. The polyglutamine tract of androgen receptor: From functions to dysfunctions in motor neurons. *Frontiers in Neuroendocrinology* 25:1–26.

Raber, J. 2008. AR, apoE, and cognitive function. *Hormones and Behavior* 53:706–715.

Rosario, E. R., Chang, L., Stanczyk, F. Z. and Pike, C. J. 2004. Age-related testosterone depletion and the development of Alzheimer disease. *Journal of American Medical Association* 292:1431–1432.

Shihan, M., Ahmed, A. B. and Scheiner-Bobis, G. 2014. Non-classical testosterone signaling is mediated by a G-protein-coupled receptor interacting with Gnα11. *Biochimica et Biophysica Acta (BBA)–Molecular Cell Research* 1843:1172–1181.

Spencer, T. E., Jenster, G., Burcin, M. M., Allis, C. D., Zhou, J., Mizzen, C. A., McKenna, N. J. et al. 1997. Steroid receptor coactivator-1 is a histone acetyltransferase. *Nature* 389:194–198.

Sullivan, S. D. and Moenter, S. M. 2003. Neurosteroids alter gamma-aminobutyric acid postsynaptic currents in gonadotropin releasing hormone neurons: A possible mechanism for direct steroidal control. *Endocrinology* 144:4366–4375.

Tan, R. S., Pu, S. J. and Culberson, J. W. 2004. Role of androgens in mild cognitive impairment and possible interventions during andropause. *Medical Hypotheses* 62: 14–18.

Thakur, M. K. 1995. Androgen receptor and the mechanism of androgen action. *Current Science* 68:806–812.

Thakur, M. K. 2009. Dementia in old age, series ed. A. Datta, 1–8. HelpAge India.

Thakur, M. K. and Kumar, R. C. 2007. 17β-Estradiol modulates age-dependent binding of 40 kDa nuclear protein to androgen receptor promoter in mouse cerebral cortex. *Biogerontology* 8:575–582.

Thakur, M. K. and Paramanik, V. 2009. Role of steroid hormone coregulators in health and disease. *Hormone Research* 71:194–200.

Yang, C. S., Xin, H. W., Kelley, J. B., Spencer, A., Brautigan, D. L. and Paschal, B. M. 2007. Ligand binding to the androgen receptor induces conformational changes that regulate phosphatase interactions. *Molecular and Cellular Biology* 27:3390–3404.

Chapter 12

Genomics of Male Infertility

Kiran Singh and Rajiva Raman

Contents

12.1 Introduction ... 299
12.2 Cytogenetics of Male Infertility ... 301
12.3 Y Chromosome Microdeletions and Male Infertility ... 302
12.4 The X Chromosome and Male Infertility ... 305
12.5 Autosomal Genes and Male Infertility ... 305
12.6 Genomic Instability in Male Factor Infertility ... 306
12.7 Endocrine Defects and Male Infertility ... 307
12.8 Epigenomics and Male Infertility ... 308
12.9 Homocysteine Pathway Genes and Susceptibility to Infertility ... 309
12.10 MicroRNAs and Male Infertility ... 311
12.11 Identification of Candidate Male Infertility Genes ... 311
12.12 Future of Genomics of Male Infertility ... 312
References ... 312

12.1 Introduction

The biological fitness of an individual or a population depends upon its reproductive potential, the numbers of progeny it leaves in the nature. Realisation of reproductive potential lies in the fertility status of the individuals contributing to the population. Fertility is established by the production of chromosomally normal (haploid), functional gametes (sperm in male and ova in female) that are capable of effecting fertilisation. In males, production of functional sperm is the outcome of a series of programmed events in testis, collectively called spermatogenesis, which starts with the interaction between the somatic and the germ cells of testis, and the induction by hormones originating in the pituitary in the brain. Spermatogenesis is completed in three specific phases: proliferation, meiosis and spermiogenesis. Primordial germ cells that differentiate into sperm, multiply in numbers through several cycles of division of the spermatogonial type A cells that later differentiate into premeiotic, type B spermatogonia. Delineation of the type B gonia into primary spermatocytes ushers the cells into meiotic division which through two successive divisions culminates into formation of haploid spermatids. Spermiogenesis is the process which oversees the differentiation of round spermatid first into elongating spermatid in the testis and then maturing into sperm in the male reproductive tracts. It is obvious that this elaborate, step-wise and well programmed process occurs under strict genetic surveillance, and an error at any of the steps may impair sperm production or its function that could lead to infertility or reduced fertility. As such, nearly 10% of human couples cannot procreate because either one (male or female) or both the partners are infertile. Infertility due to male factor contributes approximately 50% of the infertility cases in humans (World Health Organization, 1999). In more than 10% of the infertile cases, the cause is not identified and they are considered idiopathic. It is suspected that the idiopathic fraction of the infertile males may indeed be the cases caused due to genetic impairment.

A number of autosomal as well as sex chromosomal genes have been identified that are believed to play a pivotal role in the production of viable, normal and functional sperm. Studies on spermatogenesis in mouse, man and other mammals, especially those involving targeted mutations of the candidate genes have been shown to affect specific testicular cell types, resulting in male infertility or subfertility (Matzuk and Lamb, 2002, 2008). Targeted mutagenesis in mouse and infertility phenotypes in man have paved the way for identifying individual genes and their role in fertility. However, unravelling the process of spermatogenesis and the cause(s) of infertility requires understanding of the network of genes, their cross-talk and response to diverse environmental factors. The post-genomic era, especially after near-complete unravelling of the genomes of man (International Human Genome Sequencing Consortium, 2001; Venter et al., 2001) and mouse has opened a floodgate for such studies (Portin, 2007). Whole genome analyses of man and mouse (and of many others) have unleashed a mass of data and varied databases that reveal enormous possibilities of novel genes, multiple products of single genes, millions of DNA polymorphisms, vast expression data, non-coding RNAs and smaller RNA molecules that could regulate gene function. Genomic research thus offers a new opportunity for determining how diseases occur. Taking advantage of the growing array of sophisticated research tools, it is becoming increasingly possible to identify the molecular abnormalities underlying disease processes. Therefore, while understanding the genomics of male infertility it is important to decipher the candidate genes and their interactions in normal spermatogenesis, sperm maturation and sperm function. It is also desirable to know about the polymorphism or mutations resulting in infertile phenotype, and different regulators of gene expression during spermatogenesis. In the following, we will take up diverse aspects of male infertility and the role of genome.

12.2 Cytogenetics of Male Infertility

One of the common causes of male infertility is numerical or structural abnormalities in chromosomes, sex-chromosome anomaly being more frequent than the autosomal. Nearly 12% of idiopathic infertile couples harbour such a defect. Chromosomal aberrations interfere with the process of spermatogenesis, and the frequency of chromosomal abnormality increases with the decline in sperm concentrations (McLachlan and O'Bryan, 2010). Several studies have shown that Klinefelter's syndrome (47, XXY) is the most prevalent abnormality. It has been estimated that 1 in 1500 male births suffers from Klinefelter's syndrome (47, XXY), accounting for nearly 14% of all cases of non-obstructive azoospermia. Figure 12.1 represents the karyotype of a patient with Klinefelter's syndrome. In addition to the clear aneuploidy, around 10% of Klinefelters are mosaics having a karyotype, 46,XY/47,XYY/47,XXY or 47,XXY/45,XO or 47,XXY/46,XY karyotype. Obviously the mosaics would have arisen during development, and the heterogeneity in the Klinefelter phenotype could well be a reflection of the extent of mosaicism in the individual (Nakamura et al., 2001; Penna Videau et al., 2001; Tempest, 2011). Around 50%–55% of conceptions with Klinefelter's syndrome survive to term

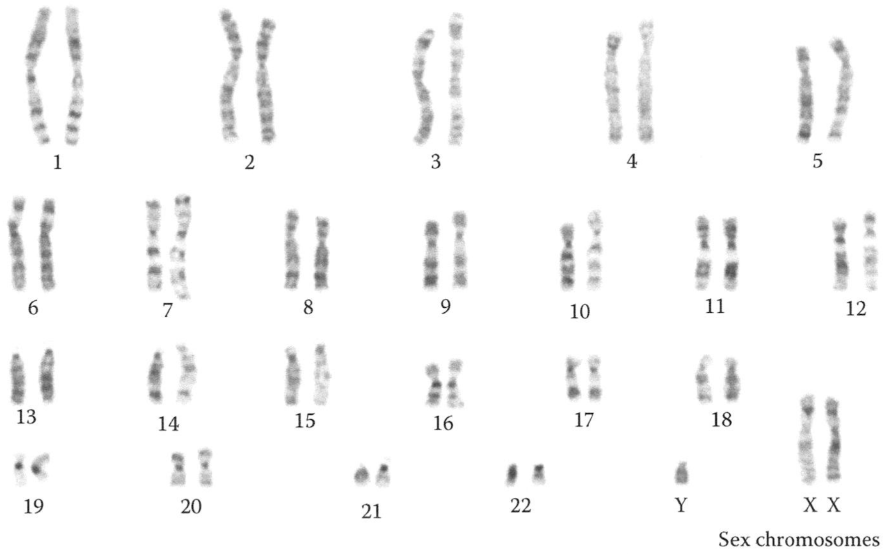

Figure 12.1 (See colour insert.) Karyotype of a Klinefelter patient 47, XXY.

(Hassold and Jacobs, 1984; Jacobs and Hassold, 1995). The aneuploidy of sex chromosome has much less deleterious effect on phenotype than autosomal aneuploidy so it often remains undiagnosed (Rimoin et al., 2006). Gross autosomal rearrangements, such as balanced or unbalanced translocations are associated with azoospermia (no sperm produced, as against various forms of oligospermia where the sperm count is lower than 20 million/mL). Heterozygotes for balanced translocations interfere with normal chromosome pairing and segregation at meiosis (De and Dao, 1991). Unequal segregation of chromosome segments resulting from recombination in translocation and inversion heterozygotes leads to loss or gain of genetic material resulting in partial aneuploidy (Carrell, 2008). Azoospermia and oligospermia are also reported in males with Robertsonian translocation t(14; 21). A few infertility cases with trisomy 7 mosaicism having skin pigmentary variation and dysmorphism have been described (Nagvenkar et al., 2005). Carriers of Robertsonian translocations may exhibit a normal phenotype but could be infertile or subfertile because of lack of gamete production (Ferlin et al., 2007). Supernumerary marker chromosome is also observed in cytogenetic investigation of infertile males. Familial pericentric inversion of chromosome 1(p34q23) in male infertile patient with stage-specific spermatogenic arrest has also been reported (Meschede et al., 1994). Presence of an extra, structurally abnormal chromosome might lead to reduced fertility in males due to meiotic arrest and instability that result in maturation arrest at the spermatocyte stage (Johnson, 1998). In recent years, comparative genomic hybridisation (CGH) at chromosomal level and CGH microarray have proved effective in tracing aneuploidy of even small parts (3–5 Mb) of chromosome while spectral karyotyping (SKY) is the more effective technique in detecting cryptic structural chromosome rearrangements associated with infertility. These technologies show much promise in detecting hitherto unidentified and unrecognised chromosome anomalies.

12.3 Y Chromosome Microdeletions and Male Infertility

The human Y chromosome (60 Mb long) which was long considered to be composed largely of repetitive DNA and heterochromatic region, is now established as one having stretches of euchromatin both in its short and long arms (Skaletsky et al., 2003). Located on the two extremities of the Y chromosome, pseudoautosomal regions are homologous with the termini of the X chromosome. Figure 12.2 illustrates the schematic representation of a typical Y chromosome. More than 95% of the euchromatic region of the Y chromosome comprises 'male-specific Y' (MSY) (Tilford et al., 2001). Among the male-determining genes, the *SRY* (sex-determining region of Y chromosome) is on its short arm while several spermatogenesis-related genes are on the long arm. The genes on its long arm also have a unique distribution pattern, occurring as direct repeats, inverted repeats and palindromes, exhibiting a high level of polymorphism. One hundred and fifty-three haplotypes

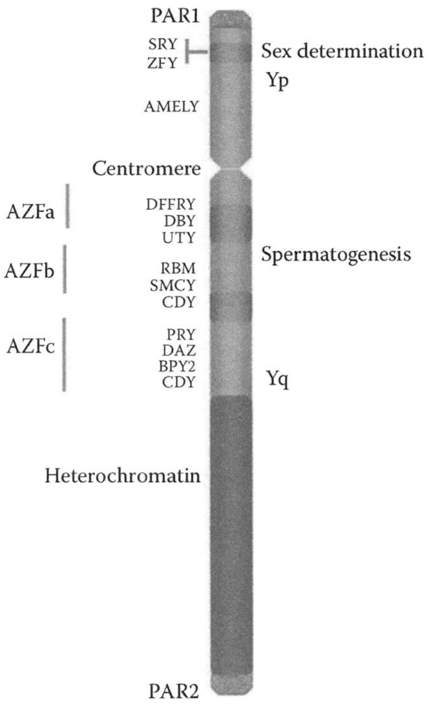

Figure 12.2 (See colour insert.) Schematic representation of the human Y chromosome.

have so far been discovered in human Y chromosome. Out of the 220 genes (104 coding genes, 111 pseudogenes and 5 other uncategorised genes) so far identified on the Y chromosome, the 104 coding genes encode a total of about 48 proteins/protein families (including putative proteins/protein families). Among them, 16 gene products have been discovered in the azoospermia factor region (AZF) and are related to spermatogenesis. Within the AZF region itself, AZFa, AZFb, AZFc and their combinations AZFbc, AZFabc and partial AZFc called AZFc/gr/gr have been identified, illustrating a high degree of polymorphism even within the euchromatic region (Nejad and Farrokhi, 2007). Because of multiplicity of gene copies in the AZF region, single gene mutations of the Y chromosome are rarely the cause of infertility. In contrast Y chromosome 'micro-deletion' is a major cause of male infertility. Tiepolo and Zuffardi (1976) were the first to report the role of Y-chromosome deletion in male infertility, through a painstaking study on karyotypes of over a thousand patients of infertility which detected small deletions in the long arm of Y chromosome of six cases. This study enabled them to identify an AZF (azoospermia factor) region in the long arm of the Y chromosome (interval 6). Subsequent molecular studies have made it possible to delineate the extent of deletion in cases of infertility with great degree of precision (Vogt, 1997; Kuroda-Kawaguchi et al.,

2001). Within AZF, the largest frequency is that of AZFc deletions. This region harbours multiple copies (at least four) of *DAZ* (deleted in azoospermia). Figure 12.3 represents the detection of *DAZ* gene clusters by fluorescence in situ hybridisation (FISH). At least 10% of idiopathic cases are attributed to the deletion in AZF, and it has been possible to draw a rough correlation with the extent of deletion and the stage of arrest in spermatogenesis (Georgiou et al., 2006). Studies on children fathered through intracytoplasmic sperm injection (ICSI) have revealed deletions in the AZF region of the Y chromosome linked to male infertility (Simoni et al., 2004). However whether different Y chromosome haplotypes/haplogroups show an association with infertility is debatable (McElreavey et al., 2000; Carvalho et al., 2003; Singh and Raman, 2009). Homologous recombination between repetitive palindromic regions in Yq explains most AZF deletions (Repping et al., 2002). But recently a novel class of sub-deletions has been reported which arises due to non-homologous recombination (NHR). Noordam et al. (2011) discovered cases with absence of only one STS, that is, sY1197 and in the same study they attributed the deletion to NHR in those cases. Analysis of ongoing Y-chromosome rearrangements and gene conversion provide a picture of the evolutionary events occurring on this reproductively relevant chromosome.

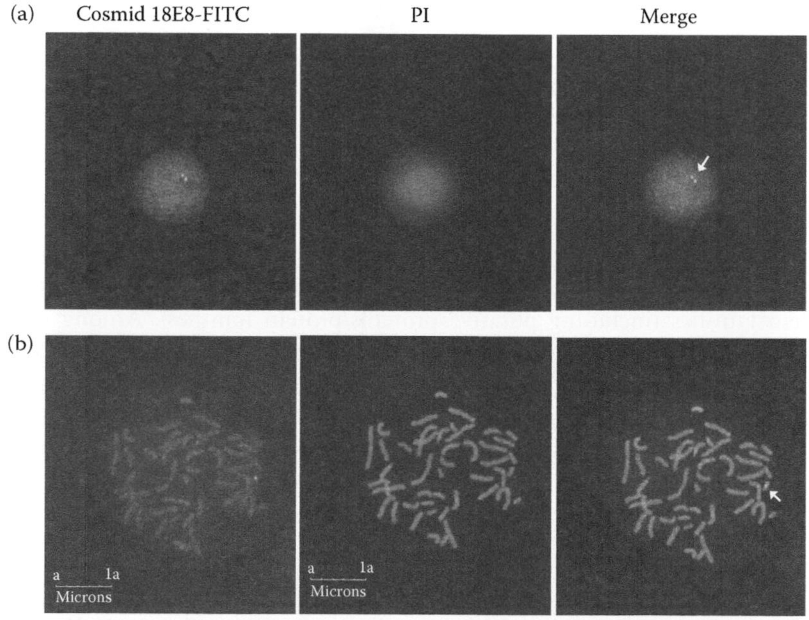

Figure 12.3 (See colour insert.) Detection of *DAZ* gene cluster by FISH. Confocal laser scanned images of representative (a) interphase nucleus and (b) metaphase chromosomes from a fertile, control individual hybridised with cosmid 18E8, showing two dots (corresponding to two *DAZ* clusters, each containing two *DAZ* genes).

12.4 The X Chromosome and Male Infertility

Idiopathic male infertility may be partially caused by mutations or other alterations in genes involved in spermatogenesis (Stouffs et al., 2009). X chromosome is of special interest since men are hemizygous for genes located on this chromosome. De novo mutations in X-linked genes will have an immediate affect since it lacks a second, normal allele. Yet, mutations can be transmitted for generations through women, without having an influence on their reproductive fitness. In males, since the X chromosome is inactivated during and after meiosis (Lifschytz and Lindsley, 1972) it is quite likely that the X-linked genes are expressed before meiosis. There are 1098 genes located on the human X chromosome (Ross et al., 2005) of which almost 10% (~99 genes) are expressed in testis.

In humans, there is enrichment in male-specific genes on the X chromosome (Lercher et al., 2003). The best studied X-linked gene so far is that of androgen receptor (AR) located at Xq11.2-q12. The phenotype of patients with mutations in AR (androgen insensitivity syndrome) varies from total feminisation in the complete androgen insensitivity syndrome (CAIS) to partial androgen insensitivity syndrome (PAIS) and minimal androgen insensitivity syndrome (MAIS) (Quigley et al., 1995). They all lead to male infertility though in more severe cases other secondary sex characters are also affected. More than 500 mutations have so far been detected in AR. Though no 'hot spot' has been identified, relatively more mutations have been recorded in exon 5 (Boehmer et al., 2001; Nagaraja et al., 2009). Involvement of the CAG repeat encoding a polyglutamine tract has also been studied in male infertile patients in different populations but its role in AIS remains equivocal (Dowsing et al., 1999; Davis-Dao et al., 2007). *SOX3* located on Xq26.3 is important for the regulation of embryonic development and shows homology with *SRY* gene. Hemizygous male null mice, although fertile, have reduced sperm counts (Weiss et al., 2003). Alteration in *USP26* gene has also been recorded (Stouffs et al., 2005). Mutations in *KAL-1* gene, which is present on the short arm of the X chromosome and codes for a neural cell adhesion molecule, results in Kallmann syndrome, one of whose manifestations is infertility in males. A more severe phenotype is seen in XY individuals having more than one copy of the *DAX1* gene which shows sex reversal towards female (pseudohermaphroditism) and infertility. Deciphering the X-chromosome mutations and variations in male infertile patients will help in understanding the importance of this chromosome in male reproductive fitness.

12.5 Autosomal Genes and Male Infertility

Autosomal genes that may be mainly involved in the regulation of spermatogenesis and sperm motility are candidates for male infertility (Saxena et al., 1996; Foresta et al., 2005). Most of these are involved in endocrine metabolism and will be discussed in a later section. One of the phenotypes in cystic fibrosis (CF), a severe and among the more common genetic disorders in Caucasians, is absence of vas deferens leading to obstructive azoospermia. CF is an autosomal recessive disorder caused due to mutation in the *CFTR* gene but lack of vas deferens is noticed even in heterozygous condition which leads to infertility without other features of CF. The AZF gene, *DAZ,* has an autosomal homologue on chromosome 3, *DAZLA*, which is involved in early stages of spermatogenesis and is suspected to be a candidate gene for male infertility (Teng et al., 2002). However, no reports yet show a causal role of *DAZLA* in infertility. Sperm morphology undergoes dramatic changes in its structure and chromatin organisation during spermiogenesis. The change in chromatin organisation leading to extreme compaction of chromatin is brought about by the replacement of histones with protamines, the low molecular weight highly basic proteins that aggregate through sulfahydral bonding. This process leads to breakdown of the nucleosomal organisation, and leads to its alignment in sheet-like structures embedded in each other. At least in the mouse, mutation in any of the protamine has been shown to lead to infertility due to inadequate compaction of chromatin (Steger, 2001). Mutations in mitochondria, which occupy the mid-piece of mature sperm and provide the energy required for sperm movement, have been linked to poor sperm motility. It raises the possibility that some types of male subfertility may be inherited only through the female line (Rovio et al., 2001). Polymorphism of apoptotic pathway genes shows significant association with infertility in Asian population (Wang et al., 2009; Jaiswal et al., 2012a,b). In addition, association of detoxification as well as immunological pathway gene polymorphism with male infertility is widely reported (Bentz et al., 2007; Jaiswal et al., 2012c; Ji et al., 2012; Wu et al., 2013). During spermiogenesis 85% of histones are replaced by protamines resulting in highly condensed chromatin that facilitates sperm motility and protects the sperm from damage. Interestingly a high expression of protamine1 (PRM1) was found in spermatid of pregnant couples when compared to non-pregnant couples (Mitchell et al., 2005). A transient expression of H1T2, a H1 histone variant is found in male haploid germ cells during spermiogenesis. *H1T2* mutant mice show delayed nuclear condensation and aberrant elongation (Martianov et al., 2005). The targeted deletion of Tnp2 gene in mice was shown to affect sperm chromatin condensation making it more vulnerable to DNA strand breaks (Zhao et al., 2001). Further functional studies will establish the significance of autosomal genes in spermatogenesis and male infertility.

12.6 Genomic Instability in Male Factor Infertility

As stated by Matzuk and Lamb (2008), spermatogenesis is a multi-step process with the primordial germ cells passing through stages of spermatogonia, spermatocyte and haploid spermatid before differentiating into sperm. In humans, this process takes about 72 days. One of the most critical events in primary spermatocytes is the synapsis of homologous chromosomes and recombination. These two steps require a repertoire of genes/proteins expressing in a precise order to achieve recombination. The prerequisite for recombination is creation of single-strand nicks or double-strand breaks and its repair as such or as a recombined molecule. It is also indicative of the high susceptibility of the germ cell DNA to damages induced by physical and/or chemical agents. Cells which suffer high degree of damage are routinely eliminated by apoptosis but less damaged, misrepaired cells may contribute to infertility. Exposure to toxins, drugs and trauma to testis is believed to reduce sperm count and/or viability (Anway and Skinner, 2006). Spermatogonia also require proper growth factor and receptor function, as well as expression of genes involved in the regulation of stem cells, mitosis and apoptosis. Thus, genes controlling genetic fidelity are of utmost importance in both mitosis and meiosis during spermatogenesis. DNA replication, mitosis and meiosis all are error prone and subject to damage. DNA repair mechanisms recognise and repair DNA damage and DNA replication errors to maintain the fidelity of the DNA sequence. Studies like targeted gene deletion and mutant mouse models to human male infertility have shown that the processes of mitosis and meiosis require proper functioning of the entire DNA repair mechanism in the cell for normal fertility including transition proteins like TP1 which play an active role in maintaining genomic integrity during spermiogenesis. Altered levels of these TP1 proteins along with compromised expression and functioning of genes like Pms2 and Msh2 of MutS family may lead to failure of the DNA repair mechanism (Agarwal and Said, 2003). Frequent chromosome rearrangements involved in infertility are translocations and Y chromosome deletions. The Y chromosome is a model of instability and all rearrangements are the results of illegitimate intra- or inter-chromosomal recombination between homologous sequences. Knockout studies in mouse have shown that genes implicated in germ cell meiosis and repair mechanism, for example, Msh4 (essential for mismatch repair) lead to male and female sterility. Damage to Spo11, critical for DSB formation, also leads to male and female sterility (Barchi and Jasin, 2003). H2Afx, having a role in DSB recognition, gives rise to male sterile/genomic sterility (Cooke and Saunders, 2002). The knowledge of these instability mechanisms is essential to assess the risk for the offspring after ICSI and to provide various treatment options and counselling to patients.

12.7 Endocrine Defects and Male Infertility

Normal spermatogenesis and sperm transport are essential to achieve fertilisation and early embryonic development. Spermatogenesis is intricately controlled by the hypothalamic–pituitary–gonadal (HPG) axis. Infertile patients in which the secretion of HPG hormones is defective are tested for serum follicle-stimulating hormone (FSH), luteinising hormone (LH), total and free testosterone, estradiol and prolactin levels. Mutations, deletions or polymorphic expansions within the regulatory genes involved in the biosynthesis of hormones, growth factors, the androgen receptors and their associated signal transduction pathways may lead to impairment of fertility. Therefore, it is essential to evaluate the possibility of defects in HPG-axis genes for possible cause and treatment of male infertility. Gonadotropin-releasing hormone (GnRH) from the hypothalamus regulates the production of FSH and LH from the pituitary gland. A low level of GnRH results in decreased levels of FSH and LH, giving rise to hypogonadotropic hypogonadism (HH) that results in low androgen secretion and impaired spermatogenesis (Seminara et al., 2000). Mutations leading to direct abnormal synthesis and function of FSH and LH as well as their receptors also play a critical role in maintaining the HPG axis function. Patients suffering from a low level of hormones can be treated with hormone replacement therapy, and normal spermatogenesis and subsequent fertility can be restored after coordinated gonadotropin stimulation (Maduro et al., 2003). The X-linked *Dax1* gene encodes an orphan nuclear hormone receptor that plays a critical role in the development of the hypothalamus, pituitary, adrenal and gonads, and in the maintenance of integrity of the testicular epithelium and spermatogenesis (Burris et al., 1996). Mutations in LH may result in variable phenotypes ranging from complete virilisation, hypogonadism and/or lack of optimal Leydig cell stimulation (Park et al., 1976; Weiss et al., 1992).

Genes involved in the testosterone biosynthesis, metabolism and action can negatively affect male sexual development and spermatogenesis. Testosterone is synthesised from the cholesterol biodegradation. Steroid and their transcriptional coactivators are also important in the regulation of reproductive functions. Conversion of testosterone to dihydrotestosterone by 5α-reductase is crucial for external genitalia and prostate growth. Variants in the *SRD5A2* gene have also been found to be associated with the quality of semen and reduced sperm motility (Zhao et al., 2012). Therefore, deficiencies in 5α-reductase are associated with male infertility as a result of underdeveloped external genitalia and the inability to effectively deliver sperm (Griffin and Wilson, 1992). The same mutation can give rise to different clinical phenotypes in the presence of modifiers (McPhaul et al., 1992). A novel 8-nucleotide deletion in 5α-reductase gene has been shown in an Indian family (Nagaraja et al., 2010). Although oestrogen receptor is also known to be required for male fertility in mouse (Eddy et al., 1996), its role in humans is less understood. A mutation in the oestrogen receptor gene exon 2 that is functionally similar to the one developed in the mouse model has been described in a male patient (Smith et al., 1994). Recently, Khattri et al. (2009) have reported mutations in the oestrogen receptor β gene in infertile Indian males. Further characterisation of gene defects will help in understanding various endocrine disturbances associated with developing an infertile phenotype.

12.8 Epigenomics and Male Infertility

The epigenetic modifications are post-replicational changes in DNA (methylation of cytosine) or post-translational modifications of histones (methylation, acetylation, phosphorylation, ubiquitylation and sumolyation) that survive cell generations and have a prolonged bearing on gene expression. Epigenetic modifications are generally part of the developmental process and thus an essential feature in regulating tissue-specific gene function. It has become increasingly clear that understanding of the epigenetic regulation of individual genes and/or genome-wide epigenetic changes (epigenomics) is essential for unravelling disparate genetic phenomena. The significance of epigenetic modifications in maintaining germ cell proliferation and differentiation indicated when 5-Aza-deoxycytidine (a hypomethylating agent) administration to neonatal mice led to failure of spermatogonial cells to differentiate into spermatocytes (Raman and Narayan, 1995). Subsequent studies on knockouts of genes of DNA methylating enzymes clearly established the role of epigenetic modifications in spermatogenesis and fertility. Certain endocrine disruptors have been identified that induce epigenetic modifications that lead to male infertility (Rakyan et al., 2003). Methylenetetrahydrofolate reductase (*MTHFR*) is a key enzyme in the folate pathway that catalyses the reduction of 5,10-methylenetetrahydrofolate to 5-methytetrahydrofolate (Chan et al., 2010). The abnormalities of *DAZL* promoter DNA methylation pattern and its expression are closely associated with spermatogenesis disorders in patients with infertility (Navarro-Costa et al., 2010; Teng et al., 2006; Davies et al., 2012). Reduced expression of DNMT3B in the germ cells of patients with bilateral spermatogenic arrest in human does not lead to changes in the global methylation status. DNA methyltransferases are expressed throughout human spermatogenesis, possibly maintaining the methylation patterns in order to avoid the transmission of imprinting errors by the male gamete (Marques et al., 2010). The embryonic period seems to be the most sensitive for chemical and environmental effects on the epigenetic marks of the male germ line, involving an altered DNA imprinting. In rat, maternal exposure to a vinclozin coincident with the sex determination phase in the embryo caused infertility in male pups, and this trait passed through four generations with high penetrance (Skinner, 2007). Studies of genomic imprints during development facilitate in understanding epigenetic stability during assisted reproductive technology (Denomme and Mann, 2012). Transgenerational effects of endocrine disruptors have been shown to cause male infertility via Igf2 DMR2 hypomethylation (Song et al., 2014). The epigenetic studies will help in establishing the variations in various epigenetic reprogramming events linked to fertilisation failure.

12.9 Homocysteine Pathway Genes and Susceptibility to Infertility

The role of gene environment interaction, the hallmark of epigenetic influence on cellular physiology, is amply demonstrated by the association of homocysteine-metabolic pathway genes and nutritional cofactors. Folate-derived one carbon units play a critical role in maintaining cellular homeostasis by participating in nucleotide synthesis and regulation of DNA transcription via methylation. Lower serum folate concentration was found to be associated with men with infertile phenotype (Murphy et al., 2011). Elevated levels of homocysteine (homocystinuria), a precursor of the amino acids cysteine and methionine, is known to be associated with several multifactorial disorders such as cardio-vascular disorders and spina bifida (van der Put et al., 1998). Elevation in its level is genetically associated with single nucleotide polymorphisms (SNPs) in the Hcy-metabolism pathway genes (*MTHFR, MTR, MTRR, CBS*). Specifically, the SNP C677T in *MTHFR* gene (methylenetetrahydrofolate reductase) has been shown to be a risk factor in these disorders (van der Put et al., 1998). A study on an Indian cohort drawn from eastern India showed a significant association of the T allele of C677T (especially T homozygosity) with male infertility (Singh et al., 2005). Figure 12.4 shows the PCR-RFLP (polymerase chain reaction–restriction fragment length polymorphism) analysis of *MTHFR*C677T SNP. Interestingly, the populations in which the nutritional intake of folic acid is optimum, incidence of mutated allele T of 677T *MTHFR* is relatively high with no significant association with male infertility. In comparison, the populations in which folic acid intake is generally low, association with infertility is statistically significant. This nutritional association with the phenotype (and Hcy level) is attributed to the fact that folic acid is an important ingredient of 5-methyltetrahydrofolate, the substrate that methylates Hcy as it converts to methionine. Such studies open a window of hope in modulating the genomic effect in disease phenotype, such as infertility in males. It has been shown in rats that the maternal vitamin B_{12} deficiency also affects spermatogenesis

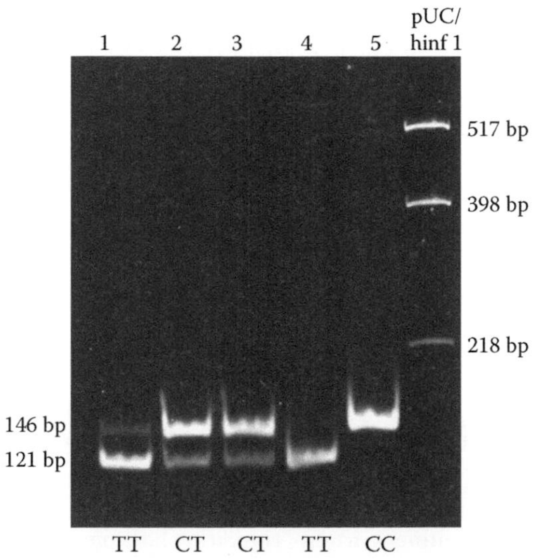

Figure 12.4 Polyacrylamide gel electrophoresis picture of C677T SNP of *MTHFR* gene.

at the embryonic and immature stages (Watanabe et al., 2007). It appears that peri-conceptional intake of adequate amount of folic acid and/or vitamin B_{12} could alleviate the risk of infertility in cases where it could be caused due to mutations in the Hcy pathway genes (Wong et al., 2002; Ebisch et al., 2006a,b).

12.10 MicroRNAs and Male Infertility

Gene expression in humans is precisely controlled in cellular, tissue-type, temporal and condition-specific manners. Complete understanding of the regulatory mechanisms of gene expression is therefore very important. Small, non-coding RNAs (sncRNAs) represent a novel important level of regulation for gene expression. MicroRNAs (miRNAs) comprise a novel class of endogenous, small, non-coding RNAs that negatively regulate gene expression via degradation or translational inhibition of their target mRNAs (Kosik, 2006). As a group, miRNAs may directly regulate 30% of the genes in the human genome (Zhang, 2008). 'microRNomics' describes a novel sub-discipline of genomics that studies the identification, expression, biogenesis, structure, regulation of expression, targets and biological functions of miRNAs on genomic scale (Zhang, 2008). miRNAs are important regulators of cell differentiation, proliferation/growth, mobility and apoptosis. These miRNAs therefore play important role in development and physiology. Consequently, dysregulation of miRNA function may lead to impairment of spermatogenesis too. Altered miRNA expression in patients with non-obstructive azoospermia has recently been established (Jehan et al., 2007; Lian et al., 2009). Interestingly the role of altered profile of seminal plasma miRNAs in patients with male infertility has been recognised. In a case-control study, altered expression of 19 miRNAs was found in patients compared to the control group (Wang et al., 2011). More studies on miRNA will establish the fine regulatory mechanism behind the pathology developing an infertile phenotype.

12.11 Identification of Candidate Male Infertility Genes

About 10% of the genes in the human genome may be related to spermatogenesis and fertility. Genomic studies in animal models have resulted in the identification of a large number of novel reproductive genes. Now it is important to determine the function of those novel and known genes and their roles in normal reproductive physiology, such as gamete production, pregnancy and fertilisation and the disease physiology such as infertility. Mouse genetics has contributed immensely to our understanding of the genetic causes of male infertility. In reproductive genomics whole genome microarray analysis of the testis has helped in identifying the candidate genes expressed exclusively in testis that play a pivotal role in spermatogenesis (Kopylow et al., 2010). Yet adequate expression and functional information is only available for less than 10% of them. Recently copy number variation studies have also been done using chromosomal analysis tools for identification of genes involved in male infertility (Stouffs et al., 2012; Jaiswal et al., 2014). High-resolution X chromosome-specific arrays have also identified novel targets involved in male infertility (Krausz et al., 2012). Functional genomics will further facilitate our knowledge of the molecular networks underlying male fertility/infertility.

12.12 Future of Genomics of Male Infertility

Though much knowledge has indeed been gained through chromosomal, genetic and molecular genetic analyses in understanding the role of genes in fertility, much remains to be learnt about numerous other genes and their interactions that affect gonadal structure, function and subsequent maturation of gametes. It is here that functional analysis of the whole genome in different cell types of the gonad and urogenital system needs to be explored. Genomics of infertility will depend on the identification of novel gene pathways and their response to environmental cues. This will enable more accurate diagnosis of the individual-specific cause of infertility and pave the path for advancement of therapeutic regimens that would address the disease in customised manner. The rapid conceptual and technological advances in genomic analysis in areas such as pharmacogenomics, metabolomics and nutrigenomics give hope that all this will be possible sooner than later.

References

Agarwal, A. and Said, T.M. 2003. Role of sperm chromatin abnormalities and DNA damage in male infertility. *Hum Reprod Update* 9: 331–345.

Anway, M.D. and Skinner, M.K. 2006. Epigenetic transgenerational actions of endocrine disruptors. *Endocrinology* 6: s43–s49.

Barchi, M. and Jasin, M. 2003. Seeking new meiotic genes. *Proc Natl Acad Sci USA* 100: 15287–15289.

Bentz, E.K., Hefler, L.A., Denschlag, D., Pietrowski, P., Buerkle, B. and Tempfer C.B. 2007. A polymorphism of the interleukin-1 beta gene is associated with sperm pathology in humans. *Fertil Steril* 88: 751–753.

Boehmer, A.L., Brinkmann, A.O., Nijman, R.M., Verleun-Mooijman, M.C., de Ruiter, P., Niermeijer, M.F. and Drop, S.L. 2001. Phenotypic variation in a family with partial androgen insensitivity syndrome explained by differences in 5-alpha dihydrotestosterone availability. *J Clin Endocrinol Metab* 86: 1240–1246.

Burris, T.P., Guo, W. and McCabe, E.R. 1996. The gene responsible for adrenal hypoplasia congenita, DAX-1, encodes a nuclear hormone receptor that defines a new class within the superfamily. *Recent Prog Horm Res* 51: 241–259.

Carrell, D.T. 2008. Contributions of spermatozoa to embryogenesis: Assays to evaluate their genetic and epigenetic fitness. *Reprod Biomed Online* 16: 474–484.

Carvalho, C.M.B., Fujisawa, M., Shirakawa, T., Gotoh, A., Kamidono, S., Paulo, T.F., Santos, S.E.B., Rocha, J. and Pena, S.D.J. 2003. Lack of association between Y chromosome haplogroups and male infertility. *Am J Med Genet* 116: 152–158.

Chan, D., Cushnie, D.W., Neaga, O.R., Lawrance, A.K., Rozen, R. and Trasler, J.M. 2010. Strain specific defects in testicular development and sperm epigenetic patterns in 5,10-methylenetetrahydrofolate reductase-deficient mice. *Endocrinology* 151: 3363–3373.

Cooke, H.J. and Saunders, P.T. 2002. Mouse models of male infertility. *Nat Rev Genet* 3: 790–801.

Davies, M.J., Moore, V.M., Willson, K.J., Van Essen, P., Priest, K., Scott, H., Haan, E.A. and Chan, A. 2012. Reproductive technologies and the risk of birth defects. *N Engl J Med* 366: 1803–1813.

Davis-Dao, C.A., Tuazon, E.D., Sokol, R.Z. and Cortessis, V.K. 2007. Male infertility and variation in CAG repeat length in the androgen receptor gene: A meta-analysis. *J Clin Endocrinol Metab* 92: 4319–4326.

De, B.M. and Dao, T.N. 1991. Cytogenetic studies in male infertility: A review. *Hum Reprod* 6: 245–250.

Denomme, M.M. and Mann, M.R.W. 2012. Genomic imprints as a model for the analysis of epigenetic stability during assisted reproductive technologies. *Reproduction* 144(4): 393–409.

Dowsing, A.T., Yong, E.L., Clark, M., McLachlan, R.I., Krester, D.M. and Trounson, A.O. 1999. Linkage between male infertility and trinucleotide repeat expansion in the androgen-receptor gene. *The Lancet* 354: 640–644.

Ebisch, I.M., Pierik, F.H., de Jong, F.H., Thomas, C.M. and Steegers-Theunissen, R.P. 2006a. Does folic acid and zinc sulphate intervention affect endocrine parameters and sperm characteristics in men? *Int J Androl* 29: 339–345.

Ebisch, I.M.W, Peters, W.H., Thomas, C.M., Wetzels, A.M., Peer, P.G. and Steegers-Theunissen, R.P. 2006b. Homocysteine, glutathione and related thiols affect fertility parameters in the subfertile couple. *Hum Reprod* 21: 1725–1733.

Eddy, E.M., Washburn, T.F., Bunch, D.O., Goulding, E.H., Gladen, B.C., Lubahn, D.B. and Korach, K.S. 1996. Targeted disruption of the estrogen receptor gene in male mice causes alteration of spermatogenesis and infertility. *Endocrinology* 137(11): 4796–4805.

Ferlin, A., Raicu, F., Gatta, V., Zuccarello, D., Palka, G. and Foresta, C. 2007. Male infertility: Role of genetic background. *Reprod Biomed Online* 14: 734–745.

Foresta, C., Garolla, A., Bartoloni, L., Bettella, A. and Ferlin, A. 2005. Genetic abnormalities among severely oligospermic men who are candidates for intracytoplasmic sperm injection. *J Clin Endocrinol Metab* 90: 152–156.

Georgiou, I., Syrrou, M., Pardalidis, N., Karakitsios, K. et al. 2006. Genetic and epigenetic risks of intracytoplasmic sperm injection method. *Asian J Androl* 8: 643–673.

Griffin, J.E. and Wilson, J.D. 1992. Disorders of sexual differentiation. *Campwell's Urology*, 2: 1509–1542.

Hassold, T.J. and Jacobs, P.A. 1984. Trisomy in man. *Annu Rev Genet* 18: 69–97.

International Human Genome Sequencing Consortium. 2001. Initial sequencing and analysis of the human genome. *Nature* 409: 860–941.

Jacobs, P.A. and Hassold, T.J. 1995. The origin of numerical chromosome abnormalities. *Adv Genet* 33: 101–133.

Jaiswal, D., Sah, R., Agrawal, N.K., Dwivedi, U.S., Trivedi, S. and Singh, K. 2012a. FAS-670 A/G and FAS-1377 G/A polymorphism in cell death pathway gene FAS and human male infertility. *Asian Pacific J Reprod* 1: 183–186.

Jaiswal, D., Sah, R., Agrawal, N.K., Dwivedi, U.S., Trivedi, S. and Singh, K. 2012b. GSTT1 and GSTM1 polymorphism's combined effect on human male infertility in North Indian Population. *Reprod Sci* 19: 312–316.

Jaiswal, D., Singh, V., Dwivedi, U.S., Trivedi, S. and Singh, K. 2014. Chromosome microarray analysis: A case report of infertile brothers with CATSPER gene deletion. *Gene* 542: 263–265.

Jaiswal, D., Trivedi, S., Singh, R., Dada, R. and Singh, K. 2012c. Association of the IL1RN gene VNTR polymorphism with human male infertility. *PLoS ONE 7*: e51899.

Jehan, Z., Vallinayagam, S., Tiwari, S., Pradhan, S., Singh, L., Suresh, A., Reddy, H.M., Ahuja, Y.R. and Jesudasan, R.A. 2007. Novel noncoding RNA in human Y distal heterochromatic block Yq12 generates testis specific chimeric CDC2L2. *Genome Res* 17: 433–440.

Ji, G., Gu, A., Wang, Y., Huang, C., Hu, F., Zhou, Y., Song, L. and Wang, X. 2012. Genetic variants in antioxidant genes are associated with sperm DNA damage and risk of male infertility in a Chinese population. *Free Radic Biol Med* 52(4): 775–780.

Johnson, M.D. 1998. Genetic risks of intracytoplasmic sperm injection in the treatment of male infertility: Recommendations for genetic counseling and screening. *Fert Steril* 70: 119–124.

Khattri, A., Pandey, R.K., Gupta, N.J., Chakravarty, B., Deenadayal, M., Singh, L. and Thangaraj, K. 2009. Estrogen receptor β gene mutations in Indian infertile men. *Mol Hum Reprod* 15: 513–520.

Kopylow, K.V., Kirchhoff, C., Jezek, D., Schulze, W., Feig, C., Primig, M., Steinkraus, V. and Spiess, A.N. 2010. Screening for biomarkers of spermatogonia within the human testis: A whole genome approach. *Hum Reprod* 25: 1104–1112.

Kosik, K.S. 2006. The neuronal microRNA system. *Nat Rev Neurosci* 7: 911–920.

Krausz, C., Giachini, C., Lo Giacco, D., Daguin, F., Chianese, C., Ars, E., Ruiz-Castane, E. et al. 2012. High resolution X chromosome-specific array-CGH detects new CNVs in infertile males. *PLoS ONE* 710: e44887.

Kuroda-Kawaguchi, T., Skaletsky, H., Brown, L.G., Minx, P.J., Cordum, H.S., Waterston, R.H., Wilson, R.H. et al. 2001. The AZFc region of the Y chromosome features massive palindromes and uniform recurrent deletions in infertile men. *Nat Genet* 29: 279–286.

Lercher, M.J., Urrutia, A.O., Pavlícek, A. and Hurst, L.D. 2003. A unification of mosaic structures in the human genome. *Hum Mol Gene* 12: 2411–2415.

Lian, J., Zhang, X., Tian, H., Liang, N., Wang, Y., Liang, C., Li, X. and Sun, F. 2009. Altered microRNA expression in patients with nonobstructive azoospermia. *Reprod Biol Endocrinol* 713: 1477–7827.

Lifschytz, E. and Lindsley, D.L. 1972. The role of X-chromosome inactivation during spermatogenesis. *Proc Natl Acad Sci USA* 69: 182–186.

Maduro, R.M., Weber, K.L., Chuang, W. and Lamb, D.J. 2003. Genes and male infertility: What can go wrong? *J Androl* 24: 485–493.

Marques, C.J., Francisco, T., Sousa, S., Carvalho, F., Barros, A. and Sousa, M. 2010. Methylation defects of imprinted genes in human testicular spermatozoa. *Fertil Steril* 94: 585–594.

Martianov, I., Brancorsini, S., Catena, R., Gansmuller, A., Kotaja, N., Parvinen, M., Sassone-Corsi, P. and Davidson, I. 2005. Polar nuclear localization of H1T2, a histone H1 variant, required for spermatid elongation and DNA condensation during spermiogenesis. *Proc Natl Acad Sci USA* 102(8): 2808–2813.

Matzuk, M.M. and Lamb, D.J. 2002. Genetic dissection of mammalian fertility pathways. *Nat Cell Bio*, 41–49.

Matzuk, M.M. and Lamb, D.J. 2008. The biology of infertility: Research advances and clinical challenges. *Nat Med* 14: 1197–1213.

McElreavey, K., Krausz, C. and Bishop, C.E. 2000. The human Y chromosome and male infertility. *Results Probl Cell Differ* 28: 211–232.

McLachlan, R.I. and O'Bryan, M.K. 2010. Clinical review: State of the art for genetic testing of infertile men. *J Clin Endocrinol Metab* 95: 1013–1024.

McPhaul, M.J., Marcelli, M., Tilley, W.D., Griffin, J.E., Isidro-Gutierrez, R.F. and Wilson, J.D. 1992. Molecular basis of androgen resistance in a family with a qualitative abnormality of the androgen receptor and responsive to high-dose androgen therapy. *J Clin Invest* 87: 1413–1421.

Meschede, D., Froster, U.G., Bergmann, M. and Nieschlag, E. 1994. Familial pericentric inversion of chromosome 1 p34q23 and male infertility with stage specific spermatogenic arrest. *J Med Genet* 31: 573–575.

Mitchell, V., Steger, K., Marchetti, C., Herbaut, J.C., Devos, P. and Rigot, J.M. 2005. Cellular expression of protamine 1 and 2 transcripts in testicular spermatids from azoospermic men submitted to TESE–ICSI. *Mol Hum Reprod* 11: 373–379.

Murphy, L.E., Mills, J.L., Molloy, A.M., Qian, C., Carter, T.C., Strevens, H., Wide-Swensson, D., Giwercman, A. and Levine, R.J. 2011. Folate and vitamin B12 in idiopathic male infertility. *Asian J Androl* 136: 856–861.

Nagaraja, M.R., Rastogi, A., Raman, R., Gupta, D.K. and Singh, S.K. 2009. Mutational analysis of the androgen receptor gene in two Indian families with partial androgen insensitivity syndrome. *J Pedia Endocrinol Metab* 22: 1–4.

Nagaraja, M.R., Rastogi, A., Raman, R., Gupta, D.K. and Singh, S.K. 2010. Molecular diagnosis of 46, XY and identification of a novel 8 nucleotide deletion in exon 1 of *SRDSA2* gene. *J Pedia Endocrinol Metab* 23: 379–385.

Nagvenkar, P., Desai, K., Hinduja, I. and Zaveri, K. 2005. Chromosomal studies in infertile men with oligozoospermia and non-obstructive azoospermia. *Indian J Med Res* 122: 34–42.

Nakamura, Y., Kitamura, M., Nishimura, K., Koga, M., Kondoh, N., Takeyama, M., Matsumiya, K. and Okuyama, A. 2001. Chromosomal variants among 1790 infertile men. *Int J Urol* 2: 49–52.

Navarro-Costa, P., Nogueira, P., Carvalho, M., Leal, F., Cordeiro, I., Calhaz-Jorge, C., Gonçalves, J. and Plancha, C.E. 2010. Incorrect DNA methylation of the DAZL promoter CpG island associates with defective human sperm. *Hum Reprod* 25: 2647–2654.

Nejad, H.S. and Farrokhi, F. 2007. Genetics of azoospermia: Current knowledge, clinical implications, and future directions. Part II: Y chromosome microdeletions. *Urol J* 3: 193–203.

Noordam, M.J., van Daalen, S.K.M., Hovingh, S.E., Korver, C.M., van der Veen, F. and Repping, S. 2011. A novel partial deletion of the Y chromosome azoospermia factor c region is caused by non-homologous recombination between palindromes and may be associated with increased sperm counts. *Hum Reprod* 26: 713–723.

Park, I.J., Burnett, L.S., Jones, H.W. Jr., Migeon, C.J. and Blizzard, R.M. 1976. A case of male pseudohermaphroditism associated with elevated LH, normal FSH and low testosterone possibly due to the secretion of an abnormal LH molecule. *Acta Endocrinol Copenh* 83: 173–181.

Penna Videau, S., Araujo, H., Ballesta, F., Ballesca, J.L. and Vanrell, J.A. 2001. Chromosomal abnormalities and polymorphisms in infertile men. *Arch Androl* 46: 205–210.

Portin, P. 2007. Evolution of man in the light of molecular genetics: A review. Part I. Our evolutionary history and genomics. *Hereditas* 144: 80–95.

Quigley, C.A., De Bellis, A., Marschke, K.B., el Awady, M.K., Wilson, E.M. and French, F.S. 1995. Androgen receptor defects: Historical, clinical, and molecular perspectives. *Endocrinol Rev* 163: 271–321.

Rakyan, V.K., Chong, S., Champ, M.E., Cuthbert, P.C., Morgan, H.D., Luu, K.V. and Whitelaw, E. 2003. Transgenerational inheritance of epigenetic states at the murine Axin Fu allele occurs after maternal and paternal transmission. *Proc Natl Acad Sci USA* 100: 2538–2543.

Raman, R. and Narayan, G. 1995. 5-Aza deoxyCytidine-induced inhibition of differentiation of spermatogonia into spermatocytes in the mouse. *Mol Reprod Dev* 423: 284–290.

Repping, S., Skaletsky, H., Lange, J., Silber, S., Van Der Veen, F., Oates, R.D., Page, D.C. and Rozen, S. 2002. Recombination between palindromes P5 and P1 on the human Y chromosome causes massive deletions and spermatogenic failure. *Am J Hum Genet* 71: 906–922.

Rimoin, D.L., Connor, J.M., Pyeritz, R.E. and Korf, B.R. 2006. *Emery and Rimoin's Principles and Practice of Medical Genetics*, 5th edn. Philadelphia, PA: Churchill Livingstone Elsevier.

Ross, J.L., Kowal, K., Quigley, C.A., Blum, W.F., Cutler, G.B. Jr., Crowe, B., Hovanes, K., Elder, F.F. and Zinn, A.R. 2005. The phenotype of short stature homeobox gene SHOX deficiency in childhood: Contrasting children with Leri-Weill dyschondrosteosis and Turner syndrome. *J Pediatr* 147: 499–507.

Rovio, A.T., Marchington, D.R., Donat, S., Schuppe, H.C., Abel, J., Fritsche, E., Elliott, D.J. et al. 2001. Mutations at the mitochondrial DNA polymerase (POLG) locus associated with male infertility. *Nat Genet* 29: 261–262.

Saxena, R., Brown, L.G., Hawkins, T., Alagappan, R.K., Skaletsky, H., Reeve, M.P., Reijo, R. et al. 1996. The DAZ gene cluster on the human Y chromosome arose from autosomal gene that was transposed, repeatedly amplified and pruned. *Nat Genet* 14: 292–299.

Seminara, S.B., Oliveira, L.M., Beranova, M., Hayes, F.J. and Crowley, W.F. Jr. 2000. Genetics of hypogonadotropic hypogonadism. *J Endocrinol Invest* 23: 560–565.

Simoni, M., Bakker, E. and Krausz, C. 2004. EAA/EMQN best practice guidelines for molecular diagnosis of Y-chromosomal microdeletions. State of the art 2004. *Int J Androl* 27: 240–249.

Singh, K. and Raman, R. 2009. Y-haplotypes and idiopathic male infertility in an Indian population. *Indian J Hum Genet* 15: 19–22.

Singh, K., Singh, S.K, Sah, R., Singh, I. and Raman, R. 2005. Mutation C677T in the methylenetetrahydrofolate reductase gene is associated with male infertility in an Indian population. *Int J Androl* 28: 115–119.

Skaletsky, H., Kuroda-Kawaguchi, T., Minx, P.J., Cordum, H.S., Hillier, L., Brown, L.G., Repping, S. et al. 2003. The male specific region of the human Y chromosome is a mosaic of discrete sequence classes. *Nature* 423: 825–837.

Skinner, M.K. 2007. Epigenetic transgenerational toxicology and germ cell disease. *Int J Androl* 30: 393–397.

Smith, E.P., Boyd, J., Frank, G.R., Takahashi, H., Cohen, R.M., Specker, B., Williams, T.C., Lubahn, D.B. and Korach, K.S. 1994. Estrogen resistance caused by a mutation in the estrogen-receptor gene in a man. *New Eng J Med* 331: 1056–1061.

Song, Y., Wu, N., Wang, S., Gao, M., Song, P., Lou, J., Tan, Y. and Liu, K. 2014. Transgenerational impaired male fertility with an Igf2 epigenetic defect in the rat are induced by the endocrine disruptor p,p′-DDE. *Hum Reprod* 11: 2512–2521.

Steger, K. 2001. Haploid spermatids exhibit translationally repressed mRNAs. *Anat Embryol* 203: 323–334.

Stouffs, K., Lissens, W., Tournaye, H., Van Steirteghem, A. and Liebaers, I. 2005. Possible role of USP26 in patients with severely impaired spermatogenesis. *Eur J Hum Genet* 13: 336–340.

Stouffs, K., Tournaye, H., Liebaers, I. and Lissens, W. 2009. Male infertility and the involvement of the X chromosome. *Hum Reprod Update* 15: 623–637.

Stouffs, K., Vandermaelen, D., Massart, A., Menten, B., Vergult, S., Tournaye, H. and Lissens, W. 2012. Array comparative genomic hybridization in male infertility. *Hum Reprod* 27: 921–929.

Tempest, H.G. 2011. Meiotic recombination errors, the origin of sperm aneuploidy and clinical recommendations. *Syst Biol Reprod Med* 57: 93–101.

Teng, Y.N., Lin, Y.M., Lin, Y.H., Tsao, S.Y., Hsu, C.C., Lin, S.J., Tsai, W.C. and Kuo, P.L. 2002. Association of a single-nucleotide polymorphism of the deleted-in-azoospermia-like gene with susceptibility to spermatogenic failure. *J Clin Endocrinol Metab* 87: 5258–5264.

Teng, Y.N., Lin, Y.M., Sun, H.F., Hsu, P.Y., Chung, C.L. and Kuo, P.L. 2006. Association of DAZL haplotypes with spermatogenic failure in infertile men. *Fertil Steril* 86: 129–135.

Tiepolo, L. and Zuffardi, O. 1976. Localisation of factors controlling spermatogenesis in the nonfluorescent portion of the human Y chromosome long arm. *Hum Genet* 34: 119–124.

Tilford, C.A., Kuroda-Kawaguchi, T., Skaletsky, H., Rozen, S., Brown, L.G., Rosenberg, M., McPherson, J.D. et al. 2001. A physical map of the human Y chromosome. *Nature* 409: 943–945.

Van der Put, N.M.J., Gabreels, F., Stevens, E.M.B., Smeitink, J.A.M., Trijbels, F.J.M., Eskes, T.K.A.B., van den Huevel, L.P. and Blom, H.J. 1998. A second common mutation in the methylenetetrahydrfolate reductase gene: An additional risk factor for neural-tube defects? *Am J Hum Genet* 62: 1044–1051.

Venter, J.C., Adams, M.D., Myers, E.W., Li, P.W., Mural, R.J., Sutton, G.G., Smith, H.O. et al. 2001. The sequence of the human genome. *Science* 291: 1304–1351.

Vogt, P.H. 1997. Genetic disorders of human spermatogenesis. In *Current Advances in Andrology,* edited by G M.H. Wailes, *Proc. 6th Int Cong Androl,* Austria: 51–75.

Wang, C., Yang, C., Chen, X., Yao, B., Zhu, C., Li, L., Wang, J. et al. 2011. Altered profile of seminal plasma microRNAs in the molecular diagnosis of male infertility. *Clin Chem* 57: 1722–1731.

Wang, W., Lu, N., Xia, Y., Gu, A., Wu, B., Liang, J., Zhang, W., Wang, Z., Su, J. and Wang, X. 2009. FAS and FASLG polymorphisms and susceptibility to idiopathic azoospermia or severe oligozoospermia. *Reprod Biomed* 18: 141–147.

Watanabe, T., Ebara, S., Kimura, S., Maeda, K., Wanatabe, Y., Wanatabe, H., Kasai, S. and Nakano, Y. 2007. Maternal vitamin B12 deficiency affects spermatogenesis at the embryonic and immature stages in rats. *Congenit Anom* 47: 9–15.

Weiss, J., Axelrod, L., Whitcomb, R.W., Harris, P.E., Crowley, W.F. and Jameson, J.L. 1992. Hypogonadism caused by a single amino acid substitution in the beta subunit of luteinizing hormone. *N Eng J Med* 326:179–183.

Weiss, J., Meeks, J.J., Hurley, L., Raverot, G., Frassetto, A. and Jameson, J.L. 2003. Sox3 is required for gonadal function, but not sex determination, in males and females. *Mol Cell Biol* 23: 8084–8091.

Wong, W.Y., Merkus, H.M., Thomas, C.M., Menkveld, R., Zielhuis, G.A. and Steegers-Theunissen, R.P. 2002. Effects of folic acid and zinc sulfate on male factor subfertility: A double blind, randomized, placebo-controlled trial. *Fertil Steril* 77: 491–498.

World Health Organization. *WHO Laboratory Manual for the Examination of Human Semen and Sperm-Cervical Mucus Interaction*, 4th edn. Cambridge: Cambridge University Press, 1999.

Wu, W., Lu, J., Tang, Q., Zhang, S., Yuan, B., Li, J., Di, Wu. et al. 2013. GSTM1 and GSTT1 null polymorphisms and male infertility risk: An updated meta-analysis encompassing 6934 subjects. *Sci Rep* 3: 2258.

Zhang, C. 2008. MicroNomics: A newly emerging approach for disease biology. *Physiol Genomics* 33: 139–147.

Zhao, M., Shirley, C.R., Yu, Y.E., Mohapatra, B., Zhang, Y., Unni, E., Deng, J.M. et al. 2001. Targeted disruption of the transition protein 2 gene affects sperm chromatin structure and reduces fertility in mice. *Mol Cell Biol* 2121: 7243–7255.

Zhao, V., Wu, W., Xu, B., Niu, Z., Cui, H., Zhang, Y., Wang, Z. and Wang, X. 2012. Variants in the SRD5A2 gene are associated with quality of semen. *Mol Med Rep* 6: 639–644.

Index

A

ABP, *see* Androgen-binding protein (ABP)
Activating transcription factor-2 (ATF-2), 104
Activation function (AF), 290
Activity of AR (ARA), 291; *see also* Androgen
AD, *see* Alzheimer's disease (AD)
Adherens junction (AJ), 108
Adjudin, 268; *see also* Male contraception
AF, *see* Activation function (AF)
AFB1, *see* Aflatoxin B1 (AFB1)
Aflatoxin B1 (AFB1), 128
AFP, *see* α-Fetoprotein (AFP)
AG, *see* Aminoguanidine (AG)
AGTR2, *see* Angiotensin II type 2 receptor (AGTR2)
AIS, *see* Androgen Insensitivity Syndrome (AIS)
AJ, *see* Adherens junction (AJ)
α-Fetoprotein (AFP), 9; *see also* Sexual dimorphism in CNS
Alzheimer's disease (AD), 286
Aminoguanidine (AG), 106
Amygdala, 15–16; *see also* Sexual dimorphism in CNS
Androgen, 4, 8, 174; *see also* Androgen receptor (AR); Epididymis; Leydig cells; Male contraception; Sexual dimorphism in CNS
- action, 175–176, 285, 294
- anti-androgen, 8, 272–273
- binding proteins, 174–175
- CAG repeat disorder, 293
- circulating, 287
- complexity of, 286–287
- coregulators, 290–292
- delivery routes, 271
- dementia, 293–294
- depression, 294
- deprivation effects, 274
- diseases related to AR coregulator, 292
- excess effects, 274
- genomic mechanisms, 287
- insensitivity syndrome, 292
- non-genomic mechanisms, 289
- receptor antagonists, 175
- -regulated genes, 176
- responsive elements, 289
- role, 286
- Rubinstein–Taybi syndrome, 292–293
- in spermatogenesis, 263
- testicular production of, 263
- testosterone and DHT in peripheral circulation, 174

Androgen-binding protein (ABP), 142, 174, 266
Androgen Insensitivity Syndrome (AIS), 292
Androgen receptor (AR), 174, 142, 285, 289–290; *see also* Androgen
- coactivators involved in, 291
- coregulators, 290
- proteins to enhance transcriptional activity of, 291
- regulation, 290

Androgen responsive elements (ARE), 289; *see also* Androgen
Angiotensin II type 2 receptor (AGTR2), 126
Anterior commissure, 19–20; *see also* Sexual dimorphism in human brain
Anti-androgen, 8
- –androgen combination regimens, 272–273; *see also* Male contraception

Anti-Mullerian hormone, 5
Apaf-1, *see* Apoptotic protease activating factor-1 (Apaf-1)
Apical cell (AC), 120, 123; *see also* Epididymal epithelim cell
- goblet-shaped, 123
- of rat epididymis, 124

Apoptosis, 97, 110–111
  activation of ERK in Sertoli cells, 107
  actvation of p38 MAPK, 105
  caspase 2, 108–110
  DNA break in germ cells, 102
  downstream caspases, 104
  Fas signalling, 103–104
  germ cell, 98–99
  pathways of, 99–103
  signal transduction pathways, 101
  testicular hyperthermia, 107
  upstream signalling pathways, 104–108, 109
Apoptotic protease activating factor-1 (Apaf-1), 99; *see also* Apoptosis
Aposomes, 131; *see also* Epididymis
AQPs, *see* Aquaporins (AQPs)
Aquaporins (AQPs), 137–138, 172
AR, *see* Androgen receptor (AR)
ARA, *see* Activity of AR (ARA)
ARE, *see* Androgen responsive elements (ARE)
ArKO, 180; *see also* Oestrogen
Aromatase, 177; *see also* Oestrogen
ART, *see* Assisted reproduction technologies (ART)
Aspermia, 256; *see also* Seminal vesicles
Assisted reproduction technologies (ART), 71
ATF-2, *see* Activating transcription factor-2 (ATF-2)
Autocrine regulation, 263
5-Aza-deoxycytidine, 309
Azoospermia, 302; *see also* Male infertility
  factor region, 303
  patients, 234

B

Basal cell (BC), 120, 126–127; *see also* Epididymal epithelim cell
Bed nucleus of the stria terminalis (BNST), 13, 15, 19; *see also* Sexual dimorphism in human brain
Behavioural masculinisation, 8
β-catenin, 5
Bioassay, 56
Blood-epididymis barrier, 130; *see also* Epididymis
BNST, *see* Bed nucleus of the stria terminalis (BNST)
Brain masculinisation, 8; *see also* Sexual dimorphism in CNS
Buffalo pituitary hormones, 47; *see also* Gonadotropins
Bulbourethral glands, *see* Cowper's glands

C

CAG repeat disorder, 293; *see also* Androgen
CAIS, *see* Complete androgen insensitivity syndrome (CAIS)
Carbonic anhydrases (CA), 172
CASA, *see* Computer-aided semen analysis (CASA)
Caspase 2, 108–110; *see also* Apoptosis
Catalytic subunit, 205
CBP, *see* CREB binding protein (CBP)
CC, *see* Clear cell (CC)
CCK, *see* Cholecystokinin (CCK)
CE 9 plasma membrane protein, 134; *see also* Epididymis
Cell electroporation, 210
CEMS, *see* Chloroethylmethanesulphonate (CEMS)
Central nervous system (CNS), 2, 286
  sexually dimorphic regions of mammalian, 3
CF, *see* Cystic fibrosis (CF)
CGH, *see* Comparative genomic hybridisation (CGH)
Chinese hamster ovary (CHO), 71
Chloroethylmethanesulphonate (CEMS), 147
CHO, *see* Chinese hamster ovary (CHO)
Cholecystokinin (CCK), 15
Cholesterol (CH), 202
Cholesterol/phospholipid ratio (CH/PL ratio), 203
CH/PL ratio, *see* Cholesterol/phospholipid ratio (CH/PL ratio)
c-jun NH2-terminal kinase (JNK), 104
Clear cell (CC), 120, 125–126; *see also* Epididymal epithelim cell
  of rat epididymis, 124, 125
Comparative genomic hybridisation (CGH), 302
Compensatory ovarian hypertrophy, 61
Complete androgen insensitivity syndrome (CAIS), 305
Computer-aided semen analysis (CASA), 200
ConA, *see* Concanavalin A (ConA)
Concanavalin A (ConA), 221
Condoms, 267–268; *see also* Male contraception
Coregulators, 286, 290; *see also* Androgen
  corepressors, 292
  types, 291

Corepressors, 292; *see also* Androgen
Corpus callosum, 20; *see also* Sexual dimorphism in human brain
Cowperitis, 249–250; *see also* Cowper's glands
Cowper's glands, 247; *see also* Male accessory sex glands
  acquired lesions, 249
  congenital lesions, 249
  location, 247
  male sexual response, 248
  neoplasms, 250
  pathology, 248
  pre-ejaculate, 248
  secretions and function, 248
  stones, 250
  structure, 247
  syringocele, 249
Cowper's stones, 250; *see also* Cowper's glands
CPA, *see* Cyproterone acetate (CPA)
CREB binding protein (CBP), 291
CRISP, *see* Cysteine-rich secretory protein (CRISP)
Cyclic AMP (cAMP), 198, 205
Cyproterone acetate (CPA), 272
Cysteine-rich secretory protein (CRISP), 149
Cystic fibrosis (CF), 305; *see also* Male infertility

D

Dangerous cells, 98
DBD, *see* DNA-binding domain (DBD)
DDRT-PCR, *see* Differential display reverse transcriptase polymerase chain reaction (DDRT-PCR)
DDs, *see* Death domains (DDs)
Death domains (DDs), 99
Decapeptide, *see* Gonadotropin-releasing hormone (GnRH)
De Graaf's thread, *see* Epididymis
Dehydroepiandrosterone (DHEA), 287
Dementia, 293–294; *see also* Androgen
De novo synthesised pre-hormone, 52
Depot medroxyprogesterone acetate (DMPA), 272
Depression, 294; *see also* Androgen
DES, *see* Diethylstilbestrol (DES)
DHEA, *see* Dehydroepiandrosterone (DHEA)
DHT, *see* Dihydrotestosterone (DHT)
DIABLO, *see* Second mitochondria-derived activator of caspases (SMAC)
Diazonium salt of sulphanilic acid (DSS), 235
Diethylstilbestrol (DES), 178
Differential display reverse transcriptase polymerase chain reaction (DDRT-PCR), 176, 183
Dihydrotestosterone (DHT), 142, 174, 253, 256, 286; *see also* Androgen; Sexual dimorphism in CNS
  5α-DHT, 5, 10
  role, 286
DLN, *see* Dorsolateral nucleus (DLN)
DMPA, *see* Depot medroxyprogesterone acetate (DMPA)
DNA-binding domain (DBD), 286
Dopamine, 7; *see also* Sexual dimorphism in CNS
Dorsolateral nucleus (DLN), 12
DSS, *see* Diazonium salt of sulphanilic acid (DSS)
Dyslexia, 21

E

ECD, *see* Extra cellulardomain (ECD)
Ecto-ATPase, 233
ecto-CIK, *see* Ecto-cyclic AMP-independent protein kinase (ecto-CIK)
Ecto-cyclic AMP-independent protein kinase (ecto-CIK), 206, 207, 212; *see also* Sperm ecto-protein kinases
  activity of, 209
  epididymal mature spermatozoa, 213
  immunoassay of, 213
  phospholipid effect on, 210
  and PPase, 217–219
  properties of, 212
  protein substrate of, 209–213
  purification, 208
  SDS PAGE of, 213
  sperm surface coupled-enzyme system, 235
  substrate specificity of, 214
  subunits, 205
Ecto-enzymes, 209; *see also* Ecto-cyclic AMP-independent protein kinase (ecto-CIK); Sperm ecto-protein kinases
Ectoplasmic specialisation (ES), 108, 268; *see also* Male contraception
Ecto-PPase, *see* Phosphoprotein phosphatase
Ecto-RCs, 205; *see also* Sperm ecto-protein kinases
EDS, *see* Ethane dimethanesulphonate (EDS)
ELSPBP1 protein, 141; *see also* Epididymis
Endocrine disorders, 45

Endocrine regulation of spermatogenesis, 78; *see also* Mammalian testis; Spermatogenesis—hormonal regulation
Endocytosis, 172; *see also* Epididymis
Endogenous opioids, 10; *see also* Sexual dimorphism in CNS
EP, *see* Epididymal plasma (EP)
EP2, *see* Epididymis protein 2 (EP2)
Epididymal epithelim cell, 120, 170; *see also* Epididymis
  apical cell, 123–124
  basal cell, 126–127
  clear cell, 125–126
  halo cells, 127–128
  mouse, 122
  narrow cell, 124–125
  organisation, 121
  pale vacuolated epithelial cell, 128–129
  principal cell, 120–123
Epididymal maturation process, 116; *see also* Epididymis
Epididymal plasma (EP), 228
Epididymal protease inhibitor (EPPIN), 149; *see also* Epididymis
Epididymis, 116, 177, 150, 168; *see also* Androgen; Epididymal epithelim cell; Oestrogen
  apocrine secretion at, 135–137
  blood-epididymis barrier, 130
  caprine, 199
  CE 9, 134
  cell culture, 144–145
  contraceptive, 147–150
  CRES, 138
  DE bound to sperm, 149
  duct, 120
  electrolyte and water transport, 137
  endocytosis, 172
  epididymal compartments, 169
  epididymosomes, 135–137
  epithelial cell types, 169–170
  fluid absorption and secretion, 171–172
  functions of, 131, 171
  growth, 173
  lumen, 130–131
  luminal compartment of, 170
  male fertility problems, 145–146
  mammalian, 198
  maturation process, 116
  mouse, 118
  organ culture, 143–144
  passage of sperm to distal parts, 141
  PH-20 protein, 132–133
  protease inhibitor, 149–150
  protection of epithelium and sperm, 139–140, 172–173
  protein secretion, 131
  of rat, 119
  reabsorption in, 138
  regulation of, 142–143, 173
  role of FSH, 188–189
  segments, 117–120
  semen coagulum proteins, 134
  sperm membrane lipids, 135
  sperm protection, 139
  sperm proteins, 131–134
  in sperm quality control, 140–141
  sperm storage and spermiophagy, 142, 173
  toxicant target, 146–147
Epididymis protein 2 (EP2), 139
Epididymis Specific Clone 42 (ESC42), 169
Epididymosomes, 131, 135–137, 234; *see also* Epididymis; Sperm cell
EPPIN, *see* Epididymal protease inhibitor (EPPIN)
Equatorial segment (ES), 133
ER Antagonists, 180; *see also* Oestrogen
ERE, *see* Estrogen response element (ERE)
ERK, *see* Extracellular signal-regulated kinase (ERK)
ES, *see* Ectoplasmic specialisation (ES); Equatorial segment (ES)
ESC42, *see* Epididymis Specific Clone 42 (ESC42)
Estradiol, 264; *see also* Leydig cells
Estrogen response element (ERE), 55
Ethane dimethanesulphonate (EDS), 147
Extra cellulardomain (ECD), 62
Extracellular signal-regulated kinase (ERK), 104

F

FADD, *see* FAS-associated death domain (FADD)
FAS-associated death domain (FADD), 99
Fas L, *see* Fas ligand (Fas L)
Fas ligand (Fas L), 103
Fas signalling system, 103; *see also* Apoptosis
Fertility, 299; *see also* Male infertility
FGF, *see* Fibroblast growth factor (FGF)
Fibrin, 252; *see also* Prostate gland
Fibrinogen, 256; *see also* Seminal vesicles

Fibrinolysin, 252; *see also* Prostate gland
Fibroblast growth factor (FGF), 266
FGF2, 84
FISH, *see* Fluorescence in situ hybridisation (FISH)
FITC, *see* Fluorescein isothiocyanate (FITC)
Fluorescein isothiocyanate (FITC), 217
Fluorescence in situ hybridisation (FISH), 304
FMP, *see* Forward motility protein (FMP)
FMSF, *see* Forward motility stimulating factor (FMSF)
Follicle stimulating hormone (FSH), 19, 31, 44, 63; *see also* Gonadotropin-releasing hormone (GnRH); Gonadotropins; Spermatogenesis—hormonal regulation
in epididymal function regulation, 188–189
in male reproduction, 66
in mice, 83–85
non-gonadal actions of, 70
in non-human primates, 85
on ovarian ascorbic acid content, 67
in rats, 83
receptor, 65
Sertoli cell, 82
in spermatogenesis, 78, 86
FORKO, *see* FSH receptor knockout (FORKO)
Forward motility, 219, 228; *see also* Sperm motility
Forward motility protein (FMP), 225, 228
Forward motility stimulating factor (FMSF), 223; *see also* Sperm motility—stimulating proteins
FMSF-HRP conjugate, 227
^{125}effect of I-labelled FMSF-I, 226
localisation of bound, 229
mechanism of action, 228, 231
MSP antibody-mediated agglutination, 226
phospho-kinase stimulating potential of, 230
effect of PKA inhibitor, 230
receptor assay, 226
source of, 225
effect of tyrosinekinase inhibitor, 230
FSH, *see* Follicle stimulating hormone (FSH)
FSH receptor knockout (FORKO), 84

G

GABA, *see* Gamma-aminobutyric acid (GABA)
Gamma-aminobutyric acid (GABA), 7, 294; *see also* Sexual dimorphism in CNS
Gamma-glutamylcysteinylglycine (GSH), 172
Gamma-glutamyl transpeptidase mRNA IV (Ggt_pr4 mRNA), 143
GDNF, *see* Glial cell line-derived neurotrophic factor (GDNF)
Germ cell death, 98; *see also* Apoptosis
Ggt_pr4 mRNA, *see* Gamma-glutamyl transpeptidase mRNA IV (Ggt_pr4 mRNA)
GL, *see* Glycolipids (GL)
Glial cell line-derived neurotrophic factor (GDNF), 84, 266
Glutathione S-transferase (GST), 126
Glutathione system, 139; *see also* Epididymis
Glycolipids (GL), 202
Glycoprotein hormone biosynthesis, 50
Glycosylphosphatidylinositol (GPI), 234
GnRH, *see* Gonadotropin-releasing hormone (GnRH)
GnRH-A, *see* Gonadotropin-releasing hormone antagonist (GnRH-A)
GnRH analogues-androgen combination regimens, 273; *see also* Male contraception
GnRHR, *see* Gonadotropin-releasing hormone receptor (GnRHR)
Gonadotropin-releasing hormone (GnRH), 19, 32
action in gonadotrope, 33–34
functions of, 36–37
receptor in humans, 37–38
release control, 32–33
in spermatogenesis, 78
stimulated gonadotropin release, 34
variants and receptor genes, 34–36
Gonadotropin-releasing hormone antagonist (GnRH-A), 99
Gonadotropin-releasing hormone receptor (GnRHR), 32; *see also* Gonadotropin-releasing hormone (GnRH)
Gonadotropins, 43; *see also* Follicle stimulating hormone (FSH); Hormone action; Human chorionic gonadotropin (hCG); Luteinising hormone (LH); Prolactin (PRL)
assay, 56–60
biopotency of prolactins, 49
bubaline LH isolation, 46
buffalo pituitary hormones, 47
clinical use, 71–73
endocrine disorders, 45

Gonadotropins (*Continued*)
enzyme activity in ovary after PMSG injection, 68
genes for gonadotropin subunit, 54
glycoprotein hormone biosynthesis, 50
GSI in animals, 67
hCGββ binding with gonadal LH receptor, 52
heterodimerisation, 50
hormone action, 60–69
hormone biosynthesis, 52
isolation and structural features, 44–51
microheterogeneity, 48
non-gonadal actions of, 69–71
oligosaccharides in pituitary and placental, 49
on ovarian ascorbic acid, 57
ovarian weight increase, 57
pharmacological receptor, 65
phenotypes of transgenic and gene knockout mouse models for, 64
pre-hormone, 52
protein glycosylation, 53
recombinant expression of, 51
secretion regulation, 51–55
signal transduction pathway of, 62
subunit of, 48
sulphation of, 50
yeast expression system, 51
GPCR, *see* G-protein-coupled receptor (GPCR)
GPI, *see* Glycosylphosphatidylinositol (GPI)
G-protein-coupled receptor (GPCR), 289
Growth factors, 266–267
GSH, *see* Gamma-glutamylcysteinylglycine (GSH)
GST, *see* Glutathione S-transferase (GST)
GSTpi gene, 139; *see also* Epididymis

H

Halo cells, 127–128; *see also* Epididymal epithelim cell
HAT, *see* Histone acetyl transferase (HAT)
hCAP18, *see* Human cationic antimicrobial protein (hCAP18)
hCG, *see* Human chorionic gonadotropin (hCG)
hCG augmentation assay, 56
HDAC, *see* Histone deacetylase (HDAC)
HDL, *see* High-density lipoprotein (HDL)
Helix bundle peptide hormones, 68
Heterodimerisation, 50
HH, *see* Hypogonadotropic hypogonadism (HH)
High-density lipoprotein (HDL), 274
Hippocampus, 18; *see also* Sexual dimorphism in human brain
Histone acetyl transferase (HAT), 291
Histone deacetylase (HDAC), 291
hLH, *see* Human LH (hLH)
Hormonal assay, 56
Hormonal contraception, 268–269; *see also* Male contraception
acceptability of, 275
side effects, 273–275
Hormone action, 60–63; *see also* Gonadotropins
enzyme activity after PMSG, 68
gonadotropin action, 63–69
GSI in animals after PMSG plus hCG, 67
hormone receptor complex, 62
pharmacological receptor, 65
prolactin, 68–69
transgenic animal models, 66
Horseradish-peroxidase (HRP), 225
HPG, *see* Hypothalamic–pituitary–gonadal (HPG)
HRP, *see* Horseradish-peroxidase (HRP)
Human brain, 2
Human cationic antimicrobial protein (hCAP18), 139
Human chorionic gonadotropin (hCG), 44; *see also* Gonadotropins
augmentation assay, 56
hCG-β, 48
thyrotropic effect, 72
Human LH (hLH), 46; *see also* Gonadotropins
Human whey acidic protein 10 (hWFDC10A), 176
hWFDC10A, *see* Human whey acidic protein 10 (hWFDC10A)
Hypogonadotropic hypogonadism (HH), 308
Hypophysectomy, 43
Hypothalamic–pituitary–gonadal (HPG), 307
Hypothalamus, 3
rat, 13

I

IAPs, *see* Inhibitor of apoptosis proteins (IAPs)
ICD, *see* Intracellular domain (ICD)
iCR, *see* Inducible carbonyl reductase (iCR)
ICSI, *see* Intracytoplasmic sperm injection (ICSI)
IELs, *see* Intraepithelial lymphocytes (IELs)

IEMs, *see* Intraepithelial macrophages (IEMs)
IGF, *see* Insulin growth factor (IGF)
IIPS, *see* International Institute for Population Sciences (IIPS)
Immunoassay, 56
Inducible carbonyl reductase (iCR), 176
Inducible nitric oxide synthase (iNOS), 104
Inhibitor of apoptosis proteins (IAPs), 99; *see also* Apoptosis
Initial segment (IS), 117
iNOS, *see* Inducible nitric oxide synthase (iNOS)
Insulin growth factor (IGF), 266
Intermediate zone (IZ), 117
International Institute for Population Sciences (IIPS), 260, 275
Interstitial cells, *see* Leydig cells
Intracellular domain (ICD), 62
Intracytoplasmic sperm injection (ICSI), 304
Intraepithelial lymphocytes (IELs), 120, 127–128; *see also* Epididymal epithelim cell
Intraepithelial macrophages (IEMs), 120, 127–128; *see also* Epididymal epithelim cell
Intra-gonadal cell types, 66; *see also* Gonadotropins
In vivo RNA interference (RNAi), 110
IS, *see* Initial segment (IS)
IZ, *see* Intermediate zone (IZ)

J

JAK tyrosine kinase, 68–69
JNK, *see* c-jun NH2-terminal kinase (JNK)

K

Klinefelter's syndrome, 301–302; *see also* Male infertility

L

Large T antigen (LTAg), 145
LBD, *see* Ligand-binding domain (LBD)
Lectin, 221; *see also* Sperm surface lectin
Leydig cells, 80, 262; *see also* Male reproduction physiology; Mammalian testis
- androgens production, 263
- autocrine and paracrine regulation, 263
- role of androgens, 263
- role of estradiol, 264

LH, *see* Luteinising hormone (LH)
Ligand-binding domain (LBD), 286
Low-density lipoprotein , 274
LTAg, *see* Large T antigen (LTAg)
Luteinising hormone (LH), 19, 31, 44, 63; *see also* Gonadotropin-releasing hormone (GnRH); Gonadotropins; Spermatogenesis—hormonal regulation
- bubaline LH isolation, 46
- cAMP responsive element, 55
- contrasting effects, 67
- cycloheximide inability, 58
- *in vitro* assay, 58
- in male reproduction, 66
- non-gonadal actions of, 70
- ovarian ascorbic acid depletion assay, 57, 67
- receptor, 65
- in spermatogenesis, 78, 86–87
- sulphation of, 50

M

MAIS, *see* Minimal androgen insensitivity syndrome (MAIS)
Major physiological protein substrate (MPS), 209; *see also* Sperm ecto-protein kinases
- immunodetection, 216
- incorporation effect on motility, 215

Male accessory sex glands, 246, 257; *see also* Cowper's glands; Prostate gland; Seminal vesicles
Male contraception, 267; *see also* Male reproduction physiology
- androgen delivery, 271
- anti-androgen–androgen combination regimens, 272–273
- contraceptive acceptability, 275
- GnRH analogues-androgen combination regimens, 273
- hormonal, 268–269
- non-hormonal agents, 268
- progestogen–androgen combination regimens, 272
- RISUG, 268
- semen coagulum proteins, 134
- side effects, 273–275
- sperm transport prevention, 267–268
- testosterone buciclate, 270
- testosterone undecanoate, 270–271
- vasectomy and vaso-occlusive devices, 268

Male-determining genes, 302; *see also* Male infertility
Male hormonal contraception (MHC), 268
Male infertility, 299; *see also* Spermatogenesis
  autosomal genes and, 305–306
  azoospermia, 302
  cytogenetics of, 301–302
  endocrine defects and, 307–308
  epigenomics and, 308–309
  genomics of, 306–307, 312
  homocysteine pathway genes, 309–311
  infertility gene identification, 311–312
  Klinefelter's syndrome, 301–302
  microRNAs and, 311
  oligospermia, 302
  X chromosome and, 305
  Y chromosome microdeletions and, 302
Male reproduction physiology, 261; *see also* Cowper's glands; Leydig cells; Male contraception; Sertoli cell; Spermatogenesis; Testis
  sexual response, 248
Male reproductive system, 168; *see also* Epididymis
  adult human, 246
Male-specific Y (MSY), 302; *see also* Male infertility
Mammalian spermatogenesis, 78; *see also* Mammalian testis; Spermatogenesis—hormonal regulation
Mammalian testis, 79; *see also* Spermatogenesis—hormonal regulation
  Leydig cells, 80
  mouse testis, 79
  peritubular myoid cells, 80
  sertoli cells, 80–81
  spermatogenic cells, 81
MAPKs, *see* Mitogen-activated protein kinases (MAPKs)
MBH, *see* Medial basal hypothalamus (MBH)
Measurement, Learning and Evaluation (MLE), 268
Medial amygdala, 15
Medial basal hypothalamus (MBH), 33
Medial preoptic area (MPOA), 7
Medroxyprogesterone acetate (MPA), 271
Membrane-bound sperm motility inhibitor, 232
Membrane-bound tyrosine kinase, 233; *see also* Sperm plasma membrane
Menopause, 70
MENT, *see* 7α-Methyl-19-nor-testosterone (MENT)
7α-Methyl-19-nor-testosterone (MENT), 271
Methylenetetrahydrofolate reductase (MTHFR), 309
MHC, *see* Male hormonal contraception (MHC)
Microheterogeneity, 48
MicroRNAs (miRNAs), 311; *see also* Male infertility
MIF, *see* Motility-inhibiting factor (MIF)
Minimal androgen insensitivity syndrome (MAIS), 305
MIP, *see* Motility initiating protein (MIP)
miRNAs, *see* MicroRNAs (miRNAs)
Mitogen-activated protein kinases (MAPKs), 104, 289
MLE, *see* Measurement, Learning and Evaluation (MLE)
MOET, *see* Multiple Ovulation and Embryo Transfer(MOET)
Monoaminergic systems in sexual dimorphism, 6; *see also* Sexual dimorphism in CNS
Motility-inhibiting factor (MIF), 232; *see also* Sperm motility; Sperm motility
  dose and action, 232
  glycosidase effect on, 233
Motility initiating protein (MIP), 228, 230; *see also* Sperm motility
  action mechanism, 231
Motility promoting protein (MSP), 225, 226
Mouse testis, 79
MPA, *see* Medroxyprogesterone acetate (MPA)
MPOA, *see* Medial preoptic area (MPOA)
MPS, *see* Major physiological protein substrate (MPS)
MSP, *see* Motility promoting protein (MSP)
MSY, *see* Male-specific Y (MSY)
MTHFR, *see* Methylenetetrahydrofolate reductase (MTHFR)
Multiple Ovulation and Embryo Transfer(MOET), 71
Myomesin 2, 175; *see also* Androgen

N

Na-H exchanger (NHE), 172
Narrow cell (NC), 120, 124–125; *see also* Epididymal epithelim cell

National Family Health Survey 2005–2006 (NFHS-3), 260
NC, *see* Narrow cell (NC)
Neoplasms, 250; *see also* Cowper's glands
Neutral lipid (NL), 202
NFHS-3, *see* National Family Health Survey 2005–2006 (NFHS-3)
NHE, *see* Na-H exchanger (NHE)
NHR, *see* Nonhomologous recombination (NHR)
NL, *see* Neutral lipid (NL)
NMS, *see* Normal monkey serum (NMS)
Non-aromatisable androgens, 8–9
Non-genetic mechanisms, 60
Nonhomologous recombination (NHR), 304
Non-hormonal agents, 268; *see also* Male contraception
Non-sexual dimorphic human behaviour, 16–17; *see also* Sexual dimorphism in human brain
Noradrenaline, 7; *see also* Sexual dimorphism in CNS
Normal monkey serum (NMS), 189
NR3C4 (nuclear receptor subfamily 3, group C, member 4), 287

O

OAAD assay, *see* Ovarian ascorbic acid depletion assay (OAAD assay)
Oestrogen, 8, 176; *see also* Epididymis; Sexual dimorphism in CNS
  ArKO, 180
  aromatase, 177
  binding with hCG, 187
  down-regulation versus up-regulation, 185
  E2 synthesis in epididymis, 184, 186
  on epididymis, 178–179
  ER antagonists, 180
  ERα, 178
  ERα expression, 182
  ERαKO, 179
  ERβ, 178, 182
  ERβKO, 179
  17-HSDI mRNA expression, 186
  ICI and fluid reabsorption, 182–183
  ICI and serum testosterone levels, 182
  ICI and sperm count, 183
  ICI and sperm motility, 183
  immunolocalisation, 181–182
  LHR expression, 186–187
  in male, 177
  P450AROM, 186, 187–188
  receptor antagonists, 180
  receptor localisation, 178
  regulated genes, 180–181, 183–184
  RT-PCR analysis, 181
  Western Blots analyses, 181
Oligospermia, 302; *see also* Male infertility
Opiates, 10; *see also* Sexual dimorphism in CNS
Orthovanadate, 233
Ovarian ascorbic acid depletion assay (OAAD assay), 57, 67
Oxytocin, 252; *see also* Prostate gland

P

PAIS, *see* Partial androgen insensitivity syndrome (PAIS)
Pale vacuolated epithelial cell (PVECs), 128–129; *see also* Epididymal epithelim cell
Paracrine regulation, 263
PARP, *see* Poly (ADP-ribose) polymerase (PARP)
Partial androgen insensitivity syndrome (PAIS), 305
PC, *see* Phosphatidylcholine (PC); Principal cell (PC)
p-chloromercuriphenylsulfonic acid (PCMPS), 234, 235
PCMPS, *see* p-chloromercuriphenylsulfonic acid (PCMPS)
PCNA, *see* Proliferating cell nuclear antigen (PCNA)
PCR-RFLP, *see* Polymerase chain reaction–restriction fragment length polymorphism (PCR-RFLP)
PE, *see* Phosphatidylethanolamine (PE)
PEMT, *see* Phosphotidylethanolamine N-methyltransferase (PEMT)
Peritubular myoid cells (PMC), 80; *see also* Mammalian testis
PH-20, 132–133
Phosphatidylcholine (PC), 135, 183, 202
Phosphatidylethanolamine (PE), 135, 183, 202
Phosphatidylinositol (PI), 203
Phosphatidylinositol 3-OH kinase (PI3K), 290
Phosphoinositide phospholipase C, 32
Phospholipids (PL), 202
Phosphoprotein phosphatase, 214; *see also* Sperm ecto-protein kinases
  activity of goat spermatozoa, 218
  purified, 217
  sperm maturation process and, 218

Phosphotidylethanolamine N-methyltransferase (PEMT), 183
PI, *see* Phosphatidylinositol (PI)
PI3K, *see* Phosphatidylinositol 3-OH kinase (PI3K)
Piceatannol, 233
PKA, *see* Protein kinase A (PKA)
PKB, *see* Protein kinase B (PKB)
PKC, *see* Protein kinase C (PKC)
PL, *see* Phospholipids (PL)
Plasma membrane (PM), 205
  protein, 134
PM, *see* Plasma membrane (PM)
PMC, *see* Peritubular myoid cells (PMC)
PMSG, *see* Pregnant mare serum gonadotropin (PMSG)
Poly (ADP-ribose) polymerase (PARP), 101
poly A, *see* Polyadenylic acid (poly A)
Polyadenylic acid (poly A), 61
Polymerase chain reaction–restriction fragment length polymorphism (PCR-RFLP), 310
Pre-ejaculate, 248; *see also* Cowper's glands
Pregnant mare serum gonadotropin (PMSG), 61; *see also* Gonadotropins
Pre-hormone, 52
Principal cell (PC), 120; *see also* Epididymal epithelim cell
  resorption, 122
  roles, 123
  ultrastructural organisation of, 121, 123
PRL, *see* Prolactin (PRL)
Progestogen–androgen combination regimens, 272; *see also* Male contraception
Programmed germ cell death, 98; *see also* Apoptosis
Prolactin (PRL), 44, 55; *see also* Gonadotropins
  anti-gonadal effects, 69
  JAK tyrosine kinase, 68–69
  radioreceptor assays for, 60
  signal transduction mechanism for, 68
Proliferating cell nuclear antigen (PCNA), 188
Prostaglandins, 256; *see also* Seminal vesicles
Prostate cancer, 254; *see also* Prostate gland
Prostate gland, 250, 251–252; *see also* Male accessory sex glands
  cancer, 254
  fibrin, 252
  fibrinolysin, 252
  oxytocin, 252
  pathologies, 252
  prostatic hypertrophy, 253–254
  prostatitis, 253
  PSA, 134, 149, 234, 248
  secretions and function, 252
  zones of, 251
Prostate-specific antigen (PSA), 134, 149, 234, 248
Prostatic hypertrophy, 253–254; *see also* Prostate gland
Prostatitis, 253; *see also* Prostate gland
Protein hormones, 44
  biosynthesis, 52
Protein kinase A (PKA), 228, 289
Protein kinase B (PKB), 290
Protein kinase C (PKC), 69, 289
PSA, *see* Prostate-specific antigen (PSA)
PVECs, *see* Pale vacuolated epithelial cell (PVECs)

Q

Quinoline-Val-asp (Ome)-CH2-O-Ph (Q-VD-OPH), 104
Q-VD-OPH, *see* Quinoline-Val-asp (Ome)-CH2-O-Ph (Q-VD-OPH)

R

Radioreceptor assay (RRA), 56
Rat protein DE, 149; *see also* Epididymis
RCA, *see Ricinus communis* agglutinin (RCA)
Regulatory subunit, 205
Reproductive physiology, 43
Reverse transcription (RT), 37
Ricinus communis agglutinin (RCA), 221
RISUG, 268; *see also* Male contraception
RNAi, *see In vivo* RNA interference (RNAi)
RRA, *see* Radioreceptor assay (RRA)
RT, *see* Reverse transcription (RT)
RTS, *see* Rubinstein–Taybi Syndrome (RTS)
Rubinstein–Taybi Syndrome (RTS), 292–293; *see also* Androgen

S

SAGA-1, *see* Sperm agglutination antigen-1 (SAGA-1)
SAPK, *see* Stress-activated protein kinase (SAPK)
SBMA, *see* Spinal and bulbar muscular atrophy (SBMA)
SCARKO, *see* Sertoli cell only knockout of androgen receptor (SCARKO)
SCN, *see* Suprachiasmatic nucleus (SCN)

SCSA, *see* Sperm chromatin structure assay (SCSA)
SDN, *see* Sexually dimorphic nucleus (SDN)
SEBS, *see* Styrene ethylene butylene styrene (SEBS)
Second mitochondria-derived activator of caspases (SMAC), 99
Self-reactive lymphocytes, 98
Semen coagulum proteins, 134; *see also* Epididymis
Seminal fluid, 254
Seminal vesicles, 254, 255; *see also* Male accessory sex glands
  fibrinogen, 256
  location, 255
  pathologies, 256–257
  prostaglandins, 256
  secretion and function, 255–256
Seminiferous epithelium, 261–262
Serotonin, 7; *see also* Sexual dimorphism in CNS
Sertoli cell, 80–81, 82, 168, 264; *see also* Follicle-stimulating hormone (FSH); Male reproduction physiology; Mammalian testis
  activation of ERK in, 107
  blood–testis barrier, 265–266
  junctional complexes, 265
Sertoli cell only knockout of androgen receptor (SCARKO), 84
Sex-determining gene, 4
Sex gland secretions, 256
Sex hormone binding globulin (SHBG), 9; *see also* Sexual dimorphism in CNS
Sexual behaviour, 5–6
Sexual differentiation timing, 7–8; *see also* Sexual dimorphism in CNS
Sexual dimorphism in CNS, 2, 23; *see also* Sexual dimorphism in human brain
  amygdala, 15–16
  androgens and oestrogens, 8
  anti-Mullerian hormone, 5
  causes of, 3
  dimorphic behaviour, 5–6
  dimorphic structures, 12
  endogenous opioids and opiates, 10
  hormonal involvement, 6
  monoaminergic systems, 6
  sex-determining gene, 4
  sex hormone binding globulin, 9
  sexual orientation and CNS, 21–23
  signalling molecule, 4
  in spinal cord, 11–12
  testosterone and 5α-dihydrotestosterone, 10
  timing of sexual differentiation, 7–8
Sexual dimorphism in human brain, 16; *see also* Sexual dimorphism in CNS
  anterior commissure, 19–20
  in anterior hypothalamus, 17–18
  bed nucleus of stria terminalis, 19
  corpus callosum, 20
  in handedness, 20–21
  hippocampus, 18
  non-sexual dimorphic behaviour, 16–17
  supraoptic and suprachiasmatic nuclei, 18
  vomeronasal organ, 19
Sexually dimorphic behaviour, 5; *see also* Sexual dimorphism in CNS
Sexually dimorphic nucleus (SDN), 13
Sexually dimorphic structures, 12; *see also* Sexual dimorphism in CNS
  in medial preoptic area and hypothalamus, 12
  MPOA, 12
  neonatally androgenised female rats, 14–15
  SDN, 13
SHBG, *see* Sex hormone binding globulin (SHBG)
Signalling molecule, 4
Signal transducer and activators of transcription (STAT), 69
Single nucleotide polymorphisms (SNPs), 310
SKY, *see* Spectral karyotyping (SKY)
SMA-4, *see* Sperm maturation antigen 4 (SMA-4)
SMAC, *see* Second mitochondria-derived activator of caspases (SMAC)
Small, non-coding RNAs (sncRNAs), 311
SNB, *see* Spinal nucleus of the bulbocavernosus (SNB)
sncRNAs, *see* Small, non-coding RNAs (sncRNAs)
SNPs, *see* Single nucleotide polymorphisms (SNPs)
SOD, *see* Superoxide dismutase (SOD)
Sodium butyrate, 286
Spectral karyotyping (SKY), 302
Spermadhesin AWN, 220; *see also* Sperm surface lectin
Sperm agglutination antigen-1 (SAGA-1), 169
Spermatogenesis, 98, 262, 299–300; *see also* Follicle stimulating hormone (FSH); Luteinising hormone (LH); Male infertility; Male reproduction physiology; Mammalian testis
  duration, 306
  germ cell apoptosis in, 98–99

Spermatogenesis (*Continued*)
hormonal regulation, 78, 81, 88–89
phases, 300
role of androgens in, 263
stem spermatogonia, 98
testosterone role in, 87–88
Spermatogenic cells, 81; *see also* Mammalian testis
Spermatogonia, 81, 261; *see also* Spermatogenesis
Spermatogonial stem cell (SSC), 266
Spermatozoa, 213, 262; *see also* Spermatogenesis; Sperm ecto-protein kinases
Sperm cell, 198; *see also* Epididymis; Male contraception; Sperm ecto-protein kinases; Sperm motility; Sperm plasma membrane; Sperm surface lectin
DHA concentration in, 203
ecto-phosphoprotein turnover, 217
epididymosomes, 234
forward motility, 219, 228
maturation process, 218
membrane lipids, 135
surface proteins, 132–134
transport prevention, 267–268
Sperm chromatin structure assay (SCSA), 84
Sperm ecto-protein kinases, 205; *see also* Ecto-cyclic AMP-independent protein kinase (ecto-CIK); Major physiological protein substrate (MPS); Phosphoprotein phosphatase; Sperm; Sperm motility; Sperm motility; Sperm plasma membrane; Sperm surface lectin
cell electroporation, 210
coupled-enzyme action, 219
ecto-cAMP-dependent protein kinase, 205–206
ecto-CIK, 206–209, 211, 212, 213, 214
ecto-PPase, 214
ecto-RCs, 205
exogenous proteins, 207
MPS, 209–214
phospholipids, 207, 210
spermatozoa immunofluorescence, 213
sperm ecto-phosphoprotein turnover, 217
substrate specificity, 206
sub-unit topography, 208
surface probe DSS, 207
surface protein dephosphorylation, 214–219
surface protein phosphorylation, 206–209, 216
Spermiogenesis, 81, 300
Spermiophagy, 173
Sperm maturation antigen 4 (SMA-4), 132
Sperm motility, 199; *see also* Forward motility stimulating factor (FMSF); Motility-inhibiting factor (MIF); Motility initiating protein (MIP); Sperm cell; Sperm ecto-protein kinases; Sperm motility; Sperm plasma membran; Sperm surface lectine
assays, 199–200
inhibitors, 233
MPS incorporation effect on, 215
stimulating proteins, 223
Sperm plasma membrane; *see also* Sperm cell; Sperm ecto-protein kinases; Sperm motility; Sperm surface lectin
absorbance vs. time plot, 201
isolation, 200–202
lipid phase transitions, 204, 205
lipids, asymmetry and fluidity, 202–205
phospholipid compositions of, 204
Sperm surface lectin, 219; *see also* Sperm cell; Sperm ecto-protein kinases; Sperm motility; Sperm plasma membrane
autoagglutination, 220
copper-dependent agglutination, 224, 225
copper-dependent sialic acid-specific lectin, 222–223
D-galactose-specific lectin, 220–221
levels with autoagglutination and motility, 222
receptor, 221–222
spermadhesin AWN, 220
SPH, *see* Sphingomyelin (SPH)
Sphingomyelin (SPH), 202
Spinal and bulbar muscular atrophy (SBMA), 293
Spinal nucleus of the bulbocavernosus (SNB), 12
SRC-1, *see* Steroid receptor coactivator 1(SRC-1)
SSC, *see* Spermatogonial stem cell (SSC)
STAT, *see* Signal transducer and activators of transcription (STAT)
Stem spermatogonia, 98
Steroid receptor coactivator 1(SRC-1), 291
Stress-activated protein kinase (SAPK), 104
Styrene ethylene butylene styrene (SEBS), 268
Superoxide dismutase (SOD), 126

Suprachiasmatic nucleus (SCN), 18
Supraoptic and suprachiasmatic nuclei, 18; *see also* Sexual dimorphism in human brain
Syringocele, 249; *see also* Cowper's glands

T

TAD, *see* Transactivation domain (TAD)
TB, *see* Testosterone buciclate (TB)
TD, *see* Testosterone decanoate (TD)
Testicular hyperthermia, 107; *see also* Apoptosis
Testis, 261; *see also* Male reproduction physiology
- estradiol role, 264
- growth factors role, 266–267
- rete testis, 261
- seminiferous epithelium, 261–262
- spermatogenesis, 262
- testicular parenchyma, 261

Testosterone, 10, 86; *see also* Androgen; Sexual dimorphism in CNS
- effect of sodium butyrate, 286

Testosterone buciclate (TB), 270; *see also* Male contraception
Testosterone decanoate (TD), 272
Testosterone undecanoate (TU), 270–271; *see also* Male contraception
TFR, *see* Total fertility rate (TFR)
TGF, *see* Transforming growth factor (TGF)
Thyroidal iodine release (TIR), 71; *see also* Gonadotropins
Thyroid-stimulating hormone (TSH), 44; *see also* Gonadotropins
Thyrotropin-releasing hormone (TRH), 55; *see also* Gonadotropins
Tip60 coactivator, 291
TIR, *see* Thyroidal iodine release (TIR)
tmAC, *see* Transmembrane adenylyl cyclase (tmAC)
TMD, *see* Transmembrane domain (TMD)
TNF, *see* Tumour necrosis factor (TNF)
TNF-related apoptosis-inducing ligand (TRAIL), 104
Total fertility rate (TFR), 260
TRAIL, *see* TNF-related apoptosis-inducing ligand (TRAIL)
Transactivation domain (TAD), 286
Transforming growth factor (TGF), 266
Transmembrane adenylyl cyclase (tmAC), 228
Transmembrane domain (TMD), 62
TRH, *see* Thyrotropin-releasing hormone (TRH)
TSH, *see* Thyroid-stimulating hormone (TSH)
TU, *see* Testosterone undecanoate (TU)
Tumour cells, 98
Tumour necrosis factor (TNF), 104

U

Urinary tract infection causatives, 249

V

Vacuolar H+-ATPase (V-ATPase), 125
Vasectomy, 268; *see also* Male contraception
Vasoactive intestinal peptide (VIP), 18
Vaso-occlusive devices, 268; *see also* Male contraception
Vasovasostomy, 268; *see also* Male contraception
V-ATPase, *see* Vacuolar H+-ATPase (V-ATPase)
VIP, *see* Vasoactive intestinal peptide (VIP)
Virus infected cells, 98
Vomeronasal organ, 19; *see also* Sexual dimorphism in human brain

W

WAP, *see* Whey acidic protein (WAP)
WGA, *see* Wheat germ agglutinin (WGA)
Wheat germ agglutinin (WGA), 221
Whey acidic protein (WAP), 132
WHO, *see* World Health Organization (WHO)
Wilms' tumour 1 (WT1), 110
World Health Organization (WHO), 269
WT1, *see* Wilms' tumour 1 (WT1)

X

X chromosome, 305; *see also* Male infertility

Y

Y chromosome, 302; *see also* Male infertility
- *DAZ* gene cluster by FISH, 304
- micro-deletion, 303

Yeast expression system, 51

T - #0086 - 061119 - C360 - 234/156/16 - PB - 9780367377458